Semantics-Oriented Natural Language Processing

Volume 27

International Federation for Systems Research International Series on Systems Science and Engineering

Series Editor: George J. Klir
Binghamton State University

IFSR was established "to stimulate all activities associated with the scientific study of systems and to coordinate such activities at international level." The aim of this series is to stimulate publication of high-quality monographs and textbooks on various topics of systems science and engineering. This series complements the Federation's other publications.

A Continuation Order Plan is available for this series. A continuation order will bring delivery of each new volume immediately upon publication. Volumes are billed only upon actual shipment. For further information please contact the publisher.

Volumes 1–6 were published by Pergamon Press.

Vladimir A. Fomichov

Semantics-Oriented Natural Language Processing

Mathematical Models and Algorithms

 Springer

Vladimir A. Fomichov
Faculty of Business Informatics
Department of Innovations and Business
  in the Sphere of Informational Technologies
State University – Higher School of Economics
Kirpichnaya Street 33, 105679 Moscow, Russia
vfomichov@hse.ru
vdrfom@aha.ru
v.fomichov@snhu.edu

Series Editor:

George J. Klir
Thomas J. Watson School of Engineering and Applied Sciences
Department of Systems Science and Industrial Engineering
Binghamton University
Binghamton, NY 13902
U.S.A

ISBN 978-1-4899-8280-3   ISBN 978-0-387-72926-8 (eBook)
DOI 10.1007/978-0-387-72926-8
Springer New York Dordrecht Heidelberg London

Mathematics Subjects Classification (2000): 03B65, 68-XX, 93A30

Printed on acid-free paper

Springer is part of Springer Science+Business Media (www.springer.com)

To my wife, Olga Svyatoslavovna Fomichova

Preface

Gluecklich, die wissen, dass hinter allen
Sprachen das Unsaegliche steht.
Those are happy who know that behind
all languages there is something unsaid
Rainer Maria Rilke

This book shows in a new way that a solution to a fundamental problem from one scientific field can help to find the solutions to important problems emerged in several other fields of science and technology.

In modern science, the term "Natural Language" denotes the collection of all such languages that every language is used as a primary means of communication by people belonging to any country or any region. So Natural Language (NL) includes, in particular, the English, Russian, and German languages.

The applied computer systems processing natural language printed or written texts (NL-texts) or oral speech with respect to the fact that the words are associated with some meanings are called *semantics-oriented natural language processing systems (NLPSs).*

On one hand, this book is a snapshot of the current stage of a research program started many years ago and called Integral Formal Semantics (IFS) of NL. The goal of this program has been to develop the formal models and methods helping to overcome the difficulties of logical character associated with the engineering of semantics-oriented NLPSs. The designers of such systems of arbitrary kinds will find in this book the formal means and algorithms being of great help in their work.

On the other hand, this book can become a source of new powerful formal tools for the specialists from several different communities interested in developing semantic informational technologies (or, shorter, semantic technologies), in particular, for the researchers developing

- the knowledge representation languages for the ontologies in the Semantic Web project and other fields;
- the formal languages and computer programs for building and analyzing the semantic annotations of Web sources and Web services;

- the formal means for semantic data integration in e-science and e-health;
- the advanced content representation languages in the field of multi agent systems;
- the general-purpose formal languages for electronic business communication allowing, in particular, for representing the content of negotiations conducted by computer intelligent agents (CIAs) in the field of e-commerce and for forming the contracts concluded by CIAs as the result of such negotiations.

During last 20 years, semantics-oriented NLPSs have become one of the main subclasses of applied intelligent systems (or, in other terms, of the computer systems with the elements of artificial intelligence).

Due to the stormy progress of the Internet, the end users in numerous countries have received technical access to NL-texts stored far away from their terminals. This has posed new demands to the designers of NLPSs. In this connection it should be underlined that several acute scientific – technical problems require the construction of computer systems being able to "understand" the meanings of arbitrary NL-texts pertaining to some fields of humans' professional activity. The collection of these problems, in particular, includes

- the extraction of information from textual sources for forming and updating knowledge bases of applied intelligent systems and the creation of a Semantic Web;
- the summarization of NL-texts stored on a certain Website or selected in accordance with certain criteria;
- conceptual information retrieval in textual databases on NL-requests of the end users;
- question answering based on the semantic-syntactic analysis of NL-texts being components of Webdocuments.

Semantics-oriented NLPSs are complex technical systems; their design is associated not only with programming but also with solving numerous questions of logical character. That is why this field of engineering, as the other fields of constructing complex technical systems, needs effective formal tools, first of all, the formal means being convenient both for describing semantic structure of arbitrary NL-texts pertaining to various fields of humans' professional activity and for representing knowledge about the world.

Systems Science has proposed a huge amount of mathematical models and methods that are useful for a broad spectrum of technical and social applications: from the design and control of airplanes, rockets, and ships to modeling chemical processes and production-sailing activity of the firms.

The principal purpose of this monograph is to open for Systems Science a new field of studies – the development of formal models and methods intended for helping the designers of semantics-oriented NLPSs to overcome numerous problems of logical character associated with the engineering of such systems.

This new field of studies can be called *Mathematical Linguocybernetics* (this term was introduced by the author in [66]).

Let's consider the informal definitions of several notions used below for describing the principal aspects of the scientific novelty of this book.

The term "semantics of Natural Language" will denote the collection of the regularities of conveying information by means of NL. *Discourses* (or narrative texts) are the finite sequences of the sentences in NL with the interrelated meanings.

If T is an expression in NL (a short word combination, a sentence, or a discourse), a *structured meaning of the expression T* is an informational structure being constructed by the brain of a person having command of the considered sublanguage of NL (Russian, English, or any other), and the construction of this structure is independent of the context of the expression T, that is, this informational structure is built on the basis of knowledge about only elementary meaningful lexical units and the rules of combining such units in the considered sublanguage of NL.

Let's agree that a *semantic representation (SR)* of an NL-expression T is a formal structure being either an image of a structured meaning of the considered NL-expression or being a reflection of the meaning (or content) of the given expression in a definite context – in a concrete situation of a dialogue, in the context of knowledge about the world, or in the context of the preceding part of the discourse.

Thus, an SR of an NL-expression T is such formal structure that its basic components are, in particular, the designations of the notions, concrete things, the sets of things, events, functions and relations, logical connectives, numbers and colors, and also the designations of the conceptual relationships between the meanings of the fragments of NL-texts or between the entities of the considered application domain.

Semantic representations of NL-texts may be, for instance, the strings and the marked oriented graphs (semantic sets).

An *algorithm of semantic-syntactic analysis* builds an SR of an NL-expression, proceeding from the knowledge about the morphology and syntax of the considered sublanguage of NL (English, Russian, etc.), from the information about the associations of lexical units with the units of conceptual level (or semantic level), and taking into account the knowledge about application domains. An SR of the text constructed by such an algorithm is interpreted by an applied computer system in accordance with its specialization, for instance, as a request to search an answer to a question, or a command to carry out an action by an autonomous intelligent robot, or as a piece of knowledge to be inscribed into the knowledge base, etc.

The scientific results stated in this monograph have been obtained by the author while fulfilling a research program started over 20 years ago. The choice of the direction of the studies was a reaction to almost complete lack in that time of mathematical means and methods that were convenient for designing semantics-oriented NLPSs.

The results of this monograph not only contribute to a movement forward but also mean a qualitative leap in the field of elaborating the formal means and methods of developing the algorithms of semantic-syntactic analysis of NL-texts. This qualitative leap is conditioned by the following main factors:

- The designers of NLPSs have received a system of the rules for constructing well-formed formulas (besides, a compact system, it consists of only ten main rules) allowing for (according to the hypothesis of the author) building semantic representations of arbitrary texts pertaining to numerous fields of humans' professional activity, i.e., SRs of the NL-texts on economy, medicine, law, technology, politics, etc. This means that the effective procedures of constructing SRs of NL-texts and effective algorithms of processing SRs of NL-texts (with respect to the context of a dialogue or of a preceding part of discourse, taking into account the knowledge about application domains) can be used in various thematic domains, and it will be possible to expand the possibilities of these procedures in case of emerging new problems.

- A mathematical model of a broadly applicable linguistic database is constructed, i.e., a model of a database containing such information about the lexical units and their interrelations with the units of conceptual level that this information is sufficient for semantic-syntactic analysis of the sublanguages of natural language being interesting for a number of applications.
- A complex and useful, strongly structured algorithm of semantic-syntactic analysis of NL-texts is elaborated that is described not by means of any programming system but completely with the help of a proposed system of formal notions, this makes the algorithm independent of program implementation and application domain.
- A possible structure of several mathematical models of the new kinds is proposed with the aim of opening for Systems Science a new field of studies high significance for Computer Science.

Informational technologies implemented in semantics-oriented NLPSs belong to the class of *Semantic Informational Technologies* (or, shorter, *Semantic Technologies*). This term was born only several years ago as a consequence of the emergence of the Semantic Web project, the use of ontologies in this project and many other projects, the elaboration of Content Representation Languages as the components of Agent Communication Languages in the field of Multiagent Systems, and of the studies on formal means for representing the records of negotiations and the contracts in the field of Electronic Commerce (E-commerce).

One of the precious features of this monograph is that the elaborated powerful formal means of describing structured meanings of NL-texts provide a broadly applicable and flexible formal framework for the development of Semantic Technologies as a whole.

Content of the Book

The monograph contains two parts. Part 1, consisting of Chaps. 1, 2, 3, 4, 5, and 6, will be of interest to a broad circle of the designers of Semantic Informational Technologies. Part 2 (Chaps. 7, 8, 9, 10, and 11) is intended for the designers of Semantics-Oriented Natural Language Processing Systems.

Chapter 1 grounds the necessity of enriching the inventory of formal means, models, and methods intended for designing semantics-oriented NLPSs. Special attention is paid to showing the necessity of creating the formal means being convenient for describing structured meanings of arbitrary sentences and discourses pertaining to various fields of humans' professional activity. The context of Cognitive

Linguistics for elaborating an appropriate approach to solving this problem is set forth. The possible structure of mathematical models of several new kinds for Systems Science is outlined.

The basic philosophical principles, history, and current composition of an original approach to formalizing semantics of NL are stated in Chap. 2; this approach elaborated by the author of this book is called Integral Formal Semantics (IFS) of Natural Language.

In Chap. 3, an original mathematical model describing a system of primary units of conceptual level used by applied intelligent systems is constructed and studied. The model defines a new class of formal objects called conceptual bases.

In Chap. 4, based on the definition of a conceptual basis, a mathematical model of a system of ten partial operations on structured meanings (SMs) of NL-texts is constructed. The essence of the model is as follows: using primary conceptual units as "blocks," we are able to build with the help of these ten partial operations the structured meanings of the texts – sentences and discourses – from a very rich sublanguage of NL (including articles, textbooks, the records of commercial negotiations, etc.) and to represent arbitrary pieces of knowledge about the world.

The model determines a new class of formal languages called standard knowledge languages (SK-languages) and can be interpreted as a formal metagrammar of a new kind. A mathematical study of the properties of SK-languages is carried out. In particular, the unambiguity of the syntactical analysis of the expressions of SK-languages is proved.

The purpose of Chap. 5 is to study the expressive possibilities of SK-languages. The advantages of the theory of SK-languages in comparison, in particular, with Discourse Representation Theory, Theory of Conceptual Graphs, Episodic Logic, and Database Semantics of Natural Language are analyzed.

Chap. 6 shows a broad spectrum of the possibilities to use the theory of SK-languages for solving a number of acute problems of Computer Science and Web Science. The possibilities of using SK-languages for (a) building semantic annotations of informational sources and of Web services; (b) constructing high-level conceptual descriptions of visual images; (c) semantic data integration in e-science, e-health, and other e-fields are indicated.

The definition of the class of SK-languages can also be used for the elaboration of formal languages intended for representing the contents of messages sent by computer intelligent agents (CIAs). It is also shown that the theory of SK-languages opens new prospects of building formal representations of contracts and records of commercial negotiations carried out by CIAs.

The broad expressive power of SK-languages demonstrated in Chaps. 4, 5, and 6 provides the possibility to propose in the final part of Chap. 6 a new, theoretically possible strategy of transforming evolutionarily, step by step, the existing Web into a Semantic Web of a new generation.

In Chap. 7, a broadly applicable mathematical model of *linguistic database* is constructed, that is, a model of a collection of semantic-syntactic data associated with primary lexical units and used by the algorithms of semantic-syntactic analysis for building semantic representations of natural language texts.

Chapter 8 sets forth a new method of transforming an NL-text (a statement, a command, or a question) into its semantic representation (SR). One of the new ideas of this method is the use of a special intermediary form of representing the results of semantic-syntactic analysis of an NL-text. This form is called a Matrix Semantic-Syntactic Representation of the introduced text. The constructed SR of an NL-text is an expression of a certain SK-language, or a K-representation of the considered NL-text. A pure syntactic representation of an analyzed text isn't used: the proposed method is oriented at directly finding the conceptual relations between the fragments of an NL-text.

Chapters 9 and 10 together describe an original, complex, and strongly structured algorithm of semantic-syntactic analysis of NL-texts; it is called the algorithm *SemSynt*1. Chapter 9 sets forth an algorithm of constructing a Matrix Semantic-Syntactic Representation of a natural language text; this algorithm is called *BuildMatr*1. The algorithm *BuildMatr*1 is multilingual: the input texts may belong to the sublanguages of English, German, and Russian languages (a Latin transcription of Russian texts is considered).

Chapter 10 describes an algorithm *BuildSem*1 of assembling a K-representation of an NL-text, proceeding from its matrix semantic-syntactic representation. The final algorithm *SemSynt*1 is defined as the composition of the algorithms *BuildMatr*1 and *BuildSem*1.

The content of Chaps. 1, 2, 3, 4, 5, 6, 7, 8, 9, and 10 can be interpreted as the principal part of *the theory of K-representations (knowledge representations)* – a new, powerful, and flexible framework for the development of semantic technologies.

The final Chap. 11 discusses two computer applications of the obtained theoretical results. The first one is a computer intelligent agent for fulfilling a semantic classification of e-mail messages. The second one is an experimental Russian-language interface implemented in the Web programming system PHP on the basis of the algorithm *SemSynt*1, it transforms NL-descriptions of knowledge pieces (in particular, definitions of concepts) first into the K-representations and then into the expressions of the ontology mark-up language OWL.

Moscow, Russia   *Vladimir Fomichov*
                                                                             December 2008

Acknowledgements

I am highly grateful to Distinguished Professor George J. Klir for the kind invitation to prepare this book and numerous constructive advices.

Many thanks to the anonymous referees of my monograph proposal and of the submitted manuscript.

The love, patience, and faith of my wife, Dr. Olga Svyatoslavovna Fomichova, and my mother-in-law, Dr. Liudmila Dmitrievna Udalova, very much helped me to obtain the results reflected in this book.

Two persons provided an outstanding help in the preparation of the electronic version of this book.

My son Dmitry, an alumnus of the Faculty of Cybernetics and Computational Mathematics of the Lomonosov Moscow State University, spent a lot of effort for translating from Russian into English the main content of Chaps. 9 and 10, in particular, for following up the tradition of stepped representation of the algorithms while preparing a LaTex file with the description of the algorithm of semantic-syntactic analysis *SemSynt*1.

I greatly appreciate the generous LaTex consulting provided by my friend, Senior Researcher in Software Development, Alexandr Mikhailovich Artyomov. Besides, special thanks to Alexandr for transforming my hand-drawn pictures into the electronic files with the figures presented in this book.

I appreciate the creative approach to the preparation of diploma work by the alumni of the Moscow State Institute of Electronics and Mathematics (Technical University), Faculty of Applied Mathematics, Inga V. Lyustig and Sergey P. Liksyutin. The applied computer systems based on the theory of K-representations and designed by Inga and Sergey are described in the final chapter of this book.

The communication with the Springer US Editors, Ms. Vaishali Damle and Ms. Marcia Bunda, during various stages of the work on the book highly stimulated the preparation of the manuscript.

Contents

Acronyms

B	conceptual basis
BuildMatr1	algorithm of constructing an MSSR of an NL-text
BuildSem1	algorithm of semantic assembly
Cb	marked-up conceptual basis
Ct, Ct(B)	concept-object system
Det	morphological determinant, M-determinant
F, F(B)	the set of functional symbols
Frp	dictionary of prepositional semantic-syntactic frames
Gen, Gen(B)	generality relation on a set of sorts
Int, Int(B)	the set of intensional quantifiers
Lec	a set of basic lexical units
Lingb	linguistic basis
Ls(B)	SK-language, standard knowledge language
Lsdic	lexico-semantic dictionary
Morphbs	morphological basis
Matr	matrix semantic-syntactic represenation
Morphsp	morphological space
P, P(B)	distinguished sort "a meaning of proposition"
R, R(B)	the set of relational symbols
Rc	classifying representation of an NL-text
ref, ref(B)	referential quantifier
Rm	morphological representation of an NL-text
Rqs	dictionary of the role interrogative word combinations
S, S(B)	sort system
Spmorph	morphological space
S, St(B)	the set of sorts
SemSynt1	algorithm of semantic-syntactic analysis
Tform	text-forming system
Tol, Tol(B)	tolerance relation on a set of sorts
X(B)	primary informational universe
V(B)	the set of variables
Vfr	dictionary of verbal – prepositional semantic-syntactic frames
W	a set of words

Part I
A Comprehensive Mathematical Framework for the Development of Semantic Technologies

Chapter 1
Mathematical Models for Designing Natural Language Processing Systems as a New Field of Studies for Systems Science

Abstract This chapter grounds the necessity of developing new mathematical tools for the design of semantics-oriented natural language processing systems (NLPSs) and prepares the reader to grasping the principal ideas of these new tools introduced in the next chapters. Section 1.1 grounds the expedience of placing into the focus of Systems Science the studies aimed at constructing formal models being useful for the design of semantics-oriented NLPSs. Sections 1.2, 1.3, and 1.4 contain the proposals concerning the structure of such mathematical models of several new types that these models promise to become a great help to the designers of semantics-oriented NLPSs. Sections 1.5 and 1.6 jointly give the rationale for the proposed structure of new mathematical models. Section 1.5 states the central ideas of Cognitive Linguistics concerning natural language (NL) comprehension. Section 1.6 describes the early stage of the studies on formal semantics of NL. Section 1.7 sets forth the idea of developing formal systems of semantic representations with the expressive power close to that of NL; this idea is one of the central ones for this monograph.

1.1 An Idea of a Bridge Between Systems Science and Engineering of Semantics-Oriented Natural Language Processing Systems

Since the pioneer works of Montague [158–160] published in the beginning of the 1970s, the studies on developing formal semantics of natural language (NL) have been strongly influenced by the look at structuring the world suggested by mathematical logic, first of all, by first-order predicate logic.

However, it appears that a rich experience of constructing semantics-oriented natural language processing systems (NLPSs) accumulated since the middle of the 1970s provides weighty arguments in favor of changing the paradigm of formalizing

V.A. Fomichov, *Semantics-Oriented Natural Language Processing*, IFSR International Series on Systems Science and Engineering 27, DOI 10.1007/978-0-387-72926-8_1, © Springer Science+Business Media, LLC 2010

NL semantics and, with this aim, placing this problem into the focus of interests of Systems Science.

The analysis allows for indicating at least the following arguments in favor of this idea:

1. The first-order logic studies the structure of propositions (in other words, the structure of statements, assertions). However, NL includes also the imperative phrases (commands, etc.) and questions of many kinds.
2. The engineering of semantics-oriented NLPSs needs, first of all, the models of transformers of several kinds. For instance, it needs the models of the subsystems of NLPSs constructing a semantic representation (SR) *Semrepr* of an analyzed sentence of an NL-discourse as the value of a function of the following arguments:

   - *Semcurrent* – a surface SR of the currently analyzed sentence *Sent* from the discourse D;
   - *Semold* – an SR of the left segment of D not including the sentence *Sent*;
   - *Lingbs* – a linguistic database, i.e., a collection of the data about the connections of lexical units with the conceptual (informational) units used by an algorithm of semantic-syntactic analysis of NL-texts; and
   - *Kb* – a knowledge base, or ontology, containing information about the world. But mathematical logic doesn't consider models of this kind.

3. NL-texts are formed as a result of the interaction of numerous mechanisms of conveying information acting in natural language. Mathematical logic doesn't possess formal means being sufficient for reflecting these mechanisms on a conceptual level. That is why mathematical logic doesn't provide sufficiently rich formal tools allowing for representing the results of semantic-syntactic analysis of arbitrary NL-texts. For instance, the first-order logic doesn't allow for building formal semantic analogues of the phrases constructed from the infinitives with dependent words, of sentences with the word "a notion," and of discourses with the references to the meanings of previous phrases and larger parts of discourses.

Due to these reasons, a hypothetical structure of several formal models of the new types for Systems Science is proposed in the next sections, expanding the content of [95]. In the examples illustrating the principal ideas of these new formal models, the strings belonging to formal languages of a new class will be used. It is the class of standard knowledge (SK) languages determined in Chapters 3 and 4 of this monograph. The strings of SK-languages will be employed for building semantic representations of word combinations, sentences, and discourses and for constructing knowledge modules. It is important to emphasize that it is not assumed that the reader of this chapter is acquainted with the definition of SK-languages. The purpose of the next sections is only to show the reasonability of undertaking the efforts for constructing the models of the proposed new types.

1.2 The Models of Types 1–4

This section proposes the structure of mathematical models of four new types promising to become useful tools of designing semantics-oriented NLPSs. In particular, this applies to the design of NL-interfaces to intelligent databases, autonomous intelligent systems (robots), advanced Web-based search engines, NL-interfaces of recommender systems, and to the design of NLPSs being subsystems of full-text databases and of the knowledge extraction systems.

1.2.1 The Models of Type 1

The models of *the first proposed class* describe a correspondence between an introduced separate sentence in NL and its semantic representation. The transformation of the inputted sentences into their semantic representations is to be carried out with respect to a linguistic database *Lingb* and a knowledge base *Kb*.

Formally, the models of the proposed type 1 (see Fig. 1.1) describe a class of the systems of the form

$$(Linp, Lingbset, Kbset, Lsemrepr, transf, Alg, Proof),$$

where

- *Linp* is an input language consisting of sentences in natural language (NL);
- *Lingbset* is a set of possible linguistic databases (each of them is a finite set of some interrelated formulas);
- *Kbset* is a set of possible knowledge bases (each of them is also a finite set of some interrelated formulas);
- *Lsemrepr* is a language of semantic representations (in other terms, a language of text meaning representations);
- *transf* is a mapping from the Cartesian product of the sets *Linp*, *Lingbset*, *Kbset* to *Lsemrepr*;
- *Alg* is an algorithm implementing the mapping (or transformation) *transf*;
- *Proof* is a mathematical text being a proof of the correctness of the algorithm *Alg* with respect to the mapping *transf*.

Example 1. If $Qs1$ is the question "How many universities in England use the e-learning platform "Blackboard" for distance education?" then $transf(Qs1)$ can be the string of the form

$$Question(x1, ((x1 \equiv Number1(S1)) \wedge Qualit - composition(S1,$$

$$university * (Location, England)) \wedge Description1(arbitrary\,university*$$

$$(Element, S1) : y1, Situation(e1, use1 * (Time, \#now\#)$$

$$(Agent1, y1)(Process, learning * (Kind1, online))$$

$$(Object1, certain\, platform3 * (Title, \,'Blackboard')))))).$$

This string includes, in particular, the following fragments: (a) a compound designation of the notion "an university in England," (b) a designation of arbitrary element of a set $S1$ consisting of some universities $arbitrary\, university * (Element, S1) : y1$, (c) a compound designation of an e-learning platform.

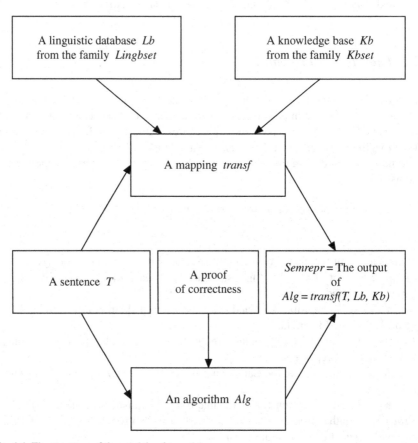

Fig. 1.1 The structure of the models of type 1

The necessity of taking into account the component Kb (a knowledge base, an ontology) while formalizing the correspondence between NL-texts and their semantic representations follows from the analysis of numerous publications on semantics-oriented NL processing: from the pioneer works of Winograd [207], Wilks [205], Schank et al. [178, 179], Hobbs et al. [126] to, in particular, the papers [13, 17, 33, 154, 165, 208].

1.2.2 The Models of Type 2

The models of the *second proposed class* describe the systems of the form

$$(Linp, Lpatterns, Lingbset, Kbset, Loutput, transf, Alg, Proof),$$

where

- *Linp* is an input language consisting of expressions in natural language (NL);
- *Lpatterns* is a language intended for indicating the patterns for extracting information from inputted NL-texts;
- *Lingbset* is the set of possible linguistic databases (each of them being a finite set of some interrelated formulas);
- *Kbset* is the set of possible knowledge bases (each of them being also a finite set of some interrelated formulas);
- *Loutput* is an output language;
- *transf* is a mapping from the Cartesian product of the sets *Linp*, *Lpattern*, *Lingbset*, *Kbset* to *Loutput*;
- *Alg* is an algorithm implementing the mapping (or transformation) *transf*;
- *Proof* is a mathematical text being a proof of the correctness of the algorithm *Alg* with respect to the mapping *transf*.

Example 2. The model describes the work of a computer-intelligent agent looking in various business texts for information about any change in the world prices of aluminum or plumbum or copper for 3 or more percent. Here the language *Linp* consists of NL-texts with commercial information, *Lpatterns* contains a semantic representation (SR) of the expression "any change of the world prices for aluminium or plumbum or copper for 3 or more percents," and *Loutput* consists of SRs of the fragments from inputted NL-texts informing about the changes of the world prices for aluminium or plumbum or copper for 3 or more percents. The string from the language *Lpatterns* can have, for instance, the form

$$change1 * (Goods1, (aluminium \lor plumbum \lor copper))$$

$$(Bottom\_border, 3/percent).$$

1.2.3 The Models of Type 3

Nowadays there are many known computer programs that are able to build semantic representations of separate short sentences in NL. However, there are a number of unsolved questions concerning the semantic-syntactic analysis of the fragments of discourses in the context of the preceding part of a dialogue or preceding part of a discourse. That is why it seems that the engineering of semantics-oriented NLPSs especially needs the models of the next proposed type.

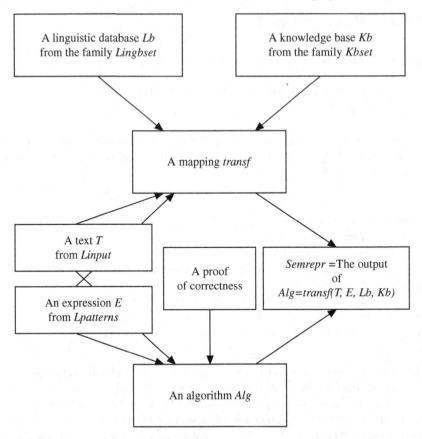

Fig. 1.2 The structure of the models of type 2

The models of the *third proposed class* are intended for designing the subsystems of NLPSs interpreting a semantic representation of the current fragment of a discourse in the context of semantic representation of the preceding part of a dialogue or preceding part of a discourse.

Formally, the models of this class describe the systems of the form

$$(Lcontext, Linp, Lingbset, Kbset, Lsem, Lreact, transf, Alg, Proof),$$

where

- *Lcontext* is a language for representing the content of the preceding part of a dialogue or a discourse;
- *Linp* is an input language consisting of underspecified or completely specified semantic representations of NL-expressions (sentences and some fragments of sentences), such NL-expressions can be, in particular, the answers to the clarifying questions of a computer system;

- *Lingbset* and *Kbset* are (as above) the sets of possible linguistic databases and knowledge bases;
- the semantic language *Lsem* is intended for representing the deep meaning of the inputs from *Linp* with respect to a semantic representation of the preceding part of a dialogue or a discourse;
- *Lreact* is a language for building semantic descriptions of the computer system's reactions to the inputted texts;
- *transf* is a mapping from the Cartesian product of the sets *Lcontext*, *Linp*, *Lingbset*, *Kbset* to the Cartesian product of the sets *Lsem* and *Lreact*;
- *Alg* is an algorithm implementing the mapping (or transformation) *transf*;
- *Proof* is a mathematical text being a proof of the correctness of the algorithm *Alg* with respect to the mapping *transf*.

Subclass 1: The models describing the work of a Recommender System.

Since the beginning of the 2000s, a new branch of E-commerce has been quickly developing called Recommender Systems (RS). The software applications of this class are intended for consulting the end users of the Internet in order to help them to take decisions about the choice of goods and/or services. The key role in the functioning of many RS is the interaction with the users by means of Natural Language (NL) – English, German, etc. [23, 24, 111, 112].

Consider a particular interpretation of the components of the models of type 3 intended for the design of NL-interfaces to RS. An input $X1$ from *Lcontext* reflects the content (in other words, the meaning) of the preceding part of a dialogue; an input $X2$ from *Linp* is an underspecified (or completely specified in particular cases) semantic representation (SR) of un utterance of the end user; an output $Y1$ from *Lsem* is a deep semantic representation of the input $X2$ in the context $X1$ with respect to a linguistic database *Lingbs* from the set *Lingbset* and to a knowledge base *Kbs* from the set *Kbset*; an output $Y2$ from *Lreact* is a semantic description of the computer system reaction to the inputted text with underspecified semantic representation $X2$ and deep semantic representation $Y1$. The knowledge base *Kbs* includes a subset of formulas *Userkbs* interpreted as a User Model.

Example 3. Suppose that an end user of an RS of an Internet shop applies to the RS with the question $Qs1$ = "What models of the cell telephones of the firm Nokia do you have, the price from 300 USD to 450 USD?" Imagine that this question is transfomed into the semantic representation $X1$ of the form

$$Question(S1, Qualitative - composition(S1, model1 * (Tech - product,$$

$$cell - telephone * (Manufacturer, \; certain \, firm1 * (Name1, 'Nokia')$$

$$(Price - diapason, 300/USD, 450/USD)))).$$

Having received an answer to this question, the user can submit the next question $Qs2$ = "And of the firm Siemens?" It is an elliptical question, and the NL-interface to the discussed RS can transform $Qs2$ into the SR $X2$ of the form

$$Question(S2, Qualitative - composition(S2, techical - object*$$

$$(Manufacturer, \ certain \ firm1 * (Name1, \ 'Siemens'))).$$

In the English language, the question $Qs2$ can have one of the following two meanings in the context of the question $Qs1$:

- Meaning 1: The end user wants to get information about all available models of cell phones produced by the firm "Siemens";
- Meaning 2: The end user wants to get information about all available models of cell phones produced by the firm "Siemens" with the price from 300 to 450 USD.

That is why the model is to be constructed in such a way that the RS asks the end user to select one of these meanings. An SR of this question, denoted by $Clarif - qs$, belongs to the language $Lreact$.

Imagine that the end user selects the second meaning. Then, according to the model, the NL-interface to the RS forms the semantic representation $Semrepr$ of the form

$$Question(S1, Qualitative - composition(S1, model1 * (Tech - product,$$

$$cell - telephone * (Manufacturer, \ certain \ firm1 * (Name1, \ 'Siemens'))$$

$$(Price - diapason, 300/USD, 450/USD)))).$$

This string expresses the deep meaning of the question $Qs2$ in the context of the questions $Qs1$ with SR $X1$ and belongs to the language $Lsem$.

Of course, this example represents one of the simplest possible dialogues of a Recommender System with the end user. With respect to the achieved level of studies on NLPSs, many people today are able to elaborate a computer system being able to function in the described way. However, the real dialogues may be much more complex. That is why the practice of designing NLPSs really needs the models of the kind.

1.2.4 The Models of Type 4

The models of *the fourth proposed class* are intended for designing computer-intelligent systems extracting knowledge from natural language sentences and complicated discourses for forming and updating a knowledge base (ontology) of an applied intelligent system. Such models describe the systems of the form

$$(Lcontext, Linp, Lingbset, Kbset, Lsem1, Lsem2, transf, Alg, Proof),$$

where

- *Lcontext* is a language for building a semantic representation of the already processed part of an NL-text,

- *Linp* is an input language consisting of underspecified or completely specified semantic representations of NL-expressions (sentences and some fragments of sentences);
- *Lingbset* and *Kbset* are (as above) the sets of possible linguistic databases and knowledge bases;
- the semantic language *Lsem*1 is intended for representing the deep meaning of the inputs from *Linp* with respect to a semantic representation of the preceeding part of an NL-text;
- *Lsem*2 is a language for representing the knowledge of the required kinds extracted from the expressions of the language *Lsem*1;
- *transf* is a mapping from the Cartesian product of the sets *Lcontext*, *Linp*, *Lingbset*, *Kbset* to the Cartesian product of the sets *Lsem*1 and *Lsem*2;
- *Alg* is an algorithm implementing the mapping (or transformation) *transf*; and
- *Proof* is a mathematical text being a proof of the correctness of the algorithm *Alg* with respect to the mapping *transf*.

1.3 The Models of Type 5

The models of *the fifth proposed class* are the models of advanced question answering systems, i.e., the models of intelligent computer systems being able to find an answer to a request in NL of an end user of a full-text database (of course, it can be Web-based) as a result of semantic-syntactic analysis of NL-texts stored in this database. Such models take into account the fact that the information enabling an intelligent system to formulate an answer to the posed question can be accumulated step by step, in the course of processing several texts in different informational sources.

The models of *the fifth proposed class* describe the systems of the form

$$(Lreq, Lct, Linp, Lbset, Kbset, Ls1, Ls2, Lans, Ind, transf, Alg, Proof),$$

where

- *Lreq* is a language for constructing semantic representations of the requests posed by the end users of an intelligent system;
- *Lct* is a language for building a semantic representation of the already processed part of an NL-text;
- *Linp* is an input language consisting of underspecified or completely specified semantic representations of NL-expressions (sentences and some fragments of sentences and discourses);
- *Lbset* and *Kbset* are (as above) the sets of possible linguistic databases and knowledge bases;
- the semantic language *Ls*1 is intended for representing the deep meaning of the inputs from *Linp* with respect to a semantic representation of the preceeding part of an NL-text;

- *Ls2* is a language for representing a piece of knowledge to be inscribed into the knowledge base *Kb* in order to be used later for formulating an answer to the request of the end user;
- *Lans* is a language for expressing the answers to the input requests, it includes the symbol *nil* called "the empty answer";
- *Ind* = {0, 1}, where 0 is interpreted as the signal to stop the search, 1 is interpreted as the signal to continue the search;
- *transf* is a mapping from the Cartesian product of the sets *Lreq, Lct, Linp, Lbset, Kbset* to the Cartesian product of the sets *Ls*1, *Ls*2, *Lans, Ind*;
- *Alg* is an algorithm implementing the mapping (or transformation) *transf*;
- *Proof* is a mathematical text being a proof of the correctness of the algorithm *Alg* with respect to the mapping *transf*.

Continuing the line of [82], let's illustrate some desirable properties of formal models reflecting the basic mechanisms of the hypothetical computer-intelligent systems of the kind.

Imagine that there is a big city D., and a user of an intelligent full-text database *Db*1 inputs the question *Qs* = "Is it true that the ecological situation in the city D. has improved during the year?" and the date of inputting *Qs* is *Date*1.

Suppose that *Qs* is transformed into the following initial semantic representation *Semreprqs*1 :

$$Question(u1, (u1 \equiv Truth - value(Better(Ecology(certain\, city*$$

$$(Name1, 'D.') : x1, Year(Date1)), Ecology(x1, Last - year(Date1)))))).$$

In the expression *Semreprqs*1, the element *Ecology* is to be interpreted as the name of the function assigning to the space object $z1$ and the year $z2$ a statement about the ecological situation in $z1$ corresponding to $z2$.

Let's assume that *Db*1 has the knowledge base *Kb*1 including a part *Objects − list*, and this part contains the string *certain city* ∗ (*Name*1, "*D*".) : $v315$. This means that the city D. is associated with the variable $v315$, playing the role of the unique system name of this city. Suppose also that *Date*1 corresponds to the year 2008. Then *Semreprqs*1 is transformed into the secondary semantic representation *Semreprqs*2 of the form

$$Question(u1, (u1 \equiv Truth - value(Better(Ecology(certain\, city*$$

$$(Name1, 'D.') : v315, 2008), Ecology(v315, 2007))))).$$

Suppose that there is the newspaper "D. News," and one of its issues published in the same month as *Date*1 contains the following fragment *Fr*1 : *"The quantity of species of birds who made their nests in the city has reached the number 7 in comparison with the number 5 a year ago. It was announced at a press-conference by Monsieur Paul Loran, Chair of the D. Association of Colleges Presidents".*

Let's consider a possible way of extracting from this fragment the information for formulating an answer to *Qs*. The first sentence *Sent*1 of *Fr*1 may have the following SR $Semrepr1a$:

$$((Quantity(certn\,species * (Compos1, bird)(Descr1, \langle S1, P1 \rangle)) \equiv 7) \wedge$$

$$(P1 \equiv (Compos1(S1, bird) \wedge Descr2(arbitrary\,bird *$$

$$(Elem, S1) : y1, \exists e1\,(sit)Is1(e1, nesting *$$

$$(Agent1, y1)(Loc, x1)(Time, 2008))))) \wedge$$

$$((Quantity(certn\,species * (Compos1, bird)(Descr1, \langle S2, P2 \rangle)) \equiv 5) \wedge$$

$$(P2 \equiv (Compos1(S2, bird) \wedge Descr2(arbitrary\,bird *$$

$$(Elem, S2) : y2, \exists e2\,(sit)Is1(e2, nesting *$$

$$(Agent1, y2)(Loc, x1)(Time, 2007)))))) : P3\,.$$

The symbol $certn$ is the informational unit corresponding to the word combination "a certain"; $Compos1$ is the designation of the binary relation "Qualitative composition of a set"; $P1, P2, P3$ are such variables that their type is the distinguished sort "a meaning of proposition".

Suppose that the second sentence $Sent2$ of $Fr1$ has the following semantic representation $Semrepr2a$:

$$\exists e3\,(sit)\,(Is(e3, announcing * (Agent1, x2)(Content1, P3)(Event,$$

$$certn\,press - conf : x3)) \wedge (x2 \equiv certain\,man * (First - name, \text{``}Paul\text{''})$$

$$(Surname, 'Loran')) \wedge (x2 \equiv Chair(certn\,association1 *$$

$$(Compos1, scholar * (Be1, President(any\,college * (Location,$$

$$certn\,city * (Name1, \text{``}D\text{''}.) : x4)))))))\,.$$

Here the element $association1$ denotes the concept "association consisting of people" (there are also the associations of universities, cities, etc.).

The analysis of the first sentence $Sent1$ shows that it is impossible to find directly in $Sent1$ the information determining the referent of the word "the city." In this case, let's take into account the knowledge about the source containing the fragment $Fr1$ and about the use of this knowledge for clarifying the referential structure of published discourses.

Imagine that the knowledge base $Kb1$ of the considered hypothetical intelligent system contains the string

$$Follows(((z1 \equiv arbitrary\,edition * (Title, z2)(Content1, Cont1)) \wedge$$

$$Associated(z2, arbitrary\,space - object : z3) \wedge Element(w, pos, Cont1) \wedge$$

$$Sem - class(w, pos, space - object) \wedge No - show - referent(w, pos, Cont1)),$$

$$Referent(w, pos, Cont1, z3))\,.$$

Let's interpret this formula as follows. Suppose that: (1) an arbitrary edition $z1$ has the title $z2$ and the content $Cont1$, its title $z2$ is associated in any way with the space-object $z3$; (2) the string $Cont1$ contains the element w in the position pos, its semantic class is $space - object$ (a city, a province, a country, etc.); (3) the text contains no explicit information about the referent of the element w in the position pos of the formula $Cont1$. Then the referent of the element w is the entity denoted by $z3$.

In order to use this knowledge item for the analysis of the fragment $Fr1$, let's remember that the list of the objects $Objects - list$ (being a part of the knowledge base $Kb1$) includes the string

$$certn\,city * (Name1, 'D.') : v315.$$

Then the system transforms the semantic representation $Semrepr1a$ of the first sentence $Sent1$ into the formula $Semrepr1b$ of the form

$$(Semr1a \wedge (x1 \equiv v315)).$$

This means that at this stage of the analysis the information extracted from $Sent1$ is associated with the city D.

Assume that the knowledge base $Kb1$ contains the knowledge items

$$\forall z1\,(person)\,\forall c1\,(concept)\,Follows(Head(z1,$$

$$arbitrary\,association1 * (Compos1, c1)), Is1(z1, c1)),$$

$$\forall z1\,(person)\,Follows((z1 \equiv$$

$$President((arbitrary\,univ : z2 \vee arbitrary\,college : z3))),$$

$$Qualification(z1, Ph.D.)),$$

and these items are interpreted as follows: (1) if a person $z1$ is the head of an association of the type 1 (associations consisting of people), the concept $c1$ qualifies each element of this association, then $z1$ is qualified by $c1$ too; (2) if a person $z1$ is the president of a university or a college, $z1$ has at least a Ph.D. degree.

Proceeding from the indicated knowledge items and from $Semrepr2a$, the system builds the semantic representation $Semrepr2b$ of the form

$$(Semr2a \wedge (x1 \equiv v315))$$

and then infers the formula $Qualification(x2, Ph.D.)$, where the variable $x2$ denotes Monsieur Paul Loran, Chair of the D. Association of Colleges Presidents.

Let $Kb1$ contain also the expression

$$Follows\,(\exists e1\,(sit)\,Is1(e1, announcing * (Agent1, arbitrary\,scholar*$$

$$(Qualif, Ph.D.))(Kind - of - event, personal - communication)$$

$$(Content1, Q1)(Time, t1)), Truth - estimation(Q1, t1, \langle 0.9, 1 \rangle)),$$

interpreted as follows: if a scholar having a Ph.D. degree announces something, and it is not a personal communication then the estimation of the truth of the announced information has a value in the interval [0.9, 1.0]. Here the substring $\exists e1 (sit) Is(e1, announcing*$ is to be read as "There is an event $e1$ of the type 'announcing' such that."

So let's imagine that, proceeding from the semantic representations $Semrepr1b$ and $Semrepr2b$ (the secondary semantic representations of the first and second sentences of the fragment $Fr1$) and the mentioned knowledge items from $Kb1$, the system infers the expression

$$((Quantity(certn\,species * (Compos1, bird)(Descr1, \langle S1, P1 \rangle))) \equiv 7) \wedge$$

$$(P1 \equiv (Compos1(S1, bird) \wedge Descr2(arbitrary\,bird*$$

$$(Elem, S1) : y1, \exists e1 (sit) Is1(e1, nesting*$$

$$(Agent1, y1)(Loc, v315)(Time, 2008))))) \wedge$$

$$((Quantity(certn\,species * (Compos1, bird)(Descr1, \langle S2, P2 \rangle))) \equiv 5) \wedge$$

$$(P2 \equiv (Compos1(S2, bird) \wedge Descr2(arbitrary\,bird*$$

$$(Elem, S2) : y2, \exists e2 (sit) Is1(e2, nesting*$$

$$(Agent1, y2)(Loc, v315)(Time, 2007)))))) : P3.$$

Suppose that the knowledge base $Kb1$ contains the knowledge unit

$$\forall z1(space - object) \,\forall t1(year) \,Follows(Better(Ecolog - sit(z1, bird, t1),$$

$$Ecolog - sit(z1, bird, t2)), Better(Ecology(z1, t1), Ecology(z1, t2)))$$

and the knowledge unit

$$\forall z1(space.ob) \forall t1(year) \forall t2(year) \,Follows((Compos1(S1, bird) \wedge$$

$$Descr2(arbitrary\,bird * (Elem, S1) : y1,$$

$$\exists e1 (sit) Is1(e1, nesting*$$

$$(Agent1, y1)(Loc, z1)(Time, t1))) \wedge (Compos1(S2, bird)$$

$$\wedge Descr2(arbitrary\,bird * (Elem, S2) : y2,$$

$$\exists e2 (sit) Is1(e2, nesting*$$

$$(Agent1, y2)(Loc, z1)(Time, t2))) \wedge$$

$$Greater(Quantity(S1), Quantity(S2))),$$

$$Better(Ecolog - sit(z1, bird, t1),$$

$$Ecolog - sit(z1, bird, t2))).$$

That is why the system finally infers the formulas

$$Better (Ecolog - sit(v315, bird, 2008), Ecolog - sit(v315, bird, 2007)),$$

$$Better (Ecology(v315, 2008), Ecology(v315, 2007)).$$

Hence the system outputs the expression of the form

$$\langle\text{``YES''}, Ground : Fr1 ('D.News', Date1)\rangle .$$

1.4 The Significance of the Models for the Design of Linguistic Processors

The analysis shows the significance of the studies aimed at constructing the formal models of the considered kinds for the engineering of natural language processing systems (NLPSs). In particular, the following factors are distinguished:

1. The algorithms being components of such formal models can be directly used as the algorithms of the principal modules of NLPSs.
2. The descriptions of the mappings $transf$, characterizing the correspondence between the inputs and outputs of the systems, can become the principal parts of the documentation of such programming modules. As a result, the quality of the documentation will considerably increase.
3. The designers of NLPSs will get the comprehensible formal means for describing the semantics of lexical units and for building semantic representations of complicated natural language sentences and discourses in arbitrary application domains. This will contribute very much to the transportability of the elaborated software of NLPSs as concerns new tasks and new application domains.

It should be underlined that even the elaboration of the *partial models* of the kind promises to be of high significance for the engineering of NLPSs. The principal difference between the complete models and partial models of the considered types consists in the lack of a proof of the correctness of the algorithm Alg with respect to the defined transformation $transf$. Besides, a partial model may include a not mathematically complete definition of a transformation $transf$ but only a description of some principal features of this transformation.

Even in case of partial models, the designers of semantics-oriented NLPSs will receive an excellent basis for the preparation of such documentation of a computer system that distinguishes the most precious or original features of the algorithms and/or data structures and creates the good preconditions of transporting the data structures and algorithms to new problems and application domains.

1.5 The Context of Cognitive Linguistics for the Formal Study of Natural Language

The first bunch of ideas underlying the proposed structure of the mathematical models intended for the design of semantics-oriented NLPSs was given by the analysis of the look of Cognitive Science at the regularities of natural language comprehension.

The problems and achievements in the field of constructing NLPSs, on the one hand, and huge difficulties in the way of formalizing regularities of NL-comprehension, on the other hand, evoked in the 1980s a considerable interest of many psychologists and linguists in investigating such regularities. A new branch of linguistics was formed called Cognitive Linguistics as a part of Cognitive Science.

Cognitive linguists consider language as "an instrument for organizing, processing, and conveying information. The formal structures of languages are studied not as if they were autonomous, but as reflections of general conceptual organization, categorization principles, processing mechanisms, and experiential and environmental influences" (see [103], p. 1).

The obtained results allowed for formulating in [66] the following now broadly accepted principles of natural language comprehension.

1. The meaning of a natural-language text (NL-text) is represented by means of a special mental language, or a language of thought [11, 20, 22, 39, 49, 50, 135, 139, 149–152, 155, 178, 203].
2. People build two different (though interrelated) mental representations of an NL-text. The first one is called by Johnson-Laird [139] the *prepositional representation (PR)*. This representation reflects the semantic microstructure of a text and is close to the text's surface structure.
   The second representation being a mental model (MM) is facultative. The MM of a text reflects the situation described in the text. Mental models of texts are built on the basis of both texts' PRs and diverse knowledge – about the reality, language, discussed situation, and communication participants [139].
3. A highly important role in building the PRs and MMs of NL-texts is played by diverse cognitive models accumulated by people during the life-semantic frames, explanations of notions' meanings, prototypical scenarios, social stereotypes, representations of general regularities and area-specific regularities, and other models determining, in particular, the use of metaphors and metonymy [39, 45, 137, 139, 149, 150, 156, 187].
4. The opinion that there exists syntax as an autonomous subsystem of language system has become out of date. Syntax should depend on descriptions of cognitive structures, on semantics of NL. Natural language understanding by people doesn't include the phase of constructing the pure syntactic representations of texts.
   The transition from an NL-text to its mental representation is carried out on the basis of various knowledge and is of integral character [20, 48, 139, 149, 151, 152, 187, 198].

5. Semantics and pragmatics of NL are inseparably linked and should be studied and described by the same means [180, 189].

A significant role in formulating the enumerated principles was played by the researches on developing computer programs being capable of carrying out the conceptual processing of NL-texts. This applies especially to the studies that can be attributed to the semantics-oriented (or semantically driven) approaches to natural language parsing.

It appears that the set of principles stated above may serve as an important reference-point for the development and comparison of the available approaches to mathematical modeling of NL-understanding.

1.6 Early Stage of Natural Language Formal Semantics

The shortcomings of the main-known approaches to the formal study of NL-semantics were felt in the 1980s by many philosophers, psychologists, and linguists. The basic philosophical ideas of model-theoretic semantics were criticized, in particular, by Putnam [170], Johnson-Laird [139], Fillmore [45], Seuren [187, 188], and Lakoff [149].

The main approaches to the formalization of NL-semantics popular in the 1980s – Montague Grammar and its extensions [141, 158–160, 164, 166], Situation Semantics [2, 15, 16, 42], Discourse Representation Theory [122, 142–144], and Theory of Generalized Quantifiers [14] – are strongly connected with traditions of mathematical logic, of model-theoretic semantics, and do not provide formal means permitting to model the processes of NL-comprehension in correspondence with the above enumerated principles of cognitive science.

In particular, these approaches do not afford effective formal tools to build (a) semantic representations of arbitrary discourses (e.g., of discourses with the references to the meaning of fragments being sentences or larger parts of texts), (b) diverse cognitive models, for instance, explanations of notions' meanings, representations of semantic frames, (c) descriptions of sets, relations and operations on sets.

Besides, these approaches are oriented toward regarding assertions. However, it is also important to study the goals, commitments, advices, commands, questions.

The dominant paradigm of describing surface structure of sentences separately from describing semantic structure (stemming from the pioneer works of Montague [158–160]) contradicts one of the key principles of cognitive linguistics – the principle assuming the dependency of syntax in semantics.

Highly emotionally the feeling of dissatisfaction with the possibilities of the main popular approaches to the formalization of NL-semantics was put into words by Seuren [187, 188]. In particular, Seuren expressed the opinion that the majority of studies on the formalization of NL-semantics was carried out by researchers interested, first of all, in demonstrating the use of formal tools possessed by them, but not in developing the formal means allowing for modeling the mechanisms of NL-comprehension.

As it is known, ecology studies the living beings in their natural environment. In [188], the need of new, adequate, ecological approaches to studying the regularities of NL-comprehension is advocated. Many considerations and observations useful for working out ecological approaches to the formalization of discourses' semantics were formulated by Seuren. In the monograph [188], a peculiar attention was given to the questions of expressing and discerning the presuppositions of discourses, and the so-called Presuppositional Propositional Calculus was introduced.

In the second half of the eighties, a number of new results concerning the formalization of NL-semantics was obtained. Let us mention here the approach of Saint-Dizier [177] motivated by the tasks of logic programming, the results of Cresswell [29] and Chierchia [26] on describing structured meanings of sentences, the theory of situation schemata [42], Dynamic Semantics in the forms of Dynamic Predicate Logic [108] and Dynamic Montague Grammar [34, 106, 107].

Unfortunately, the restrictions pointed above in this section apply also to these new approaches. It should be added that Chierchia [26] describes structured meanings of some sentences with infinitives. But the expressive power of semantic formulae corresponding to such sentences is very small in comparison with the complexity of real discourses from scientific papers, text books, encyclopedic dictionaries, legal sources, etc.

Thus, the approaches mentioned in this section do not provide effective and broadly applicable formal tools for modeling NL-understanding in accordance with stated principles of cognitive psychology and cognitive linguistics.

The lack in the eighties of such means for modeling NL-understanding can be seen also from the substantial text book on mathematical methods in linguistics by Partee et al. [167].

One can't say that all approaches to the formalization of NL-semantics mentioned in the precedent subsection are not connected with the practice of designing NLPSs. There are publications, for example, on using for the design of NLPSs Montague Grammar in modified forms [28, 125, 185], Situation Semantics [213], and Discourse Representation Theory [124].

The language of intentional logic provided by Montague Grammar is used also in Generalized Phrase Structure Grammars [102] for describing semantic interpretations of sentences. Such grammars have found a number of applications in natural language processing.

Nevertheless, these and other approaches mentioned in this section possess a number of important shortcomings as concerns applying formal methods to the design of NLPSs and to developing the theory of NLPSs.

The demands of diverse application domains to the formal means for describing natural language may differ. That is why we distinguish for further analysis the following groups of application domains:

- Natural-language interfaces to databases, knowledge bases, autonomous robots.
- Full-text databases; computer systems automatically forming and updating knowledge bases of artificial intelligence systems by means of extracting information from scientific papers, text books, etc., in particular, intelligent text summarization systems.

- Such subsystems of automatized programming systems that are determined for transforming the NL-specifications of tasks into the formal specifications for the further synthesis of computer programs; such similar subsystems of computer-aided design systems that are determined for transforming the NL-specifications of technical objects to be designed into the formal specifications of these objects.

Obviously, the enumerated application domains represent only a part of all possible domains, where the development and use of NLPSs are important. However, for our purpose it is sufficient to consider only the mentioned important domains of applying NLPSs.

The analysis of formal means for the study of NL needed for these domains will allow us to get a rather complete list of demands to the formal theories of NL which should be satisfied by useful for practice and broadly applicable mathematical tools of studying NL-semantics and NL-pragmatics.

Let us regard for each distinguished group of applications the most significant restrictions of the approaches to the formal study of NL-semantics mentioned in this section.

Group 1

Semantics-oriented, or semantically driven NL-interfaces work in the following way. They transform an NL-input (or at first its fragment) into a formal structure reflecting the meaning of this input (or the meaning of a certain input's fragment) and called a semantic representation (SR) or a text meaning representation of the input or input's fragment. Then the SR is used (possibly, after transforming into a problem-oriented representation) for working out a plan for the reaction to the input with respect to a knowledge base, and after this a certain reaction is produced. The reactions may be highly diverse: applied intelligent systems can pose questions, fulfill calculations, search for required information, and transport things.

For constructing NL-interfaces in accordance with these principles, the following shortcomings of MG and its extensions, including Dynamic Montague Grammar, of Situation Semantics, Discourse Representation Theory, Theory of Generalized Quantifiers, Dynamic Predicate Logic, and of other approaches mentioned in this section are important:

1. The effective formal means for describing knowledge fragments and the structure of knowledge bases are not provided; in particular, this applies to the formal means for building semantic representations of complex definitions of notions.
2. There are no sufficiently powerful and flexible formal means to describe surface and semantic structures of questions and commands expressed by complicated NL-utterances.
3. There are no sufficiently powerful and flexible formal means to represent surface and semantic structures of the goals of intelligent systems expressed by complex NL-utterances.
4. The possibilities of intelligent systems to understand the goals of communication participants and to use the information about these goals for planning the reaction to an NL-input are not modeled.

5. The enumerated approaches do not give the flexible and powerful formal means for describing structured meanings of NL-discourses (including the discourses from scientific papers, legal sources, patents, etc.). The means of describing structured meanings of discourses are extremely restricted and unsatisfactory from the viewpoint of practice. In particular, the discourses with the references to the meaning of sentences and larger fragments of texts are not considered.

6. The existence of sentences of many types broadly used in real life is ignored. For instance, the structure of the following sorts of sentences is not studied: (a) containing expressions built from the descriptions of objects, sets, notions, events, etc., by means of logical connectives ("Yves has bought a monograph on mathematics, a text-book on chemistry, and a French-Russian dictionary"), (b) describing the operations on sets ("It will be useful to include Professor A. into the Editorial Board of the journal B."), (c) with the words "notion" or "term" (the latter in the meaning "a notion").

7. The models of the correspondences between the texts, knowledge about the reality, and semantic representations of texts are not built, and adequate means for developing models of the kind are not provided.

8. The inputs of NLPSs may be incomplete phrases, even separate words (for instance, the answers to questions in the course of a dialogue). The interpretation of such inputs is to be found in the context of precedent phrases and with respect to the knowledge about the reality and about the concrete discussed situation. However, such a capability of NL-interfaces isn't studied and isn't formally modeled.

9. The structure of metaphors and incorrect but understandable expressions from input texts, the correspondences between metaphors and their meanings are not investigated by formal means.

Wilks ([206], p. 348) writes that many NLPSs (in particular, the systems of machine translation) do not work as explained by the "official" theories in publications about these systems and function "in such a way that it cannot be appropriately described by the upper-level theory at all, but requires some quite different form of description." The analysis carried out above shows that the approaches mentioned in this section do not afford the opportunities to adequately describe the main ways of processing information by semantic components of NLPSs.

Group 2

Obviously, the restrictions 1, 3–6, and 8–9 are also important from the viewpoint of solving tasks like the development of intelligent full-text databases. The restriction 8 should be replaced by a similar restriction, since the fragments of discourses pertaining to business, technology, science, etc., may be incomplete, elliptical phrases.

The following restriction is to be pointed out additionally: the semantic structure of discourses with the proposals, commitments (the protocols, contracts often include such discourses), the interrelations between surface and semantic structures are not studied and modeled.

Over thirty years ago Wilks ([205], p. 116) noted that "any adequate logic must contain a dictionary or its equivalent if it is to handle anything more than terms with naive denotations such as 'chair'."

However, all approaches to the formalization of NL-semantics mentioned in this section do not take into account the existence and the role of various conceptual dictionaries. Due to this, in particular, reason, there is no opportunity to model the correspondence between texts, knowledge about the reality, and semantic representations of texts.

At first sight, the demands to the means of describing structured meanings of discourses and to the models of the correspondences between texts, knowledge, and SRs of texts are much stronger for the second group of applications than for the first one.

Nevertheless, it is not excluded that the joint future work of philosophers, linguists, specialists on computer science, and mathematicians will show that such demands are in fact very similar or the same for these two groups of NLPSs' applications.

Group 3

In addition to the shortcomings important for the groups 1 and 2, the following restriction should be mentioned: there are no effective formal means to represent structured meanings of NL-discourses describing the algorithms, the methods of solving diverse tasks. In particular, there are no adequate formal means to describe on semantic level the operations with the sets.

It appears that the collection of restrictions stated above provides a useful reference-point for enriching the stock of the means and models for the mathematical study of NL-communication.

1.7 The Significance of Highly Expressive Formal Systems of Semantic Representations

The collection of the tasks faced by the theory of semantics-oriented linguistic processors (LPs) in the beginning of the 1980s proved to be extremely complicated. As a consequence, the development of the theory of LPs in the 1980s slowed down. Though many projects of designing LPs were fulfilled in a number of countries, a substantial progress in this field was not achieved.

The principal cause of this deceleration is as follows. In natural language, numerous mechanisms of coding and decoding information interact in an intricate manner. That is why in order "to understand" even rather simple (for the human being) phrases and discourses, a computer system very often has to use the knowledge about the regularities of different levels of NL (morphological, syntactical, semantic) as well as the knowledge about thematic domains and the concrete situation of communication.

For instance, for making the decision about the referent of the pronoun "them", it may be necessary to apply common sense and/or to carry out a logical reasoning. Similar situations take place for the problem of reconstructing the meaning of an elliptical phrase (i.e., a phrase with some omitted words and word combinations) in the context of a discourse or a communicative situation.

That is why, while trying to formalize the understanding by a computer system even of rather simple NL-texts, the researchers quickly came to the conclusion that for solving their particular tasks, it is necessary to first find the theoretical decisions pertaining to arbitrary texts from a group of natural languages (for example, English, German, French). As a consequence, in the 1980s, in the scientific publications on NL processing even the metaphor "the theory bottleneck" emerged as a reflection of the considerable obstacles to be overcome for creating an adequate theory of understanding NL by computer systems.

Fortunately, several groups of the researchers from different countries (including the author of this monograph) proposed an idea that allowed finding a way out of the described deadlock situation. The essence of this idea is as follows. It is necessary to elaborate such formal languages for representing knowledge about the world and for building semantic representations (SRs) of NL-texts that these languages provide the possibility to construct SRs as the formal expressions reflecting many structural peculiarities of the considered NL-texts.

In other words, it is necessary to develop the formal languages (or formal systems, since the set of well-formed expressions given by the definition of a formal system is a formal language) for describing structured meanings of NL-texts with the expressive possibilities being rather close to the expressive possibilities of NL.

In this case it will be possible to carry out the semantic-syntactic analysis of an NL-text T from the considered sublanguage of NL in two stages reflected by the scheme

$$A\ NL-text\ T \Rightarrow Underspecified\ semantic\ representation\ of\ T$$

$$\Rightarrow Final\ (completely\ specified)\ semantic\ representation\ of\ T.$$

This scheme is to be interpreted in the following way. First, an intermediary semantic representation of the analyzed text T is to be constructed, it is called an underspecified semantic representation (USR) of T. Most often, this expression will reflect only partially the meaning of the considered text T. For instance, an USR of the input text T may indicate no referent of the pronouns "her" or "them" from the text T or may indicate no concrete meaning of the word "station" from T but only the set of all possible meanings of this word.

However, an USR of an NL-text T is an formal expression in contrast to the text T. That is why during the second stage of processing T it will be possible for eliminating an uncertainty to call and apply one of numerous specialized procedures being "experts" on concrete questions. Such procedures can be developed with the use of formal means of representing information, because the databases and knowledge bases of linguistic processors store the expressions of formal knowledge representation languages, and the USR to be analyzed and the final SR of the input text T are formal expressions.

It must be noticed that this idea was formulated for the first time in the author's works [52–56], and the new classes of formal languages of semantic representations with very high expressive power were defined in the mentioned works (see next sections).

Since the end of the 1980s, the idea of employing the formal systems of semantic representations with high expressive possibilities in the design of semantics-oriented LPs has been the central one for the development of the theory of understanding NL by computer-intelligent systems. The growth of the popularity of this idea was stimulated in the 1980s and 1990s by

- the series of publications on Episodic Logic (EL) [130–134, 181–184], the use of EL as the theoretical basis for the implementation of the project TRAINS, aimed at the formalization of problem-oriented dialogue in natural (English) language [4];
- the realization of the machine translation project Core Language Engine (CLE) in the Cambridge division (England) of the Stanford Research Institute [8, 9];
- the implementation in the 1980s–1990s of the project SnepS in USA [192, 193];
- the publications on the Theory of Conceptual Graphs [37, 195–197].

The general feature of the major part of the proposed approaches to constructing formal systems of semantic representations with high expressive power can be characterized as enhancing the expressive possibilities of first-order predicate logic by means of adding a number of new possibilities reflecting (on the formal level) some expressive mechanisms functioning in natural language.

Example 1. We often encounter in discourses the fragments "due to this event," "this caused," and the like. The referent of the word combination "this event" (first fragment) and of the pronoun "this" (second fragment) is a situation which took place at some moment in the past. NL allows for using in discourses such short designations of the situations in case a previous fragment of the considered discourse contains a complete description of this situation. It is a manifestation of one of the mechanisms of compactly coding information in NL.

However, the first-order predicate logic provides no possibility to associate an arbitrary formula F being a part of a formula H with a mark (being a variable or a constant) and then use only this compact mark instead of all other occurrences of F in H.

Episodic Logic overcomes this restriction of first-order predicate logic, and it is one of EL's distinguished features. Suppose, for instance, that $T1$ = "A predatory animal attacks a nearby creature only when it is hungry or feeling nasty." Then $T1$ may have the following semantic representation [130]:

$$(\forall x : [x((attr\ predatory)\ animal)]((\forall y : [y\ creature]$$

$$(\forall e1 : [[y\ near\ x] * *e1]$$

$$(\forall e2 : [e2\ during\ e1]$$

$$[[[x\ attack\ y] * *e2]$$

$$(\exists e3 : [e3\ same - time\ e2]$$

$$[[[x\ hungry] * *e3] \vee [[x\ feel - nasty] * *e3]])])))))).$$

In this formula of EL, the string ** designates the episodic operator; this operator connects a formula with the mark of the situation (or episode) it describes. The introduction of the episodic operator provides the possibility to model the mechanism of compactly encoding information in NL manifested in discourses due to the word combinations "this event," "this situation," "this caused," etc.

Example 2. Sentences and discourses in NL often contain compound designations of sets. For instance, commercial contracts may contain the expressions "5 containers," "a party of containers with bicycles," etc. The first-order predicate logic gives no convenient means for building formal analogues of the compound designations of sets.

In the project Core Language Engine (CLE), the formal expressions used for constructing underspecified semantic representations of the sentences are called quasilogical forms (QLF). In particular, the expression "the three firms" can be associated with the QLF (see [6])

$$q\_term(\langle t = quant, n = plur, l = all \rangle),$$

$$S,$$

$$[subset, S,$$

$$q\_term(\langle t = ref, p = def, l = the, n = number(3) \rangle),$$

$$X, [firm, X])]).$$

Thus, the language of quasilogical forms allows for building formal analogues of the compound natural language designations of sets.

Analysis shows that none of the approaches to the formalization of semantic structure of NL-texts mentioned above in this section is convenient for modeling (on the formal semantic level) *every* mechanism of encoding information in NL manifested in the structure of NL-expressions of the following kinds:

- texts with direct and indirect speech;
- texts containing compound designations of goals formed from the infinitives or gerunds with dependent words by means of the conjunctions "and," "or" and the particle "not" (such texts may express commands, advices, wishes, obligations, commitments);
- texts containing compound designations of notions;
- texts containing compound designations of sets;
- discourses with references to the meanings of phrases and larger parts of a discourse;
- texts with the word "a notion" ("This notion is used in chemistry and biology," etc.).

That is why it seems that the demands of practice, in particular, the demand to have formal means being convenient for building semantic annotations of arbitrary Web-documents with NL-components show that it is necessary to continue the studies aimed at the elaboration of formal systems of semantic representations with the expressive power being very close to the expressive power of natural language.

One of the possible broadly applicable approaches to this problem is the principal subject of Chaps. 2, 3, 4, 5, and 6.

Problems

1. What are the principal restrictions of first-order predicate logic from the standpoint of building mathematical models useful for the designers of semantics-oriented natural language processing systems?
2. Explain the term "the Cartesian product of the sets X and Y" without mathematical designations, continuing the phrase "The Cartesian product of the sets X and Y is the set consisting of."
3. What is the difference between the structure of the models of type 1 and type 2?
4. What proposed kinds of models could be of use for the design of NL-interfaces to (a) recommender systems, (b) autonomous intelligent robots?
5. What kinds of the proposed models could be of use for the design of machine translation systems?
6. How does the structure of the models of type 5 reflect the fact that the information for formulating an answer to a request posed by an end user of an intelligent full-text database can be accumulated step by step, as a result of analyzing not one but several informational sources?
7. Why can the partial models of types 1–5 contribute to increasing the quality of documentation of semantics-oriented NLPSs?
8. What are the main principles of Cognitive Linguistics concerning the study of natural language comprehension?
9. Explain the term "an underspecified semantic representation of a NL-text."
10. What is the purpose of introducing the episodic operator?

Chapter 2
Introduction to Integral Formal Semantics of Natural Language

Abstract This chapter sets forth the basic ideas and components of Integral Formal Semantics (IFS) of Natural Language – a many-component branch both of formal semantics of NL and Computer Science developed by the author of this book. Section 2.1 describes the basic principles of IFS and introduces the notion of a broadly applicable conceptual metagrammar. Section 2.2 shortly characterizes the principal components of IFS. Sections 2.3, 2.4, 2.5, 2.6, and 2.7 describe a number of the principal components of IFS. These sections contain numerous examples reflecting the different stages of elaborating powerful and flexible formal means for describing semantic structure of NL-texts – sentences and discourses.

2.1 The Basic Principles of Integral Formal Semantics of Natural Language

Integral Formal Semantics of Natural Language (IFS) is a many-component branch both of formal semantics of NL and of the theory of natural language processing systems as a part of Computer Science. It consists of several theories, mathematical models, and algorithms developed by the author of this monograph since the beginning of the 1980s.

2.1.1 Basic Principles

The basic principles of IFS stated below correspond very well to the requirements of Cognitive Linguistics and Computer Science concerning the formal study of the regularities of conveying information by means of NL. IFS proposes, first of all, a new class of formal systems for building semantic representations of sentences and discourses with high expressive power being close to the expressive power of NL.

V.A. Fomichov, *Semantics-Oriented Natural Language Processing*, IFSR International Series on Systems Science and Engineering 27, DOI 10.1007/978-0-387-72926-8_2, © Springer Science+Business Media, LLC 2010

The total content of the next chapters of this book can be considered as the kernel of the current configuration of IFS.

The basic principles of IFS are as follows:

1. The main goal of the researches on the formalization of NL-semantics is to be the construction of formal models of Natural Language Processing Systems (NLPSs) and of such subsystems of NLPSs which belong to the so-called semantic components of NLPSs. This means that the accent in the researches is to be on modeling the regularities of the communication of intelligent systems by means of NL.

2. The studies are to be oriented toward considering not only the assertions but also the commands, questions, and discourses which may be the inputs of NLPSs.

3. The basis of the studies is to be a formal model reflecting many peculiarities of semantic structures of sentences and discourses of arbitrary big length and providing a description of some class *Langsem* of formal languages being convenient for building semantic representations (SRs) of NL-texts in a broad spectrum of applications and on different levels of representation.

4. The central roles in the development of formal models for the design of NLPSs must play the models of the following correspondences:

   • "NL-text or its special representation (e.g., a marked-up representation of a text) + Knowledge ⇔ Semantic representation of a text" (both for the analyzers and generators of NL);

   • "An NL-text or its special (marked-up) representation + Knowledge → Semantic representation of a text + Plan of the reaction" for designing NL-interfaces to the recommender systems, expert systems, personal robots, etc. (the reactions may be questions, movements, calculations, etc.);

   • "Text of a request or its special (marked-up) representation + Text of an information source (or a semantic representation of the latter text) + Knowledge → A textual or semantic representation of retrieved information or Negative answer" for designing full-text databases and the systems which automatically form and update the knowledge bases of applied intelligent systems.

5. The model-theoretical semantics of NL is to play the auxiliary roles. The first to third sections of this chapter and the papers [58, 64, 65] contain the proposals concerning the formal structure of models of the listed kinds.

6. Semantics and pragmatics of NL should be studied jointly by means of the same formal techniques. It should be noted that this principle underlies the works [52, 53, 55, 56]. Hence this principle was formulated several years before the publication of the works [180, 189, 190].

7. A formal description of the surface structure of any NL-text T is to be based on a formal description of the structured meaning of T and on a formal description of the semantic – syntactic structure of T. Purely syntactic descriptions of texts' structures may be useful, but are not necessary. Such syntactic descriptions are to be the derivatives of the descriptions of semantic and semantic–syntactic

structures of NL-texts. This point of view is directly opposite to the approach used, in particular, in Montague Grammar and in Generalized Phrase Structure Grammars. However, it seems that the suggested viewpoint is (a) more similar to the processes realizing in the course of human thinking; (b) more practically effective; and (c) the only useful as concerns describing surface structure of scientific articles, books, etc. The stated principle may be considered as a possible formulation of one of the key ideas of Cognitive Linguistics. This idea set forth, in particular, in [20, 149, 152, 187] is the dependency of syntax on semantics.

8. The semantic interpretation of a phrase being a fragment of a published discourse is to depend on the knowledge about reality, on the source where the discourse is published, and on the meanings of precedent fragments of the discourse (in some cases – on the meanings of some next fragments too).

9. The semantic interpretation of an utterance in the course of a dialogue is to depend in general case on the knowledge about reality, about dialogue participants (in particular, about their goals), about the discussed situation, and about the meanings of previous utterances.

10. The languages from the class *Langsem* are to provide the possibility to represent knowledge about the reality and, in particular, to build formal descriptions of notions and regularities and also the descriptions of the goals of intelligent systems and of the destinations of things.

11. The languages from the class *Langsem* are to give the opportunity to represent the knowledge modules (blocks, chunks in other terms) as the units having some external characteristics (Authors, Date, Application domains, etc.) or metadata.

12. The languages from the class *Langsem* are to allow for building the models of structured hierarchical conceptual memory of applied intelligent systems, the frame-like representations of knowledge and are to be convenient for describing the interrelations of knowledge modules.

2.1.2 The Notion of a Broadly Applicable Conceptual Metagrammar

Let's call a formal model of the kind described above in the principle 3 a Broadly Applicable Metagrammar of Conceptual Structures or a Broadly Applicable Conceptual Metagrammar (BACM).

A Broadly Applicable Conceptual Metagrammar should enable us to build formal semantic analogues of sentences and discourses; hence the expressive power of formal languages determined by the model may be very close to the expressive power of NL (if we take into account the surface semantic structure of NL- texts). Besides, a BACM is to be convenient for describing various knowledge about the world [52, 54–56, 62, 63, 65, 67, 68].

If a model is convenient for describing arbitrary conceptual structures of NL-texts and for representing arbitrary knowledge about the world, we say about a Universal Metagrammar of Conceptual Structures or a Universal Conceptual Metagrammar (UCM) .

The reason to say about a metagrammar but not about a grammar is as follows: A grammar of conceptual structures is to be a formal model dealing with the elements directly corresponding to some basic conceptual items (like "physical object," "space location")

An example of such semi-formal grammar is provided by the known Conceptual Dependency theory of Schank. On the contrary, a metagrammar of conceptual structures is to postulate the existence of some classes of conceptual items, to associate in a formal way with arbitrary element from each class certain specific information, and to describe the rules to construct arbitrarily complicated structured conceptual items in a number of steps in accordance with such rules (proceeding from elementary conceptual items and specific information associated with arbitrary elements of considered classes of items).

The most part of the known approaches to the formalization of NL-semantics practically doesn't give the cues for the construction of an UCM. This applies, in particular, to Montague Grammar, Discourse Representation Theory (DRT), Theory of Generalized Quantifiers, Situation Theory, Dynamic Montague Grammar, Dynamic Predicate Logic, Theory of Conceptual Graphs, and Episodic Logic.

For instance, it is difficult to not agree with the opinion of Ahrenberg that "in spite of its name, DRT can basically be described as formal semantics for short sentence sequences rather than as a theory of discourse" [3]. This opinion seems to be true also with respect to the content of the monograph [143].

Happily, a considerable contribution to outlining the contours of a Universal Conceptual Metagrammar has been made by Integral Formal Semantics of NL.

2.2 The Components of Integral Formal Semantics of Natural Language

In order to list the principal components of IFS, we need the notion of a formal system, or a calculus. In discrete mathematics, the development and investigation of formal systems, or calculuses, is the main manner of studying the structure of strings belonging to formal languages.

Following [194], by *a formal system*, or *a calculus*, we'll mean any ordered triple

$$F = (L, L_0, R),$$

where L is a formal language in an alphabet, $L_0 \subset L$, R is a finite set of rules enabling us to obtain from the strings of L another strings of L. The rules from R are called *the inference rules*.

The strings of L that one can obtain as a result of applying the rules from R and starting with the strings from L_0 are called the formulas of the system F. We will be interested in what follows in such an interpretation of a calculus when formulas are considered not as theorems but as expressions of some language (in applications – as semantic representations (SRs) of texts and as parts of SRs).

It should be mentioned in this connection that for every context-free grammar generating a language L, one can easily define such calculus that the set of its formulas will be L.

The principal components of IFS are as follows:

1. The theory of S-calculuses and S-languages (the SCL-theory) developed in the first half of 1982 and proposed the new formal means for describing both separate sentences and complex discourses in NL of arbitrary big length (see Sect. 2.3).

2. A mathematical model of a correspondence between the NL-texts (sentences and discourses expressing the commands to a dynamic intelligent device or the commands to draw the geometric figures) and their semantic representations being the strings of restricted S-languages (see Sect. 2.4).

3. The theory of T-calculuses and T-languages (the TCL-theory) studying the semantic structure of discourses introducing a new notion or a new designation of an object (see Sect. 2.5).

4. The initial version of the theory of K-calculuses (knowledge calculuses) and K-languages (knowledge languages), or the KCL-theory, is a new step (in comparision with the SCL-theory) on the way of creating the formal means convenient for describing semantic structure of both sentences and complex discourses in NL (see Sect. 2.6).

5. The current version of the theory of K-calculuses and K-languages (its kernel is the theory of SK-languages – standard knowledge languages) set forth in [85, 91] and in Chaps. 2, 3, 4 and 5 of this book.

6. The analysis of the possibilities to use the theory of SK-languages for solving a number of significant problems of modern Computer Science and Web Science (see Chap. 6 of this monograph).

7. A broadly applicable mathematical model of a linguistic database, that is, a model of a collection of semantic-syntactic data associated with primary lexical units and used by the algorithms of semantic-syntactic analysis for building semantic representations of natural language texts (see Chap. 7 of this book).

8. A new method of transforming an NL-text (a statement, a command, or a question) into its semantic representation (see Chap. 8 of this book).

9. Two complex, strongly structured algorithms of semantic-syntactic analysis of NL-texts (they possess numerous common features). The first one is described in the book [85] and the second one is proposed in Chaps. 9 and 10 of this monograph.

10. The proposals concerning the structure of formal models being useful for the design of semantics-oriented NLPSs (Chap. 1 of this book and [64, 65]).

The components 5–10 of Integral Formal Semantics of Natural Language form the theory of K-representations (knowledge representations). The principal part of the theory of K-representations is set forth in this monograph.

2.3 The Theory of S-Calculuses and S-Languages

The theory of S-calculuses and S-languages (the SCL-theory) is set forth in the publications [52, 53, 55, 56] and in the Ph.D. dissertation [54]. This theory proposed already in 1981–1983 is a really ecological approach to the formalization of NL-semantics, providing powerful and convenient mathematical means for representing both structured meanings of NL-texts and knowledge about the reality.

The basic ideas of the SCL-theory were presented, in particular, at the First symposium of the International Federation of Automatic Control (IFAC) on Artificial Intelligence, which was held in 1983 in Sankt-Petersburg, Russia, and were published in the proceedings of this symposium; it should be noted that the paper [56], published by Pergamon Press, is a considerably abridged version of the publication [55].

The principal part of the SCL-theory is a new formal approach (new for the beginning of the 1980s) to describing conceptual (or semantic) structure of sentences and discourses in NL. The paper [52] for the first time in the world stated the task of developing mathematical models destined for describing structured meanings not only of sentences but also of complicated discourses in NL. Besides, this paper proposed the schemas of 16 partial operations on the finite sequences consisting of conceptual structures associated with NL-texts.

Example 1. Let T1 be the discourse "Sergey and Andrey are friends of Igor and are the physicists. He had told them that he didn't want to work as a programmer. Sergey believed that it would be useful for Igor to have a talk with the Associate Professor Somov and advised him to act in such a way."

In the paper [52], it was proposed to associate the discourse T1 with the following semantic representation $Semrepr1$:

$$((((((\{\downarrow man * Name(\Delta 1, Sergey) : x1,$$

$$\downarrow man * Name(\Delta 1, Andrey) : x2\} = M1)\wedge$$

$$Subset(M1, Friends(\downarrow man * Name(\Delta 1, Igor) : x3)))\wedge$$

$$(Profession((x1 \wedge x2)) = physicist))\wedge$$

$$((P1 = \neg Want[time - gramm, past](Subject, x3)(Goal,$$

$$Work(Qualification, programmer)(Institution,$$

$$\downarrow res - inst * Name((\Delta 1, Plastics Research Institute)))))\wedge$$

$$Say[time - gramm, past](Subject, x3)(Addressees, M1)$$

$(Proposition, P1) :: e1)) \wedge ((Believe[time - gramm, past](Subject, x1)$

$(Proposition, Useful[time - gramm, future](Person1, x3)$

$(Goal, Talk(Person2, \downarrow man * (Title((\Delta1, Assoc - Professor)$

$\wedge Surname((\Delta1, Somov))) : G1)) :: e2 \wedge Cause(el, e2)) \wedge$

$(Advise[time - gramm, past](Subject, x1)(Addressees, x3)$

$(Goal, G1) :: e3 \wedge Cause(e2, e3)))).$

Let's pay attention to the peculiarities of this formal expression being new at the time of publishing the paper [52]. We can find in this formal expression the following original features:

- the compound designations of the notions $man * Name(\Delta1, Sergey)$, $man * Name(\Delta1, Andrey)$, $man * Name(\Delta1, Igor)$;
- the compound designations of the concrete persons with the names $Sergey$, $Andrey$, $Igor$;
- a description of a set

$$(\{\downarrow man * Name(\Delta1, Sergey) : x1,$$

$$\downarrow man * Name(\Delta1, Andrey) : x2\} - M1);$$

- a formal representation of the meaning of a sentence with indirect speech;
- the substring $(x1 \wedge x2)$, where the logical connective $\wedge$ (conjunction, and) joins the designations of the persons $X1$ and $X2$ (but not the formulas representing propositions as in first-order logic);
- the substrings of the forms $Expression1 :: e1$, $Expression2 :: e2$, and $Expression3 :: e3$, where $Expression1$, $Expression2$, and $Expression3$ are the descriptions of some events, and $e1$, $e2$, $e3$ are the marks of these events;
- a compound designation of a goal to have a talk with the Associate Professor Somov

$$Talk(Person2, \downarrow man * (Title(\Delta1, Assoc - Professor)$$

$$\wedge Surname(\Delta1, Somov))) : G1;$$

- the compact representations of causal relationships $Cause(e1, e2)$ and $Cause(e2, e3)$, constructed due to the association of the marks $e1$, $e2$, $e3$ with the descriptions of concrete events in the left fragments of the semantic representation $Semrepr1$.

Example 2. Let T2 be the question "What did Igor say, and to whom did he tell it ?", Then, according to [52], the formula

$$?Transfer - information[time, past](Subject, \downarrow man * Name(\Delta1, Igor))$$

$$(Mode1, voice)(Addressees, ?y1)(Proposition, ?p1)$$

may be regarded as a possible semantic representation of T2.

The ideas of the papers [52, 53] received a mathematical embodiment in the Ph.D. dissertation [54]. This dissertation contains a complete mathematical model of a system consisting of 14 partial operations on the finite sequences consisting of conceptual structures associated with NL-texts.

Example 3. Let $Seq1$ be the sequence consisting of the informational units $\vee$, *airplane, helicopter, dirigible, glider, deltaplane*. Then one of these partial operations allows for constructing the formal expression

$$(airplane \vee helicopter \vee dirigible \vee glider \vee deltaplane),$$

considered as the value of this operation on the sequence $Seq1$.

The mathematical model constructed in the Ph.D. dissertation [54] defines the formal systems (or calculuses) of four new kinds (the S-calculuses of types 1–4) and, as a consequence, the formal languages of four new kinds (the restricted S-languages of types 1–4). The S-calculuses of types 1–3 and the restricted S-languages of types 1–3 were determined as preliminary results in order to achieve the final goal: the definition of the class of restricted S-languages of type 4.

Some denotations introduced in [54] are different from the denotations used in [52]. In particular, the expressions of the form $\{d_1, \ldots, d_n\}$, used in [52] for denoting the sets consisting of the objects $d_1, \ldots, d_n$, are not employed in [54].

For instance, the expression

$$\{\downarrow man * Name(\Delta 1, Sergey) : x1,$$

$$\downarrow man * Name(\Delta 1, Andrey) : x2\},$$

being a substring of the string $Semrepr1$ in the Example 1 is to be replaced by the string

$$(\downarrow group * Elements(\Delta 2, (\downarrow man * Name(\Delta 1, Sergey) : \{x1\},$$

$$\wedge \downarrow man * Name(\Delta 1, Andrey) : \{x2\}))$$

Let's illustrate some expressive possibilities of restricted S-languages of type 4 defined in [54].

Example 4. Let T3 be the discourse "Peter said that he had studied both in the Moscow Institute of Civil Engineering (MICE) and in the Moscow Institute of Electronic Engineering (MIEE). It was new for Somov that Peter had studied in the Moscow Institute of Electronic Engineering." The discourse T4 is associated in the Ph.D. dissertation [54] with the following semantic representation $Semrepr2$:

$$(Say[time - gramm, past](Subject, \downarrow person(Name, Peter) : \{x1\})$$

$$(Proposition1, Study1[time - gramm, past](Subject, x1)$$

$$(Learn - institution, (\downarrow techn - univer(Title, MICE) : \{x2\}\wedge$$

$$(\downarrow techn - univer(Title, MIEE) : \{x3\})) : P1)$$

$$\wedge(New(\downarrow person(Surname, Somov) : \{x4\}, P1)$$

$$\equiv Study1[time - gramm, past](Subject, x1)(Learn - institution,$$

$$(\downarrow techn - univer(Title, MICE) : \{x2\}))).$$

The analysis of this formal expression enables us to notice that the distinguished features of the proposed approach to modeling communication in NL are the possibilities listed below:

- to build (on the semantic level) the formal analogues of the phrases with indirect speech;
- to construct the compound designations of the notions and, as a consequence, the compound designations of concrete objects;
- to associate the marks with the compound descriptions of the objects (the substrings : $\{x1\}$, : $\{x2\}$, : $\{x3\}$, : $\{x4\}$);
- to associate the marks with the semantic representations of the phrases and larger fragments of a discourse (the indicator of an association of the kind in the string *Semrepr2* is the substring : *P1*);
- to build the semantic representations of the discourses with the references to the meanings of phrases and larger fragments of the considered discourse.

The class of restricted S-languages of type 4 introduced in [54] allows also for building an improved SR of the discourse T3 which the following formula *Semrepr3* :

$$(Say[time - gramm, past](Subject, \downarrow person(Name, Peter) : \{x1\})$$

$$(Time, t1)(Proposition1, (Study1[time - gramm, past](Subject, x1)(Time, t2)$$

$$(Learning - institution, (\downarrow techn - univer(Title, MICE) : \{x2\})$$

$$\wedge Study1[time - gramm, past](Subject, x1))(Time, t3)$$

$$(Learn - institution, \downarrow techn - univer(Title,$$

$$MIEE) : \{x3\})) : P1)$$

$$\wedge(New(\downarrow person(Surname, Somov : \{x4\}, P1)$$

$$\equiv Study1[time - gramm, past](Subject, x1)(Learn - institution,$$

$$x2))(Time, t2) \wedge Precedes1(t2, t1) \wedge Precedes1(t3, t1)$$

$$\wedge Precedes1(t1, current - moment)).$$

In comparison with SR *Semrepr2*, the representation *Semrepr3* is more exact, because it introduces the mark $t1$ for the short time interval of speaking by Peter, the marks $t2$ and $t3$ for time intervals when Peter had studied in the first and second university, respectively, and shows that $t2$ and $t3$ precede $t1$, and $t1$ precedes the current moment.

Example 5. Let T4 = "Somebody didn't turn off a knife-switch. This caused a fire." Then the string *Semrepr4* of a restricted S-language of type 4

$$(\neg Switch - off[time - gramm, past](Subject, \downarrow person)(Object1,$$

$$\downarrow knife - switch) :: \{e1\} \wedge Cause(e1, \downarrow fire : \{e2\}))$$

may be interpreted as a semantic representation of T4 [54]. In this string, the substrings *e1, e2* denote the events, and the symbol :: is used for associating events with semantic representations of assertions. The formula

$$Switch - off[time - gramm, past](Subject, \downarrow person)$$

$$(Object1, \downarrow knife - switch)$$

is built from the components *Switch − off[time, past]* (called a predicator element in the SCL-theory), *Subject, ↓ person, Object1, ↓ knife − switch* by means of applying to these components exactly one time one of the inference rules introduced in [54]. The items *Subject* and *Object1* are the designations of thematic roles (or conceptual cases, or semantic cases, or deep cases).

The papers [55, 56] contain the detailed proposals aimed at making more compact the complicated structure of the mathematical model constructed in [54]. It must be noted that these proposals modify a little the structure of formulas built in accordance with some rules of constructing semantic representations of NL-texts.

Example 6. In the paper [55], the discourse T4 = "Somebody didn't turn off a knife-switch. This caused a fire" is associated with the following semantic representation *Semrepr5* :

$$(\neg Switch - off[time - gramm, past](Subject, \downarrow person)$$

$$Object1, \downarrow knife - switch) :: e1 \wedge Cause(e1, \downarrow fire : e2)).$$

We can see that, in accordance with the proposals from [55, 56], we use as the marks of subformulas describing the events not the strings $\{e1\}$ and $\{e2\}$ but the strings *e1* and *e2*. In the semantic representations constructed in the Examples 1, 4, and 6, the symbol :: is used for associating the marks of events with the semantic images of statements describing these events.

It is easy to see that the symbol :: is used in the SCL-theory with the same purpose as the episodic operator ** in Episodic Logic [130–132, 183]. However, it was done in the year 1982, i.e., 7 years before the publication of the paper [183], where the episodic operator ** was introduced.

Example 7. Let T5 be the discourse "Victor said that he had lived in Kiev and Moscow. It was new for Rita that Victor had lived in Kiev." Then we can associate with T5 a semantic representation *Semrepr6* [55] being the formula

$$(Say[time - gramm, past](Subject, \downarrow person(Name, Victor) : x1)$$

$$(Proposition, Live[time - gramm, past](Subject, x1)(Location,$$

$$(Kiev \wedge Moscow)) \; : \; P1) \wedge (New(\downarrow person(Name, Rita) \; : \; x2, P1)$$

$$\equiv Live[time - gramm, past](Subject, \downarrow person(Name, Victor) \; : \; x1)$$

$$(Location, Kiev))).$$

The string *Semrepr6* contains the substrings : $x1$, : $x2$ but not the substrings : $\{x1\}$ and : $\{x2\}$ which could be expected by us as a consequence of our acquaintance with the semantic representation *Semrepr2* in Example 4.

Example 8. Let T6 = "How many students are there in the Lomonosov Moscow State University?". Then the formula

$$??(Number1(all person * Study1[time - gramm, present](Subject, \Delta 1)$$

$$(Learning - institution, Lomonosov - Moscow - State - Univ) = ?x1)$$

may be considered as a semantic representation of the question T6 [55].

It must be added that the proposals formulated in [55, 56] concern not only 14 partial operations of building semantic representations of NL-texts stated in [54] but also two more operations schematically outlined in [53].

2.4 A Model of a Correspondence Between NL-Texts and Their Semantic Representations

The next component of Integral Formal Semantics of NL is a mathematical model elaborated in [54] and describing a correspondence between the NL-texts (sentences and discourses) and their semantic representations being the strings of restricted S-languages of type 4. This model proposes a unified description of at least two different sublanguages of NL. The first one is a collection of the texts in Russian, English, German, and some other languages expressing the commands to fulfill certain actions.

For instance, the first sublanguage contains the command C1 = "Turn to the left. The radius – 3 m." The expression C1 can be interpreted as a command to a radio-controlled model of ship. The second sublanguage contains, for example, the command C2 = "Draw two circles. The centers are the points (9, 14) and (12, 23). The diameters – 8 and 12 cm."

The general feature of these sublanguages is that they contain the commands to create one or several entities of a particular kind. The model contains the parameter *entity − sort*, its value is a semantic unit (called a sort) qualifying the class of entities to be created in accordance with the input sequence of commands.

If *entity − sort = event*, the model describes the commands to fulfill certain actions. If *entity − sort = geom − object*, the model describes the commands to draw certain geometrical figures on the plane.

The correspondence between NL-texts and their semantic representations determined by the model is based on the following central idea: the command "Turn to the left" is replaced by the statement "It is necessary to turn to the left," the command "Draw two circles" is replaced by the statement "It is necessary to draw two circles," and so on. This approach provides the possibility to construct a semantic representation of an input discourse as a conjunction of the semantic representations of the statements.

Example 1. Let C3 = "Turn to the left and give the light signal. The radius of the turn is 3 m." Then, following [54], we can associate C3 with the following semantic representation:

$$(necessary[time - gramm, present](Goal, (turn(Orientation, left) :: \{e1\})$$

$$\land produce1(Result1, \downarrow signal\,(Kind - signal, light) : \{e2\})))$$

$$\land(Radius(e1) \equiv 3/m)).$$

Example 2 [54]. Consider C4 = "Draw two circles. Diameters – 8 cm and 12 cm." Then the following string is a possible semantic representation of C4:

$$(necessary[time - gramm, present]\,(Goal, draw(Object - geom,$$

$$\downarrow circle : \{x1, x2\})) \land (Diameter((x1 \land x2)) \equiv (8/cm \land 12/cm))).$$

The constructed mathematical model was later used as the theoretical basis for designing the software of a prototype of an NL-interface to a computer training complex destined for acquiring (by ship captains and their deputies) the skills necessary for preventing the collisions of ships [58, 59, 61].

2.5 The Theory of T-Calculuses and T-Languages

The theory of T-calculuses and T-languages (the TCL-theory) is an expansion of the theory of S-calculuses and S-languages (the SCL-theory). The outlines of the TCL-theory can be found in [55, 56]. This part of IFS studies in a formal way the semantic structure of the discourses defining a new notion or introducing a new designation of an object and, as a consequence, playing the role of an order to an intelligent system to include a new designation of a notion or of an object into the inner conceptual system. T-languages allow for describing semantic structure of sentences and discourses.

Example 1 [55]. Let T1 = "A tanker is a vessel for carrying liquid freights." Then there is a T-language containing the string *Semrepr*1 of the form

$$(tanker \Longleftarrow\uparrow transp) \land (tanker \equiv vessel*$$

$$Destination((\Delta2, Carry1(Objects, diverse\,freight * Kind(\Delta1, liquid)))))$$

being a possible semantic representation of the definition T1.

Let's presume that an applied intelligent system can semantically analyze the strings of T-languages. Then the substrings with the symbols $\Longleftarrow$ or $\longleftarrow$ are to be interpreted as the commands to up date the considered knowledge base (KB). In particular, the substring $(tanker \Longleftarrow\uparrow transp)$ is to signify that an intelligent system should include in its knowledge base the notion's designation *tanker*, qualifying a transport means.

Example 2. Consider the text T2 = "Let M be the intersection of lines AH and BE, P be the intersection of lines CD and FK. Then it is necessary to prove that the line MP is a tangent to the circle with the center N and the radius 12 mm."

One can define such a T-language L_t that L_t will include the string *Semrepr2* of the form

$$(M \longleftarrow geom - ob) \wedge (P \longleftarrow geom - ob) \wedge Intersection(AH, BE, M) \wedge$$

$$Intersection(CD, FK, P) \wedge Necessary(Prove(Proposition,$$

$$Tangent(MP, \downarrow circle(Center, N)(Radius, 12/mm)))).$$

In the string *Semrepr2*, the substrings $(M \longleftarrow geom - ob)$ and $(P \longleftarrow geom - ob)$ indicate that an intelligent system will include in its knowledge base the constants *M* and *P*, denoting some geometrical objects.

The rules allowing us to construct the formulas *Semrepr1* and *Semrepr2* are explained in [55]. Thus, the theory of S-calculuses and S-languages and the theory of T-calculuses and T-languages provided already in 1983 a broadly applicable variant of discourses' dynamic semantics.

The examples considered above show that the expressive power of S-languages and T-languages is very high and essentially exceeds, in particular, the expressive power of Discourse Representation Theory.

2.6 The Initial Version of the Theory of K-Calculuses and K-Languages

The SCL-theory and the TCL-theory became the starting point for developing the theory of K-calculuses, algebraic systems of conceptual syntax, and K-languages (the KCL-theory) that are nowadays the central component of IFS. The first variant of this theory elaborated in 1985 is used in [58] and is discussed in [60, 61].

The second variant is set forth in the textbook [62] and in [63, 65, 67, 68] (see also the bibliography in [65]). We'll discuss below the second variant of the KCL-theory. The basic model of the KCL-theory describes a discovered collection consisting of 14 partial operations on the conceptual structures associated with NL-texts and destined for building semantic representations of sentences and discourses.

The KCL-theory provides much more powerful formal means for describing the sets and n-tuples, where $(n > 1)$, than the SCL-theory. It should be noted that the KCL-theory allows for regarding the sets containing the sets and the n-tuples with components being sets. This enables us to consider the relationships between the sets, untraditional functions with arguments and/or values being sets, etc.

The KCL-theory gives the definition of a class of formulas providing the possibility to (a) describe structured meanings of complicated sentences and discourses and (b) build the representations of diverse cognitive structures.

Example. Let D1 be the discourse "The chemical action of a current consists in the following: for some solutions of acids (salts, alkalis), by passing an electrical current across such a solution one can observe isolation of the substances contained in the solution and laying aside these substances on electrodes plunged into this solution. For example, by passing a current across a solution of blue vitriol ($CUSO4$) pure copper will be isolated on the negatively charged electrode. One uses this to obtain pure metals" [65].

Then D1 may have a semantic representation $Semreprdisc1$ of the form

$$(Description(action * (Kind, chemical), current1,$$

$$\exists x1 \, (solution1 * (Subst, (acid \lor salt \lor alkali)))$$

$$If - then(Pass(\langle Agent1, . : current1 : y1 \rangle,$$

$$\langle Envir, x1 \rangle), Observe((. : isolation2 * (Agent2,$$

$$diverse \, substance * Contain(x1, \#) : z1) \land$$

$$. : laying - aside * (Agent, z1)(Loc1,$$

$$certain \, electrode * (Plunge, x1)))) : P1 \land$$

$$Example(P1, If - then(Pass(\langle Agent1, . : current1 : y2 \rangle,$$

$$\langle Envir, . : solution1 * (Subst,$$

$$blue - vitriol * (Formula, CuSO4)))),$$

$$Isolate2(\langle Agent2, . : matter1 * (Is, copper * (Kind, pure)) \rangle,$$

$$\langle Loc1, . : electrode * (Charge, neg) \rangle)))) \land$$

$$Use(. : phenomenon * (Charact, P1),$$

$$Obtain(*, diverse \, metal * (Kind, pure)))).$$

Here the referential structure of D1 is reflected with the help of variables $x1$, $y1$, $y2$, $z1$, $P1$; the symbol . : is interpreted as the referential quantifier, i.e., as the informational unit corresponding to the word *certain*.

The text D1 is taken from the textbook on physics destined for the pupils of the eighth class in Russia (the initial class – 6-year-old children – has the number 1, the last class – the number 11). This textbook was written by A. Pyoryshkin and N. Rodina and published in Moscow in 1989. This information is reflected by the K-string $Semreprdisc2$ of the form

$$. : text * (Content, Semreprdisc1)(Source, . : text - book *$$

$$(Educ - inst, any\, school * (Country, Russia)(Grade1, 8))$$

$$(Area, physics)(City, Moscow)(Year, 1989))$$

$$(Authors, (A.Pyoryshkin \wedge N.Rodina)) : inf218,$$

where the string $inf218$ is a mark of a concrete informational object. So we see that K-languages allow for building the formulas reflecting both the content of an informational object and its metadata – the data about the informational object as a whole.

Numerous examples of K-strings are adduced in [62, 65, 67, 68]. Hence the expressive power both of standard K-languages and of S-languages of type 5 considerably exceeds the expressive possibilities of other approaches to the formalization of NL-semantics discussed above.

In [65], some opportunities of recording NL-communication by means of standard K-languages are explained. That is, it is shown how it is possible to represent in a formal manner the actions carried out by intelligent systems in the course of communication.

The paper [65] also shows how to use standard K-languages for describing semantic-syntactic information associated with words and fixed word combinations.

2.7 The Theory of K-Representations as the Kernel of the Current Version of Integral Formal Semantics

The theory of K-representations is an expansion of the theory of K-calculuses and K-languages (the KCL-theory). The basic ideas and results of the KCL-theory are reflected in numerous publications in both Russian and English, in particular, in [65–100].

The first basic constituent of the theory of K-representations is the theory of SK-languages (standard knowledge languages), stated, in particular, in [70–94]. The kernel of the theory of SK-languages is a mathematical model describing a system of such 10 partial operations on structured meanings (SMs) of natural language texts (NL-texts) that, using primitive conceptual items as "blocks," we are able to build SMs of arbitrary NL-texts (including articles, textbooks) and arbitrary pieces of knowledge about the world. The outlines of this model can be found in two papers published by Springer in the series "Lecture Notes in Computer Science" [83, 86].

A preliminary version of the theory of SK-languages – the theory of restricted K-calculuses and K-languages (the RKCL-theory) – was set forth in [70].

The analysis of the scientific literature on artificial intelligence theory, mathematical and computational linguistics shows that today the class of SK-languages opens the broadest prospects for building semantic representations (SRs) of NL-texts (i.e., for representing structured meanings of NL-texts in a formal way).

The expressions of SK-languages will be called below the K-strings. If T is an expression in natural language (NL) and a K-string E can be interpreted as an SR of T, then E will be called a K-representation (KR) of the expression T.

The second basic constituent of the theory of K-representations is a broadly applicable mathematical model of a linguistic database (LDB). The model describes the frames expressing the necessary conditions of the existence of semantic relations, in particular, in the word combinations of the following kinds: "Verbal form (verb, participle, gerund) + Preposition + Noun," "Verbal form + Noun," "Noun1 + Preposition + Noun2," "Noun1 + Noun2," "Number designation + Noun," "Attribute + Noun," "Interrogative word + Verb."

The third basic constituent of the theory of K-representations is formed by two complex, strongly structured algorithms carrying out semantic-syntactic analysis of texts from some practically interesting sublanguages of NL. The first algorithm is described in Chapters. 8 and 9 of the book [85]. The second algorithm being a modification of the first one is set forth in Chapters 9 and 10 of this monograph. Both algorithms are based on the elaborated formal model of an LDB.

The other components of the theory of K-representations are briefly characterized in Sect. 2.2.

Problems

1. What are the components of Integral Formal Semantics of Natural Language?
2. What is a formal system or a calculus?
3. What are the new features of the theory of S-calculuses and S-languages (the SCL-theory) in comparison with the first-order predicate logic?
4. Discover the new ways of using the logical connectives $\land$ and $\lor$ in the SCL-theory in comparison with the first-order predicate logic.
5. What are the main ideas of building compound representations of notions (concepts) in the SCL-theory?
6. What is the purpose of using the symbols ":" and "::" in the formulas of the SCL-theory?
7. What is common for the SCL-theory and Episodic Logic?
8. What is the purpose of using the symbols $\leftarrow$ and $\Leftarrow$ in the formulas of the theory of T-calculuses and T-languages?

Chapter 3
A Mathematical Model for Describing a System of Primary Units of Conceptual Level Used by Applied Intelligent Systems

Abstract The first section of this chapter formulates a problem to be solved both in Chaps 3 and 4: it is the problem of describing in a mathematical way the structured meanings of a broad spectrum both of sentences and discourses in natural language. The second section states a subproblem of this problem, it is the task of constructing a mathematical model describing a system of primary units of conceptual level and the information associated with such units and needed for joining the primary units with the aim of building semantic representations of arbitrarily complicated Natural Language texts. A solution to this task forms the main content of this chapter. From the mathematical standpoint, the proposed solution is a definition of a new class of formal objects called conceptual bases.

3.1 Global Task Statement

Let's formulate a problem to be solved in Chapters 3 and 4.

In the situation when the known formal methods of studying the semantics of natural language proved to be ineffective as regards solving many significant tasks of designing NLPSs, a number of researchers in diverse countries have pointed out the necessity to search for new mathematical ways of modeling NL-communication.

Habel [113] noted the importance of creating adequate mathematical foundations of computational linguistics and underlined the necessity to model the processes of NL-communication on the basis of formal methods and theories of cognitive science.

Fenstad and Lonning ([44], p. 70) posed the task of working out adequate formal methods for Computational Semantics – "a field of study which lies at the intersection of three disciplines: linguistics, logic, and computer science." Such methods should enable us, in particular, to establish the interrelations between pictorial data and semantic content of a document.

A.P. Ershov, the prominent Russian theoretician of programming, raised in [38] the problem of developing a formal model of the Russian language. It is very

V.A. Fomichov, *Semantics-Oriented Natural Language Processing*, IFSR International Series on Systems Science and Engineering 27, DOI 10.1007/978-0-387-72926-8_3,
© Springer Science+Business Media, LLC 2010

interesting how close the ideas of Seuren [188] about the need of ecological approaches to the formal study of NL are to the following words of A. P. Ershov published in the same 1986: "We want as deeply as it is possible to get to know the nature of language and, in particular, of Russian language. A model of Russian language should become one of manifestations of this knowledge. It is to be a formal system which should be adequate and equal-voluminous to the living organism of language, but in the same time it should be anatomically prepared, decomposed, accessible for the observation, study, and modification" ([38], p. 12).

Having analyzed the state of the researches on formalizing semantics of NL, Peregrin [168] drew the conclusion that the existing logical systems didn't allow for formalizing all the aspects of NL-semantics being important for the design of NLPSs. He wrote that we couldn't use the existing form of logic as such molding form that it is necessary to squeeze natural language into this form at any cost. That is why, according to this scholar, in order to create an adequate formal theory of NL-semantics, it is necessary to carry out a full-fledged linguistic analysis of all components of NL and to establish the connections between the logical approaches to the formalization of NL-semantics and linguistic models of meaning.

In essence, the same conclusion but considerably earlier, in the beginning of the 1980s (see [52, 55, 56]), was drawn by the author of this monograph. This conclusion became the starting point for the elaboration of the task statement below.

It seems that the borders of mathematical logic are too narrow for providing an adequate framework for computer-oriented formalization of NL-semantics. That is why the problem of creating mathematical foundations of designing intellectually powerful NLPSs requires not only an expansion of the first-order logic but rather the development of new mathematical systems being compatible with first-order logic and allowing for formalizing the logic of employing NL by computer-intelligent systems.

Taking this into account, let's pose the task of developing a mathematical model of a new kind for describing structured meanings of NL sentences and complicated discourses. Such a model is to satisfy two groups of requirements: the first group consists of several very general requirements to *the form* of the model, and the second group consists of numerous requirements concerning the reflection (on semantic level) of concrete phenomena manifested in NL.

The first group of requirements to a model to be constructed is as follows:

1. A model is to define a new class of formal objects to be called *conceptual bases* and destined for (a) explicitly indicating the primary units of conceptual level used by an applied intelligent system; (b) describing in a formal way the information associated with primary units of conceptual level and employed for joining primary and compound conceptual units into complex structures interpreted as semantic representations of NL-texts.

2. For each conceptual basis B, a model is to determine a formal language (in other words, a set of formulas) $Ls(B)$ in such a way that the class of formal languages $\{ Ls(B) \mid B$ is a conceptual basis $\}$ is convenient for building semantic representations both of separate sentences and complicated discourses in NL.

3. Let *Semunits*(*B*) be the set of primary units of conceptual level defined by the conceptual basis *B*. Then a model is to determine the set of formulas *Ls*(*B*) simultaneously with a finite collection of the rules $R_1, \ldots, R_k$, where $k > 1$, allowing for constructing semantic representations both of sentences and complicated discourses step by step from the elements of the set *Semunits*(*B*) and several special symbols.

For elaborating *the second group of requirements* to a model, a systemic analysis of the structure of natural language expressions and expressions of some artificial languages has been carried out; it has been aimed at distinguishing lexical and structured peculiarities of (a) the texts in Russian, English, German, and French languages; (b) a number of artificial languages used for constructing semantic representations of NL-texts by linguistic processors; (c) the expressions of artificial knowledge representation languages, in particular, of terminological (or KL-ONE-like) knowledge representation languages.

There are some important aspects of formalizing NL-semantics which were underestimated or ignored until recently by the dominant part of researchers. First, this applies to formal investigation of the structured meanings of (a) narrative texts including descriptions of sets; (b) the discourses with references to the meaning of sentences and larger fragments of texts; (c) the phrases where logical connectives "and," "or" are used in non-traditional ways and join not the fragments expressing assertions but the descriptions of objects, sets, concepts; (d) phrases with attributive clauses; (e) phrases with the lexical units "a concept," "a notion."

Besides, the major part of the most popular approaches to the mathematical study of NL-semantics (mentioned in Chap. 2) practically doesn't take into account the role of knowledge about the world in NL comprehension and generation and hence does not study the problem of formal describing knowledge fragments (definitions of concepts, etc.).

It should be added that NL-texts have authors, may be published in one or another source, may be inputted from one or another terminal, etc. The information about these external ties of a text (or, in other terms, about its metadata) may be important for its conceptual interpretation. That is why it is expedient to consider a text as a structured item having a surface structure *T*, a set of meanings *Senses* (most often, *Senses* consists of one meaning) corresponding to *T*, and some values $V_1, \ldots, V_n$ denoting the author (authors) of *T*, the date of writing (or of pronouncing) *T*, indicating the new information in *T*, etc. The main popular approaches to the mathematical study of NL-semantics provide no formal means to represent texts as structured items of the kind.

We'll proceed from the hypothesis that there is only one mental level for representing meanings of NL-expressions (it may be called the conceptual level) but not the semantic and the conceptual levels. This hypothesis is advocated by a number of scientists, in particular, by Meyer [157].

Let's demand that the formal means of our model allow us:

1. To build the designations of structured meanings (SMs) of both phrases expressing assertions and of narrative texts; such designations are called usually *semantic representations (SRs)* of NL-expressions.

2. To build and to distinguish the designations of items corresponding to (a) objects, situations, processes of the real world and (b) the notions (or concepts) qualifying these objects, situations, processes.

3. To build and to distinguish the designations of (a) objects and sets of objects, (b) concepts and sets of concepts, (c) SRs of texts and sets of SRs of texts.

4. To distinguish in a formal manner the concepts qualifying the objects and the concepts qualifying the sets of objects of the same kind.

5. To build compound representations of the notions (concepts), e.g., to construct formulas reflecting the surface semantic structure of NL-expressions such as "a person graduated from the Stanford University and being a biologist or a chemist".

6. To construct the explanations of more general concepts by means of less general concepts; in particular, to build the strings of the form $(a \equiv Des(b))$, where a designates a concept to be explained and $Des(b)$ designates a description of a certain concretization of the known concept b.

7. To build the designations of ordered n-tuples of objects $(n > 1)$.

8. To construct: (a) formal analogues of complicated designations of the sets such as the expression "this group consisting of 12 tourists being biologists or chemists," (b) the designations of the sets of n-tuples $(n > 1)$, (c) the designations of the sets consisting of sets, etc.

9. To describe set-theoretical relationships.

10. To build the designations of SMs of phrases containing, in particular: (a) the words and word combinations "arbitrary," "every," "a certain," "some," "all," "many," etc.; (b) the expressions formed by means of applying the connectives "and", "or" to the designations of things, events, concepts, sets; (c) the expressions where the connective "not" is located just before a designation of a thing, event, etc.; (d) indirect speech; (e) the participle constructions and the attributive clauses; (f) the word combinations "a concept", "a notion".

11. To build the designations of SMs of discourses with references to the mentioned objects.

12. To explicitly indicate in SRs of discourses causal and time relationships between described situations (events).

13. To describe SMs of discourses with references to the meanings of phrases and larger fragments of a considered text.

14. To express the assertions about the identity of two entities.

15. To build formal analogues of formulas of the first-order logic with the existential and/or universal quantifiers.

16. To consider nontraditional functions (and other nontraditional relations of the kind) with arguments and/or values being: (a) sets of things, situations (events); (b) sets of concepts; (c) sets of SRs of texts.

17. To build conceptual representations of texts as informational objects reflecting not only the meaning but also the values of external characteristics of a text: the author (authors), the date, the application domains of the stated results, etc.

This task statement develops the task statements from [52, 54–56, 62, 65, 70] and coincides with the task statements in [81, 82, 85].

3.2 Local Task Statement

The analysis shows that the first step in creating a broadly applicable and domain-independent mathematical approach to represent structured meanings of NL-texts is the development of a formal model enumerating primary (i.e. not compound) units of conceptual level employed by an applied intelligent system and, besides, describing the information associated with such units and needed for combining these units into the compound units reflecting structured meanings of arbitrarily complicated NL-texts.

For constructing a formal model possessing the indicated property, first an analysis of the lexical units from Russian, English, German, and French languages was fulfilled. Second, the collections of primary informational units used in artificial knowledge representation languages were studied, in particular, the collections of units used in terminological (or KL-ONE-like) knowledge representation languages.

Proceeding from the fulfilled analysis, let's state the task of developing such a domain-independent mathematical model for describing (a) a system of primary units of conceptual level employed by an NLPS and (b) the information of semantic character associated with these units that, first, this model constructively takes into account the existence of the following phenomena of natural language:

1. A hierarchy of notions is defined in the set of all notions; the top elements in this hierarchy are most general notions. For instance, the notion "a physical object" is a concretization of the notion "a space object".
2. Very often, the same thing can be qualified with the help of several notions, where none of these notions is a particular case (a concretization) of another notion from this collection. It is possible to metaphorically say that such notions are "the coordinates of a thing" on different "semantic axes." For example, every person is a physical object being able to move in the space. On the other hand, every person is an intelligent system, because people can solve problems, read, compose verses, etc.
3. The English language contains such words and word groups as "a certain", "definite", "every", "each", "any", "arbitrary", "all", "some", "a few", "almost all", "the majority" and some other words and word combinations that these words and word groups are always combined in sentences with the words and word combinations designating the notions. For instance, we can construct the expressions "every person", "a certain person", "any car", "arbitrary car", "all people", "several books", etc. The Russian, German, and French languages contain similar words and word combinations.

Second, the model is to allow for distinguishing in a formal way the designations of the primary units of conceptual level corresponding to

- the objects, situations, processes of the real world and the notions (the concepts) qualifying these objects, situations, processes;
- the objects and the sets of objects;
- the notions qualifying objects and the notions qualifying the sets of objects of the same kind ("a ship" and "a squadron", etc.);

- the sets and the finite sequences (or the ordered n-tuples, where $n > 1$) of various entities.

Third, the model is to take into account that the set of primary units of conceptual level includes

- the units corresponding to the logical connectives "not", "and", "or" and to the logical universal and existential quantifiers;
- the names of nontraditional functions with the arguments and/or values being (a) the sets of things, situations; (b) notions; (c) the sets of notions; (d) semantic representations of NL-texts; (e) the sets of semantic representations of NL-texts;
- the unit corresponding to *the words* "a notion", "a concept" and being different from the conceptual unit "a concept"; the former unit contributes, in particular, to forming the meaning of the expression "an important notion used in physics, chemistry, and biology."

A mathematical model for describing a system of primary units of conceptual level is constructed in Sect. 3.3, 3.4, 3.5, 3.6, 3.7, 3.8, and 3.9.

3.3 Basic Denotations and Auxiliary Definitions

3.3.1 General Mathematical Denotations

$x \in Y$ the element x belongs to the set Y

$x \neg \in Y$ the element x doesn't belong to the set Y

$X \subset Y$ the set X is a subset of the set Y

$Y \cup Z$ the union of the sets Y and Z

$Y \cap Z$ the intersection of the sets Y and Z

$Y \setminus Z$ the set-theoretical difference of the sets Y and Z, that is the collection of all such x from the set Y that x doesn't belong to the set Z

$Z_1 \times \ldots \times Z_n$ the Cartesian product of the sets $Z_1, \ldots, Z_n$, where $n > 1$

$\emptyset$ empty set

$\exists$ existential quantifier

$\forall$ universal quantifier

$\Rightarrow$ implies

$\Leftrightarrow$ if and only if

3.3.2 The Preliminary Definitions from the Theory of Formal Grammars and Languages

Definition 3.1. An arbitrary finite set of symbols is called *alphabet*. If A is an arbitrary alphabet, then A^+ is the set of all sequences $d_1, \ldots, d_n$, where $n \geq 1$, for $i = 1, \ldots, n, d_i \in A$.

Usually one writes $d_1 \ldots d_n$ instead of $d_1, \ldots, d_n$. For example, if $A = \{0,1\}$, then the sequences of symbols 011, 11011, 0, 1 $\in A^+$.

Definition 3.2. If A is an arbitrary alphabet, the elements of the set A^+ are called the *non-empty strings in the alphabet A* (or over the alphabet A).

Definition 3.3. Let A be an arbitrary alphabet, d be a symbol from A then $d^1 = d$; for $n > 1$, $d^n = d \ldots d$ (n times).

Definition 3.4. Let $A^* = A^+ \cup \{e\}$, where A is an arbitrary alphabet, e is the empty string. Then the elements of the set A^* are called *the strings in the alphabet A* (or *over the alphabet A*).

Definition 3.5. For each $t \in A^*$, where A is an arbitrary alphabet, the value of the function $length(t)$ is defined as follows: (1) $length(e) = 0$; if $t = d_1 \ldots d_n$, $n \geq 1$, for $i = 1, \ldots, n$ the symbol d_i belongs to A, then $length(t) = n$.

Definition 3.6. Let A be an arbitrary alphabet. Then *a formal language in the alphabet A* (or over the alphabet A) is an arbitrary subset L of the set A^*, i.e. $L \subseteq A^*$.

Example 1. If $A = \{0, 1\}$, $L_1 = \{0\}$, $L_2 = \{e\}$, $L_3 = \{0^{2k} 1^{2k} \mid k \geq 1\}$, then L_1, L_2, L_3 are the formal languages in the alphabet A (or over the alphabet A).

3.3.3 The Used Definitions from the Theory of Algebraic Systems

Definition 3.7. Let $n \geq 1$, Z be an arbitrary non-empty set. Then the Cartesian n-degree of the set Z is called (and denoted by Z^n) the set Z in case $n = 1$ and the set of all ordered n-tuples of the form $(x_1, x_2, \ldots, x_n)$, where $x_1, x_2, \ldots, x_n$ are the elements of the set Z in case $n > 1$.

Definition 3.8. Let $n \geq 1$, Z be an arbitrary non-empty set. Then an *n-ary relation on the set Z* is an arbitrary subset R of the set Z^n. In case $n = 1$ one says about *an unary relation* and in case $n = 2$ we have *a binary relation*. The unary relations are interpreted as the distinguished subsets of the considered set Z.

Example 2. Let Z_1 be the set of all integers, and Odd be the subset of all even numbers. Then Odd is an unary relation on Z_1. Let $Less$ be the set of all ordered pairs of the form (x, y), where x, y are the arbitrary elements of Z_1, and $x < y$. Then $Less$ is a binary relation on the set Z_1.

Very often one uses a shorter denotation bRc instead of the denotation $(b, c) \in R$, where R is a binary relation on the arbitrary set Z.

Definition 3.9. Let Z be an arbitrary non-empty set, R be a binary relation on Z. Then

- if for arbitrary $a \in Z$, $(a, a) \in R$, then R is *a reflexive relation* ;

- if for arbitrary $a \in Z$, $(a, a) \neg \in R$, then R is *an antireflexive relation* ;
- if for every $a, b, c \in Z$, it follows from $(a, b) \in R$, $(b, c) \in R$ that $(a, c) \in R$, then R is *a transitive relation* ;
- if for every $a, b \in Z$, it follows from $(a, b) \in R$ that (b, a) belongs to R, then R is *a symmetric relation* ;
- if for every $a, b \in Z$, it follows from $(a, b) \in R$ and $a \neq b$ that $(b, a) \neg \in R$, then R is an *antisymmetric relation* ;
- if R is a reflexive, transitive, and antisymmetric relation on Z, then R is called *a partial order on Z* [140] .

Example 3. The binary relation *Less* from the previous example is antireflexive, transitive, and antisymmetric relation on $Z1$.

Example 4. Let Z_1 be the set of all integers, and *Eqless* be the set of all ordered pairs of the form (x, y), where x, y are arbitrary elements of Z_1, and the number x is either equal to the number y or less than y. Then *Eqless* is a binary relation on the set Z_1. This relation is reflexive, transitive, and antisymmetric. Thus, the relation *Eqless* is a partial order on Z_1.

Example 5. Let Z_2 be the set of all notions denoting the transport means, and *Genrel* is the the set of all ordered pairs of the form (x, y), where x, y are arbitrary elements of the set Z_2, and the notion x either coincides with the notion y or is a generalization of y. For example, the notion *a ship* is a generalization of the notion *an ice-breaker*; hence the pair *(ship, ice-breaker)* belongs to the set *Genrel*. Obviously, *Genrel* is a binary relation on the set Z_2. This relation is reflexive, transitive, and antisymmetric. That is why the relation *Genrel* is a partial order on Z_2.

3.4 The Basic Ideas of the Definition of a Sort System

Let us start to solve the posed task. Assuming that it is necessary to construct a formal description of an application domain, we'll consider the first steps in this direction.

Step 1. Let's consider a finite set of symbols denoting the most general notions of a selected domain: a space object, a physical object, an intelligent system, an organization, a natural number, a situation, an event (i.e. a dynamic situation), etc. Let's agree that every such notion qualifies an entity that is not being regarded as either a finite sequence (a tuple) of some other entities or as a set consisting of some other entities. Denote this set of symbols as St and call the elements of this set *sorts*.

Step 2. Let's distinguish in the set of sorts St a symbol to be associated with the semantic representations (SRs) of NL-texts either expressing the separate assertions or being the narrative texts. Denote this sort as P and will call it *the sort* "a meaning of proposition ." For instance, the string *prop* may play the role of the distinguished sort P for some applications. A part of the formulas of a new kind considered in this monograph can be represented in the form $F \& t$, where F is an SR of an NL-expression, and t is a string qualifying this expression. Then, if $t = P$, then the formula F is interpreted as an SR of a simple or compound assertion (or a statement, a proposition). In particular, the formula

$$(Weight(certain\,block1 : x1) \equiv 4/tonne) \,\&\, prop$$

can be regarded as a formula of the kind.

Step 3. A hierarchy of the concepts on the set of sorts St with the help of a binary relation Gen on St is defined, that is a certain subset Gen of the set $St \times St$ is selected. For instance, the relationships $(integer, natural)$, $(real, integer)$, $(phys.object, dynamic.phys.object)$, $(space.object, phys.object) \in Gen$ may take place.

Step 4. Many objects can be characterized from different standpoints; metaphorically speaking, these objects possess "the coordinates" on different "semantic axes."

Example 1. Every person is both a dynamic physical object (we can run, spring, etc.) and an intelligent system (we can read and write, a number of people are able to solve mathematical problems, to compose poems and music). That is why, metaphorically speaking, each person has the coordinate $dyn.phys.ob$ on one semantic axis and the coordinate $intel.system$ on another semantic axis (Fig. 3.1).

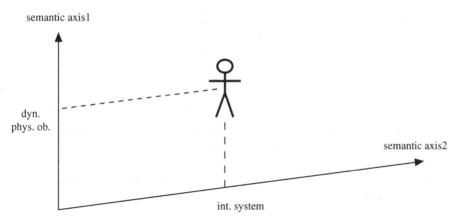

Fig. 3.1 "Semantic coordinates" of arbitrary person

Example 2. One is able to drive or to go to a certain university, that is why every university has the "semantic coordinate" $space.object$. For each university, there is a person who is the head (the rector) of this university, so the universities possess the "semantic coordinate" $organization$. Finally, a university is able to elaborate certain technology or certain device, hence it appears to be reasonable to believe that the universities have the "semantic coordinate" $intelligent.system$.

Taking into account these considerations, let's introduce a binary *tolerance relation* Tol on the set St. The interpretation of this relation is as follows: if $(s, u) \in Tol \subset St \times St$, then in the considered domain such entity x exists that it is possible to associate with x the sort s as one "semantic coordinate" and the sort u as the second

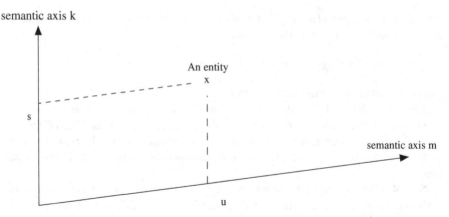

Fig. 3.2 Illustration of the metaphor of semantic axes: the case of two "semantic coordinates" of an entity x

"semantic coordinate"; besides, the sorts s and u are not comparable for the relation *Gen* reflecting a hierarchy of the most general concepts (see Fig. 3.2).

For instance, the sets *St* and *Tol* can be defined in such a way that *Tol* includes the ordered pairs

$$(space.object, organization), \ (space.object, intelligent.system),$$

$$(organization, intelligent.system), \ (organization, space.object),$$

$$(intelligent.system, space.object), \ (intelligent.system, organization).$$

The considered organization of the relation *Tol* implies the following properties: (1) $\forall u \in St$, $(u, u)\neg \in Tol$, i.e. *Tol* is an antireflexive relation; (2) $\forall u, t \in St$, it follows from $(u, t) \in Tol$ that $(t, u) \in Tol$, i.e. *Tol* is a symmetric relation.

A sort system will be defined below as an arbitrary four-tuple S of the form (St, P, Gen, Tol) with the components satisfying certain conditions.

3.5 The Formal Definition of a Sort System

Definition 3.10. A sort system is an arbitrary four-tuple S of the form

$$(St, P, Gen, Tol),$$

where *St* is an arbitrary finite set of symbols, $P \in St$, *Gen* is a non-empty binary relation on *St* being a partial order on *St*, *Tol* is a binary relation on *St* being antireflexive and symmetric, and the following conditions are satisfied:

1. *St* doesn't include the symbols $\uparrow$, {, }, (,), [$\uparrow$ *entity*], [$\uparrow$ *concept*], [$\uparrow$ *object*], [*entity*], [*concept*], [*object*];

2. If *Concr*(*P*) is the set of all such *z* from the set *St* that $(P, z) \in Gen$, then $St \setminus Concr(P) \neq \emptyset$, and for every $u \in St \setminus Concr(P)$ and every $w \in Concr(P)$, the sorts *u* and *w* are incomparable both for relation *Gen* and for the relation *Tol*;
3. for each $t, u \in St$, it follows from $(t, u) \in Gen$ or $(u, t) \in Gen$ that *t*, *u* are incomparable for the relation *Tol*;
4. for each $t1, u1 \in St$ and $t2, u2 \in St$, it follows from $(t1, u1) \in Tol$, $(t2, t1) \in Gen$, $(u2, u1) \in Gen$, that $(t2, u2) \in Tol$.

The elements of the set *St* are called sorts, *P* is called the sort "a meaning of proposition," the binary relations $Gen \subset St \times St$ and $Tol \subset St \times St$ are called respectively *the generality relation* and *the tolerance relation*. If $t, u \in St$, $(t, u) \in Gen$, then we often use an equivalent notation $t \rightarrow u$ and say that *t is a generalization of u*, and *u is a concretization of t*. If $(t, u) \in Tol$, we use the denotation $t \perp u$ and say that the sort *t is tolerant to the sort u*.

The symbols ↑, {, }, (,), [↑ *entity*], [↑ *object*], [↑ *concept*], [*entity*], [*object*], [*concept*] play special roles in constructing (from the sorts and these symbols) the strings called types and being the classifiers of the entities considered in the selected application domain.

The requirement 4 in the definition of a sort system is illustrated by Fig. 3.3 and 3.4. Suppose that the following situation takes place:

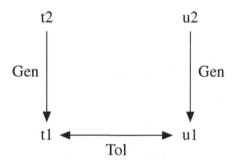

Fig. 3.3 A visual representation of the presupposition in the requirement 4 of the definition of a sort system

Then this situation implies the situation reflected in Fig. 3.4.

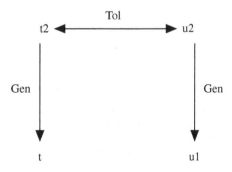

Fig. 3.4 A visual representation of the implication in the requirement 4 of the definition of a sort system

Example 1. Suppose that the sort *ints* (intelligent system) and the sort *dyn.phys.ob* (dynamic physical object) are associated by the tolerance relation *Tol*, i.e.

$$(ints, dyn.phys.ob) \in Tol,$$

and the sort *dyn.phys.ob* is a concretization of the sort *phys.ob* (physical object), i.e.,

$$(phys.ob, dyn.phys.ob) \in Gen.$$

Due to reflexivity of the generality relation *Gen*,

$$(ints, ints) \in Gen.$$

Using the denotation from the item 4 in the definition of a sort system, we have

$$t1 = ints, t2 = ints, (t2, t1) \in Gen,$$

$$u1 = dyn.phys.ob, u2 = phys.ob, (u2, u1) \in Gen.$$

This situation is illustrated by Fig. 3.5:

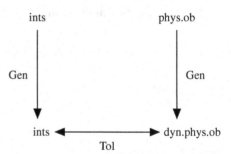

Fig. 3.5 A particular case of the presupposition in the requirement 4 of the definition of a sort system

Then, according to the requirement 4 in the definition above, $(t2, u2) \in Tol$ (see Fig. 3.6), that is,

$$(ints, phys.ob) \in Tol.$$

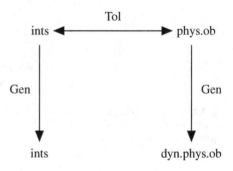

Fig. 3.6 A particular case of the situation mentioned in the implication in the requirement 4 of the definition of a sort system

Example 2. Let's construct a sort system S_0. Let

$$St_0 = \{nat, int, real, weight.value, space.ob, phys.ob, dyn.phys.ob,$$

$$imag.ob, ints, org, mom, sit, event, prop\}.$$

The elements of St_0 designate the notions (the concepts) and are interpreted as follows: *nat* – "natural number," *int* – "integer," *real* – "real number," *weight.value* – "value of weight," *space.ob* – "space object," *phys.ob* – "physical object," *dyn.phys.ob* – "dynamic physical object," *imag.ob* – "imaginary space object," *ints* – "intelligent system," *org* – "organization," *mom* – "moment," *sit* – "situation," *event* – "event" ("dynamic situation"), *prop* – "semantic representation of an assertion or of a narrative text."

Let $P_0 = prop$, the sets Ge_1, Ge_2, Gen_0, T_1, T_2 be defined as follows:

$$Ge_1 = \{(u, u) \mid u \in St_0\},$$

$$Ge_2 = \{(int, nat), (real, nat), (real, int),$$

$$(space.ob, phys.ob), (space.ob, imag.ob), (phys.ob, dyn.phys.ob),$$

$$(space.ob, dyn.phys.ob), (sit, event)\},$$

$$Gen_0 = Ge_1 \cup Ge_2,$$

$$T_1 = \{(ints, dyn.phys.ob), (ints, phys.ob), (ints, space.ob), (org, ints),$$

$$(org, phys.ob), (org, space.ob)\},$$

$$T_2 = \{(u, s) \mid (s, u) \in T_1\},$$

$$Tol_0 = T_1 \cup T_2.$$

Let S_0 be the four-tuple $(St_0, prop, Gen_0, Tol_0)$. Then it is easy to verify that S_0 is a sort system, and the sort *prop* is its distinguished sort "a meaning of proposition."

With respect to the definition of the set Gen_0, the following relationships take place:

$$real \rightarrow nat, int \rightarrow nat, space.ob \rightarrow phys.ob, space.ob \rightarrow imag.ob,$$

$$phys.ob \rightarrow dyn.phys.ob, ints \perp phys.ob, ints \perp dyn.phys.ob,$$

$$ints \perp org, phys.ob \perp ints, dyn.phys.ob \perp ints.$$

3.6 Types Generated by a Sort System

Let us define for any sort system S a set of strings $Tp(S)$ whose elements are called *the types of the system S* and are interpreted as the characteristics of the entities which are considered while reasoning in a selected domain. Let's agree that if in a reasoning about an entity z it is important that z is not a concept (a notion), we say that z is an object.

Suppose that the strings $[\uparrow entity]$, $[\uparrow concept]$, $[\uparrow object]$ are associated with the terms "an entity," "a concept" ("a notion"), "an object," respectively, or, in other words, these strings are the types of semantic items corresponding to the expressions "an entity," "a concept" ("a notion"), "an object."

For formalizing a considered domain, let's agree to proceed from the following recommendations. If the nature of an entity z considered in a reasoning is of no importance, we associate with z the type $[entity]$ in the course of reasoning. If all that is important concerning z is that z is an object, we associate with z the type $[object]$. If, to the contrary, all that is important as concerns z is that z is a concept (a notion), we associate with z the type $[concept]$. The purpose of introducing the types $[entity]$, $[concept]$, $[object]$ can be explained also by means of the following examples.

Let E_1 and E_2 be, respectively, the expressions "the first entity mentioned on page 12 of the issue of the newspaper 'The Moscow Times' published on October 1, 1994" and "the first object mentioned on page 12 of the issue of 'The Moscow Times' published on October 1, 1994." Then we may associate the types $[entity]$ and $[object]$ with the entities referred in E_1 and E_2 respectively in case we haven't read page 12 of the indicated issue.

However, after reading page 12 we'll get to know that the first entity and the first object mentioned on this page is the city Madrid. Hence, we may associate now with the mentioned entity (object) a more informative type $popul.area$ (a populated area).

Let E_3 be the expression "the notion with the mark AC060 defined in the Longman Dictionary of Scientific Usage (Moscow, Russky Yazyk Publishers, 1989)." Not seeing this dictionary, we may associate with the notion mentioned in E3 only the type $[concept]$. But after finding the definition with the mark AC060, we get to know that it is the definition of the notion "a tube" (a hollow cylinder with its length much greater than its diameter). Hence we may associate with the notion mentioned in E_3 a more informative type $\uparrow phys.ob$ (designating the notion "a physical object").

Let's consider the strings

$$[\uparrow entity], [\uparrow concept], [\uparrow object], [entity], [concept], [object]$$

as symbols in the next definitions.

Definition 3.11. Let S be a sort system of the form (St, P, Gen, Tol), and

$$Spectp = \{[\uparrow entity], [\uparrow concept], [\uparrow object]\},$$

$$Toptp = \{[entity], [concept], [object]\}.$$

Then the set of types $Tp(S)$ is the least set M satisfying the following conditions:

1. $Spect\,p \cup Topt\,p \cup St \cup \{\uparrow s \mid s \in St\} \subset M$. The elements of the sets $Spect\,p$ and $Topt\,p$ are called *special types* and *top types* respectively.
2. If $t \in M \setminus Spect\,p$, then the string of the form $\{t\}$ belongs to M.
3. If $n > 1$, for $i = 1, \ldots, n$, $t_i \in M \setminus Spect\,p$, then the string of the form $(t_1, \ldots, t_n)$ belongs to M.
4. If $t \in M$ and t has the beginning $\{$ or $($, then the string $\uparrow t$ belongs to M.

Definition 3.12. If S is a sort system, then

$$Mtp(S) = Tp(S) \setminus Spect\,p;$$

the elements of the set $Mtp(S)$ are called *the main types* .

Let's formulate the principles of establishing the correspondence between the entities considered in a domain with a sort system S of the form (St, P, Gen, Tol) and the types from the set $Mtp(S)$.

The types of notions (or concepts), as distinct from the types of objects, have the beginning $\uparrow$. So if a notion is denoted by a string s from St, we associate with this notion the type $\uparrow s$.

The type $\{t\}$ corresponds to any set of entities of type t. If $x_1, \ldots, x_n$ are the entities of types $t_1, \ldots, t_n$, then the type $(t_1, \ldots, t_n)$ is assigned to the n-tuple $(x_1, \ldots, x_n)$.

Example 1. We may assign the types from $Mtp(S_0)$ to some concepts and objects (including relations and other sets) with the help of the following table:

ENTITY	TYPE
The notion "a set"	$\uparrow \{[entity]\}$
The notion "a set of objects"	$\uparrow \{[object]\}$
The notion "a set of notions"	$\uparrow \{[concept]\}$
The notion "a person"	$\uparrow ints * dyn.phys.ob$
Tom Soyer	$ints * dyn.phys.ob$
The concept "an Editorial Board"	$\uparrow \{ints * dyn.phys.ob\}$
The Editorial Board of	
"Informatica" (Slovenia)	$\{ints * dyn.phys.ob\}$
The notion "a pair of integers"	$\uparrow (int, int)$
The pair (12, 144)	(int, int)

We can also associate with the relation "Less" on the set of integers the type $\{(int, int)\}$, with the relation "To belong to a set" the type

$$\{([entity], \{[entity]\})\},$$

with the relation "An object Y is qualified by a notion C" the type

$$\{([object], [concept])\},$$

and with the relation "A notion D is a generalization of a notion C" the type

$$\{([concept], [concept])\}.$$

3.7 The Concretization Relation on the Set of Types

The purpose of this section is to determine a transitive binary relation $\vdash$ on the set $Tp(S)$, where S is an arbitrary sort system; this relation will be called *the concretization relation* on the set $Tp(S)$. The basic ideas of introducing this relation are stated in the first subsection of this section; a formal definition of the concretization relation can be found in the second subsection.

It is worthwhile to note that the second subsection contains a rather tedious series of definitions. That is why in case you are reading this book not with the aim of modifying the described formal means but in order to apply them to the elaboration of semantic informational technologies, it is recommended to skip the second subsection of this section while reading this chapter for the first time.

3.7.1 Basic Ideas

Suppose that $S = (St, P, Gen, Tol)$ is an arbitrary sort system. The *first, simplest requirement* to the relation $\vdash$ on the set of types $Tp(S)$ is that the relation $\vdash$ coincides on the set of sorts St with the generality relation Gen.

Hence, for instance, if *phys.ob* and *dyn.phys.ob* are the sorts "a physical object" and "a dynamic physical object" and $(phys.ob, dyn.phys.ob) \in Gen$ (the equivalent denotation is $phys.ob \rightarrow dyn.phys.ob$), then $phys.ob \vdash dyn.phys.ob$.

The second requirement (also very simple) is as follows: Each of the basic types *[concept]*, *[object]* is a concretization of the basic type *[entity]*, that is

$$[entity] \vdash [concept], \quad [entity] \vdash [object].$$

The concretizations of the type *[concept]* are to be, in particular, the types $\uparrow phys.ob, \uparrow dyn.phys.ob, \uparrow ints * dyn.phys.ob$, where *ints* is the sort "intelligent system." In general, the types with the beginning $\uparrow$ (they are interpreted as the types of notions) are to be the concretizations of the type *[concept]*.

The concretizations of the type *[object]* are to be, in particular, the types

$$phys.ob, \ dyn.phys.ob, \ ints * dyn.phys.ob, \ \{ints * dyn.phys.ob\},$$

$$\{(ints * dyn.phys.ob, ints * dyn.phys.ob)\}.$$

Taking this into account, the following relationships are to take place:

$$[concept] \vdash \uparrow ints, \ [concept] \vdash \uparrow ints * dyn.phys.ob,$$

$$[concept] \vdash \uparrow \{ints * dyn.phys.ob\},$$

$$[object] \vdash phys.ob, \ [object] \vdash dyn.phys.ob,$$

$$[object] \vdash ints * dyn.phys.ob,$$

$$[object] \vdash (real, real), \ [object] \vdash \{(real, real)\}.$$

Consider now a more complex requirement to the relation $\vdash$. Let $n > 1$, $C_1, \ldots, C_n$ be some classes of entities, and R be a relation on the Cartesian product $C_1 \times \ldots \times C_n$, that is, let R be a set consisting of some n-tuples with the elements from the sets $C_1, \ldots, C_n$ respectively.

If $n = 2$, one says that R is a relation from C_1 to C_2. If $C_1 = \ldots = C_n$, the set R is called an n-ary relation on the set C_1. If in the latter case $n = 2$, R is called *a binary relation on the set C_1*.

Suppose that for every $k = 1, \ldots, n$ it is possible to associate with every entity $z_k \in C_k$ a certain type $t_k \in Tp(S)$. Then we'll believe that semantic restrictions of the attributes of the relation R are given by the type $\{(t_1, \ldots, t_n)\}$.

Example 1. Let U be the class of all real numbers, and $R1$ be the binary relation "Less" on U. Then the semantic restrictions of the attributes of $R1$ can be expressed by the type $\{(real, real)\}$.

Example 2. Let C_1 be the class of all physical objects having a definite, stationary shape, and C_2 be the class of all values of the distance in the metrical measurement system. Then the function "The diameter of a physical object" can be defined as follows: if $X \in C_1$, then the diameter of X is the maximal length of a line connecting some two points of the physical object X. We can interpret this function as a relation *Diameter* from C_1 to C_2.

Then the semantic requirements to the attributes of *Diameter* can be expressed by the type $\{(phys.ob, length.value)\}$ in case the considered set of sorts St contains the sorts *phys.ob* and *length.value* denoting the notions "a physical object" and "the value of length."

Let's continue to consider the idea of introducing the concretization relation on a set of types. Suppose that for $k = 1, \ldots, n$, we've distinguished a subclass D_k in the class C_k, and the following condition is satisfied: it is possible to associate with every entity $Z \in D_k$ not only a type t_k but also a type u_k conveying more detailed information about the entity Z. Then we would like to define a binary relation $\vdash$ on the set of types $Tp(S)$ in such a way that the relationship $t_k \vdash u_k$ takes place.

Example 3. Let's expand the previous example. Suppose that D_1 is the set of all dynamic physical objects, and the type *dyn.phys.ob* is associated with every object from the set D_1. Let E_1 be the set of all people, and the type *ints * dyn.phys.ob* is associated with each person, where *ints* is the sort "an intelligent system." Naturally, if $x \in C_1$, $y \in D_1$, $z \in E_1$, $w \in C_2$, then the following expressions are well-formed (according to our common sense):

$$Diameter(x) = w, \, Diameter(y) = w, \, Diameter(z) = w.$$

That is why let's demand that the following relationships take place:

$$phys.ob \vdash dyn.phys.ob,$$

$$dyn.phys.ob \vdash ints * dyn.phys.ob,$$

$$phys.ob \vdash ints * dyn.phys.ob.$$

These ideas can be formulated also in the following way. Let *conc* be a denotation of a notion; in particular, it is possible that $conc \in St$, where St is the considered set of sorts. Then let $Dt(conc)$ be the designation of all entities which can be qualified (in other words, characterized) by the notion with the denotation *conc*; we'll say that $Dt(conc)$ is *the denotat of the notion conc*.

Suppose that S is a sort system, *Rel* is the designation of a certain n-ary relation on a certain set Z, and a certain mapping tp assigns to R a description of the semantic requirements to the attributes of R of the form $(t_1, \ldots, t_n)$, i.e. $tp(Rel) = \{(t_1, \ldots, t_n)\}$, where $n > 1$, $t_1, \ldots, t_n \in Tp(S)$.

We'll believe that $(x_1, \ldots, x_n) \in Rel$ if and only if there exist such types $u_1, \ldots, u_n \in Tp(S)$ that for every $k = 1, \ldots, n$, the type u_k is a concretization of t_k (we'll use in this case the denotation $t_k \vdash u_k$), and x_k belongs to the denotat of u_k; this means that x_k is an entity qualified by the type u_k.

Example 4. The academic groups of university students are the particular cases of sets. The relationship

$$tp(Number - of - elem) = \{(\{[entity]\}, nat)\}$$

can be interpreted as a description of semantic requirements to the arguments and value of the function "The number of elements of a finite set" denoted by the symbol $Number - of - elem$.

Suppose that the list of all identifiers being known to an intelligent database used by the administration of a university includes the element $Mat08 - 05$, and that the mapping tp associates with this element the type $\{ints * dyn.phys.ob\}$. This means that the hypothetical intelligent database considers the object with the identifier $Mat08 - 05$ as a certain set of people (because each person is both an intelligent system and a dynamic physical object).

Let $tp(14) = nat$. Since $nat \to nat$, it follows from the relationship (if it takes place) $[entity] \vdash \{ints * dyn.phys.ob\}$ that the expression $Number - of - elem(Mat08 - 05, 14)$ is well-formed.

3.7.2 Formal Definitions

Definition 3.13. Let S be an arbitrary sort system with the set of sorts St. Then *elementary compound types* are the strings from $Tp(S)$ of the form $s_1 * s_2 * \ldots * s_k$, where $k > 1$, for $i = 1, \ldots, k$, $s_i \in St$.

Example 5. The string $ints * dyn.phys.ob$ is an elementary compound type for the sort system S_0.

Definition 3.14. Let S be a sort system with the set of sorts St. Then $Elt(S)$ is the union of the set of sorts St with the set of all elementary compound types. The elements of the set $Elt(S)$ will be called *elementary types* .

Definition 3.15. If S is a sort system of the form (St, P, Gen, Tol), $t \in Elt(S)$, then *the spectrum of the type t* (denoted by $Spr(t)$) is (a) the set $\{t\}$ in case $t \in St$; (b) the set $\{s_1, \ldots, s_k\}$ in case the type t is the string of the form $s_1 * \ldots * s_k$, where $k > 1$, for every $i = 1, \ldots, k$, $s_i \in St$.

Example 6. If S_0 is the sort system constructed in Sect. 3.5, then the following relationships are valid:

$$Spr(phys.ob) = \{phys.ob\},$$

$$Spr(ints * dyn.phys.ob) = \{ints, dyn.phys.ob\}.$$

Definition 3.16. Let S be any sort system of the form (St, P, Gen, Tol), $u \in St$, t be an elementary compound type from $Tp(S)$. Then the type t is called *a refinement of the sort* $u \Leftrightarrow$ the spectrum $Spr(t)$ contains such sort w that $u \rightarrow w$ (i.e. $(u, w) \in Gen$).

Example 7. Let $u = phys.ob$, $t = ints * dyn.phys.ob$. Then the spectrum $Spr(t) = \{ints, dyn.phys.ob\}$.

That is why it follows from $phys.ob \rightarrow dyn.phys.ob$ that the type t is a refinement of the sort u.

Let's remember that the sorts are considered in this book as symbols, i.e. as indivisible units.

Definition 3.17. Let S be any sort system of the form (St, P, Gen, Tol), $u \in St$, t is a type from $Tp(S)$, and t includes the symbol u. Then an occurrence of u in the string t is *free* $\Leftrightarrow$ either $t = u$ or this occurrence of u in t is not an occurrence of u in any substring of the form $s_1 * s_2 \ldots * s_k$, where $k > 1$, for every $i = 1, \ldots, k$, $s_i \in St$, and there exists such m, $1 \leq m \leq k$, that $u = s_m$.

Example 8. It is possible to associate with the function "Friends" the type

$$t1 = \{(ints * dyn.phys.ob, \{ ints * dyn.phys.ob\})\}.$$

Both the first and the second occurences of the symbol (a sort) $dyn.phys.ob$ in the string $t1$ are not the free occurrences. The function "The weight of a set of physical objects" can be associated with the type

$$t2 = \{(\{phys.ob\}, (real, kg))\};$$

the occurrences of the symbol $phys.ob$ in the string $t2$ and in the string $t3 = \uparrow phys.ob$ (a possible type of the concept "a physical object") are the free occurrences.

Definition 3.18. Let S be a sort system of the form (St, P, Gen, Tol). Then

$$Tc(S) = \{t \in Tp(S) \setminus (Spectp \cup Toptp) \mid \text{the symbol } \uparrow \text{ is the beginning of } t\},$$

where

$$Spectp = \{[\uparrow entity], [\uparrow concept], [\uparrow object]\},$$

$$Toptp = \{[entity], [concept], [object]\},$$

$$Tob(S) = Tp(S) \setminus (Spectp \cup Toptp \cup Tc(S)).$$

Definition 3.19. Let S be a sort system of the form (St, P, Gen, Tol). Then the transformations $tr_1, \ldots, tr_6$, being partially applicable to the types from the set $Tp(S)$, are defined in the following way:

1. If $t \in Tp(S)$, t includes the symbol $[entity]$, then the transformations tr_1 and tr_2 are applicable to the type t. Let w_1 be the result of replacing in t an arbitrary occurrence of the symbol $[entity]$ by the type $[concept]$; w_2 be the result of replacing in t an arbitrary occurrence of the symbol $[entity]$ by the type $[object]$. Then w_1 and w_2 are the possible results of applying to the type t the transformations tr_1 and tr_2, respectively.
2. If $t \in Tp(S)$, t includes the symbol $[concept]$, the type $u \in Tc(S)$, then the transformation tr_3 is applicable to the type t, and the result of replacing an arbitrary occurrence of the symbol $[concept]$ by the type u is a possible result of applying the transformation tr_3 to the type t.
3. If $t \in Tp(S)$, t includes the symbol $[object]$, the type $z \in Tob(S)$, then the transformation tr_4 is applicable to the type t, and the result of replacing an arbitrary occurrence of the symbol $[object]$ by the type z is a possible result of applying the transformation tr_4 to the type t.
4. If $t \in Tp(S)$, t includes the symbol $s \in St$, $u \in St$, $s \neq u$, $(s, u) \in Gen$, and the type w is the result of replacing in t an arbitrary free occurrence of the sort s by the sort u, then w is a possible result of applying the transformation tr_5 to the type t.
5. If $t \in Tp(S)$, $u \in St$, z is an elementary compound type from $Tp(S)$ being a refinement of the sort u, and w is obtained from t by replacing in t an arbitrary free occurrence of the sort u by the string z, then w is a possible result of applying the transformation tr_6 to the type t.

Example 9. If S_0 is the sort system built above, $t_1 = [object]$, $t_2 = space.ob$, $w_1 = ints * dyn.phys.ob$, $w_2 = dyn.phys.ob$, then w_1 and w_2 are the possible results of applying the transformations tr_4 and tr_5 to the types t_1 and t_2, respectively.

If $t_3 = \{phys.ob\}$, $w_3 = \{ints * dyn.phys.ob\}$, then w_3 is a possible result of applying the transformation tr_6 to the type t_3 (since the sort $ints * dyn.phys.ob$ is a concretization of the sort $phys.ob$, because dynamic physical objects form a subclass of the set of all physical objects). Here the element $ints$ is the distinguished sort "intelligent system."

Definition 3.20. Let S be a sort system of the form (St, P, Gen, Tol), and $t, u \in Tp(S)$. Then the type u is called *a concretization of the type* t, and the type t is called *a generalization of the type* u (the designation $t \vdash u$ is used) $\Leftrightarrow$ either t coincides with u or there exist such types $x_1, x_2, \ldots, x_n \in Tp(S)$, where $n > 1$, that $x_1 = t$, $x_n = u$, and for each $j = 1, \ldots, n-1$, there exist such $k[j] \in \{1, 2, \ldots, 6\}$ that the transformation $tr_{k[j]}$ can be applied to the string x_j, and the string x_{j+1} is a possible result of applying the transformation $tr_{k[j]}$ to x_j.

Example 10. It is easy to verify for the sort system S_0 that the following relationships take place:

$$[entity] \vdash [object], \; [entity] \vdash [concept],$$

$$[object] \vdash ints, \; [object] \vdash phys.ob, \; phys.ob \vdash dyn.phys.ob,$$

$$ints \vdash ints * dyn.phys.ob, \; phys \vdash ints * dyn.phys.ob,$$

$$\{[object]\} \vdash \{phys.ob\}, \; \{[object]\} \vdash \{ints * dyn.phys.ob\},$$

$$[object] \vdash real, \; [object] \vdash (nat, nat), \; [object] \vdash \{nat\},$$

$$[concept] \vdash\uparrow ints, \; [concept] \vdash\uparrow \{ints\},$$

$$[concept] \vdash\uparrow ints * dyn.phys.ob,$$

$$[concept] \vdash\uparrow \{ints * dyn.phys.ob\}.$$

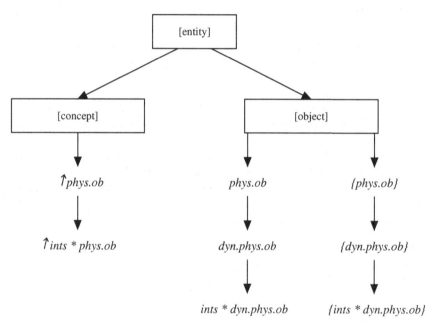

Fig. 3.7 A fragment of the hierarchy on the set of main types Mtp(S) induced by the concretization relation on the set of types Tp(S), where S is a sort system

Theorem 3.1. *Let S be an arbitrary sort system. Then the concretization relation $\vdash$ is a partial order on the set of types $Tp(S)$.*

Proof

The reflexivity and transitivity of the relation $\vdash$ immediately follow from its definition. Let's show that the antisymmetricity of the relation $\vdash$ follows from the properties of the transformations $tr_1, tr_2, \ldots, tr_6$.

As a result of applying the transformation tr_1 or tr_2, the number of occurrences of the symbol $[entity]$ is reduced, the difference is 1. After applying the transformations

tr_3 or tr_4, the number of the occurrences of the symbol $[concept]$ or $[object]$ is $n - 1$, where n is the initial number of occurrences of the considered symbol.

If $t1, t2 \in Tp(S)$, and the type $t2$ has been obtained from the type $t1$ as a result of applying just one time the transformation tr_5, then this means that there exist such sorts $s, u \in St$, that $s \neq u$, $(s, u) \in Gen$, $t1$ includes the symbol s, and $t2$ has been obtained by means of replacing an occurrence of the symbol s in the string $t1$ by the symbol u. It follows from the antisymmetricity of the relation Gen on the set of sorts St that the reverse transformation of $t2$ into $t1$ is impossible.

If the type $t2$ has been obtained from the type $t1$ by means of applying just one time the transformation tr_6, the number of the symbols in $t2$ is greater than the number of the symbols in $t1$.

3.8 Concept-Object Systems

Let's proceed from the assumption that for describing an application domain on the conceptual level, we should choose some sets of strings X and V. The first set is to contain, in particular, the designations of notions (concepts), physical objects (people, ships, books, etc.), events, n-ary relations ($n \geq 1$). The set V should consist of variables which will play in expressions of our knowledge representation language the roles of marks of diverse entities and, besides, will be used together with the quantifiers $\exists$ and $\forall$.

We'll distinguish in X a certain subset F containing the designations of diverse functions. Each function f with n arguments will be considered as a certain set consisting of $n + 1$-tuples $(x_1, \ldots, x_n, y)$, where $y = f(x_1, \ldots, x_n)$. Besides, we'll introduce a mapping tp assigning to each element $d \in X \cup V$ a certain type $tp(d)$ characterizing the entity denoted by d.

Definition 3.21. Let S be any sort system of the form (St, P, Gen, Tol). Then a four-tuple Ct of the form

$$(X, V, tp, F)$$

is called *a concept-object system (c.o.s.) coordinated with the sort system S* (or a *concept-object system for S*) $\Leftrightarrow$ the following conditions are satisfied:

- X and V are countable non intersecting sets of symbols;
- tp is a mapping from $X \cup V$ to the set of types $Tp(S)$;
- F is a subset of X; for each $r \in F$, the string $tp(r)$ has the beginning $/\{(/$ and the ending $/)\}/$;
- St is a subset of X, and for each $s \in St$, $tp(s) = \uparrow s$;
- the set $\{u \in V \mid tp(v) = [entity]\}$ is countable.

The set X is called *the primary informational universe*, the elements of V and F are called *the variables* and *functional symbols*, respectively. If $d \in X \cup V$, $tp(d) = t$, then we'll say that t is *the type* of the element d.

Example. Let's construct a concept-object system Ct_0 coordinated with the sort system S_0, determined in Sect. 3.5. Let *Nat* be the set of all such strings *str* formed from the ciphers 0, 1, ..., 9 that the first symbol of *str* is distinct from 0 in case *str* contains more than one symbol. Let

$$U1 = \{ person, chemist, biologist, stud.group, tour.group, J.Price,$$

$$R.Scott, N.Cope, P.Somov, Friends, Numb, Less, Knows, Isa1,$$

$$Elem, Subset, Include1, Before, \#now\#, concept \}.$$

The strings of $U1$ denote respectively the notions "a person," "a chemist," "a biologist," "a student group," "a tourist group;" some concrete persons with the initial and surname J. Price, R. Scott, N. Cope, P. Somov; the function "Friends" assigning to a person Z the set of all friends of Z; the function "Numb" assigning to a set the number of elements in it; the relations "Less" (on the set of real numbers), "Knows" ("The memory of an intelligent system Z_1 at the moment Z_2 contains a semantic representation of an assertion (in other words, of a statement, a proposition) Z_3"), "Isl" ("An object Z_1 is qualified by a notion Z_2"; an example of a phrase: "P. Somov is a chemist"), "Element" ("An entity Z_1 is an element of the set Z_2"), "Subset" ("An entity Z_1 is a subset of the set Z_2"), Include1" ("An intelligent system Z_1 includes an entity Z_2 at the moment Z_3 into a set of objects Z_4"), "Before" on the set of the moments of time.

The symbol $\#now\#$ will be used in semantic representations of texts for denoting a current moment of time. The symbol *concept* will be interpreted as the informational unit corresponding to *the word groups* "a notion," "a concept," and, besides, the word group "a term" in the meaning "a notion," "a concept."

Let's define a mapping $t1$ from $U1$ to the set $Tp(S_0)$ by the following table:

d	t1(d)
person, chemist, biologist	$\uparrow ints * dyn.phys.ob$
stud.group, tour.group	$\uparrow \{ints * dyn.phys.ob\}$
J.Price, R.Scott	$ints * dyn.phys.ob$
N.Cope, P.Somov	$ints * dyn.phys.ob$
Friends	$\{(ints * dyn.phys.ob, \{ints * dyn.phys.ob\})\}$
Numb	$\{(\{[entity]\}, nat)\}$
Less	$\{(real, real) \}$
Knows	$\{(int, mom, prop)\}$
Isa1	$\{([object], [concept])\}$
Elem	$\{([entity], \{[entity]\})\}$
Subset	$\{(\{[entity]\}, \{[entity]\})\}$
Include1	$\{(ints, [entity], mom, \{[entity]\})\}$
Before	$\{(mom, mom)\}$
#now#	mom
concept	$[\uparrow concept]$

Let's believe that *Firm_Ocean*, *Firm_Rainbow*, *Firm_Sunrise* are the designations of the firms; *Suppliers*, *Staff*, *Director* are the designations of the functions "The set of all suppliers of an organization," "The set of all persons working at an organization," and "The director of an organization." Let

$$U2 = \{Firm\_Ocean, Firm\_Rainbow, Firm\_Sunrise,$$

$$Suppliers, Staff, Director\},$$

and the mapping $t2$ from $U2$ to $Tp(S)$ is determined by the following relationships:

$$t2(Firm\_Ocean) = t2(Firm\_Rainbow) = t2(Firm\_Sunrise)$$

$$= org * space.ob * ints,$$

where *org* is the sort "organization", *space.ob* is the sort "space object", and *ints* is the sort "intelligent system";

$$t2(Suppliers) = \{(org, \{org\})\},$$

$$t2(Staff) = \{(org, \{ints * dyn.phys.ob\})\},$$

$$t2(Director) = \{(org, ints * dyn.phys.ob)\}.$$

Let $Vx = \{x1, x2, \ldots\}$, $Ve = \{e1, e2, \ldots\}$, $Vp = \{P1, P2, \ldots\}$,

$$Vset = \{S1, S2, \ldots\},$$

$$V_0 = Vx \cup Ve \cup Vp \cup Vset,$$

where the elements of the sets Vx, Ve, Vp, $Vset$ will be interpreted as the variables for designating respectively (a) arbitrary entities, (b) situations (in particular, events), (c) semantic representations of statements (assertions, propositions) and narrative texts, (d) sets.

Let $X_0 = St_0 \cup Nat \cup U1 \cup U2 \cup Weights$, where $Weights = \{x/y/x \in Nat, y \in \{kg, tonne\}\}$, and the mapping $tp_0 : X_0 \cup V_0 \rightarrow Tp(S_0)$ is defined by the following relationships:

$$d \in St_0 \Rightarrow tp_0(d) = \uparrow d;$$

$$d \in Nat \Rightarrow tp_0(d) = nat;$$

$$d \in Weights \Rightarrow tp_0(d) = weight.value;$$

$$d \in U1 \Rightarrow tp_0(d) = t1(d);$$

$$d \in U2 \Rightarrow tp_0(d) = t2(d);$$

$$d \in Vx \Rightarrow tp_0(d) = [entity];$$

$$d \in Ve \Rightarrow tp_0(d) = sit;$$

$$d \in Vp \Rightarrow tp_0(d) = prop;$$

$$d \in V\,set \Rightarrow tp_0(d) = \{[entity]\}.$$

Let's define the set of functional symbols

$$F_0 = \{\,Friends, Numb, Suppliers, Staff, Director\,\},$$

and let

$$Ct_0 = (\,X_0, V_0, tp_0, F_0\,).$$

Then it is easy to verify that the system Ct_0 is a concept-object system coordinated with the sort system S_0.

3.9 Systems of Quantifiers and Logical Connectives: Conceptual Bases

Presume that we define a sort system S of the form (St, $P\,Gen$, Tol) and a concept-object system Ct of the form (X, V, tp, F) coordinated with S in order to describe an application domain. Then it is proposed to distinguish in the primary informational universe X two non intersecting and finite (hence non void) subsets Int_1 and Int_2 in the following manner: we distinguish in St two sorts int_1 and int_2 and suppose that for $m = 1, 2$,

$$Int_m = \{q \in X \mid tp(x) = int_m\}.$$

The elements of Int_1 correspond to the meanings of the expressions "every," "a certain," "any," "arbitrary," etc. (and, may be, "almost every," etc.); these expressions are used to form the word groups in singular. The elements of Int_2 are interpreted as semantic items corresponding to the expressions "all," "several," "almost all," "many," and so on; the minimal requirement is that Int_2 contains a semantic item corresponding to the word "all."

Let Int_1 contain a distinguished element ref considered as an analogue of the word combination "a certain" in the sense "quite definite" (but, possibly, unknown). If Ct is a concept-object system of the form (X, V, tp, F), $d \in X$, d denotes a notion, and a semantic representation of a text includes a substring of the form $ref\,d$ (e.g., the substring $certain\,chemist$, where $ref = certain$, $d = chemist$), then we suppose that this substring denotes a certain concrete entity (but not an arbitrary one) that is characterized by the concept d.

Let X contain the elements $\equiv$, $\neg$, $\wedge$, $\vee$, interpreted as the connectives "is identical to," "not," "and," "or," and contain the elements $\forall$ and $\exists$, interpreted as the universal and existential quantifiers.

Figure 3.8 is intended to help grasp the basic ideas of the definition below.

Definition 3.22. Let S be any sort system of the form (St, P, Gen, Tol), where St be the set of sorts, P be the distinguished sort "a meaning of proposition"; let $Concr(P)$ be the set of all such z from St that $(P, z) \in Gen$ (i.e. $Concr(P)$ be the set of all sorts being the concretizations of the sort P), Ct be any concept-object system of

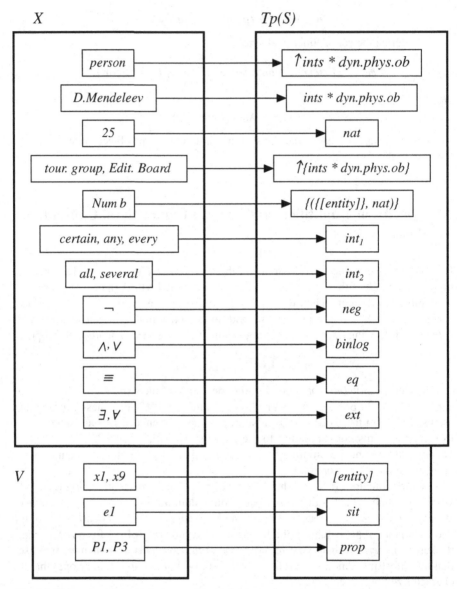

Fig. 3.8 Illustration of the basic ideas of the definitions of a concept-object system and a system of quantifiers and logical connectives

the form (X, V, tp, F) coordinated with S, ref be the intensional quantifier from X, the different elements int_1, int_2, eq, neg, $binlog$, ext be some distinguished sorts from $St \setminus Concz(P)$, and each pair of these sorts be incomparable with respect to the generality relation Gen and incomparable with respect to the tolerance relation Tol.

Then the seven-tuple Ql of the form

$$(int_1, int_2, ref, eq, neg, binlog, ext)$$

is called *a system of quantifiers and logical connectives (s.q.l.c.) coordinated with S and Ct (or a s.q.l.c. for S and Ct)* ⇔ the following conditions are satisfied:

1. For each $m = 1, 2$, the set $Int_m = \{ d \in X \mid tp(d) = int_m \}$ is a finite set; $ref \in Int_1$, the sets Int_1 and Int_2 don't intersect.
2. The primary informational universe X includes the subset

$$\{ \equiv, \neg, \wedge, \vee, \forall, \exists \};$$

besides, $tp(\equiv) = eq, tp(\neg) = neg, tp(\wedge) = tp(\vee) = binlog, tp(\forall) = tp(\exists) = ext$.

3. There are no such $d \in X \setminus (Int_1 \cap Int_2 \cap \{ \equiv, \neg, \wedge, \vee, \forall, \exists \})$ and no such $s \in \{ int_1, int_2, eq, neg, binlog, ext \}$ that $tp(d)$ and s are comparable with respect to the relation *Gen* or are comparable with respect to the relation *Tol*.
4. For each sort $u \in \{ int_1, int_2, eq, neg, binlog, ext \}$ and each sort $w \in Concr(P)$, where P is the distinguished sort "a meaning of proposition," the sorts u and w are incomparable with respect to the relation *Gen* and are incomparable with respect to the relation *Tol*.

The elements of Int_1 and Int_2 are called *intensional quantifiers*, the element ref is called *the referential quantifier*, the symbols $\forall$ and $\exists$ are called *extensional quantifiers*.

Example 1. Let $S_0 = (St_0, prop, Gen_0, Tol_0)$ be the sort system built above, $Ct_0 = (X_0, V_0, tp_0, F_0)$ be the concept-object system coordinated with the sort system S_0; the system Ct_0 was defined in Sect. 3.8. Suppose that

$$Str = \{ sort.qr.int_1, sort.qr.int_2, sort.eqvt, sort.not, sort.bin.log, sort.ext.qr \};$$

$$Gen_1 = Gen_0 \cup \{ (s, s) \mid s \in Str \};$$

$$St_1 = St_0 \cup Str;$$

$$S_1 = (St_1, prop, Gen_1, Tol_0).$$

Then, obviously, the four-tuple S_1 is a sort system.

Let's define now a concept-object system Ct_1 and a system of quantifiers and logical connectives Ql_1. Let

$$Z = \{ certn, all, \equiv, \wedge, \vee, \forall, \exists \},$$

where the string *certn* is interpreted as the semantic item "a certain," and

$$X_1 = X_0 \cup Str \cup Z.$$

Let's determine a mapping tp_1 from $X_1 \cup V_0$ to the set of types $Tp(S_1)$ in the following way:

$$u \in Str \Rightarrow tp_1(u) = \uparrow u;$$

$$d \in X_0 \Rightarrow tp_1(d) = tp_0(d);$$

$$tp_1(certn) = sort.qr.int1, \; tp_1(all) = sort.qr.int2,$$

$$tp_1(\equiv) = sort.eqvt, \; tp_1(\neg) = sort.not,$$

$$tp_1(\wedge) = tp_1(\vee) = sort.bin.log,$$

$$tp_1(\forall) = tp_1(\exists) = sort.ext.qr.$$

Let the systems Ct_1 and Ql_1 be defined by the relationships

$$Ct_1 = (X_1, V_0, tp_1, F_0),$$

$$Ql_1 = (\, sort.qr.int_1, \; sort.qr.int_2, \; certn, \; sort.eqvt,$$

$$sort.not, \; sort.bin.log, \; sort.ext.qr \,).$$

Then it is easy to verify that Ct_1 is a concept-object system coordinated with the sort system S_1, and Ql_1 is a system of quantifiers and logical connectives coordinated with the sort system S_1 and the concept-object system Ct_1. In the system Ql_1, the element *certn* is the informational item interpreted as the *referential quantifier* *ref* (that is as a semantic unit corresponding to the word combination "a certain").

Definition 3.23. An ordered triple B of the form

$$(\, S, Ct, Ql \,)$$

is called *a conceptual basis (c.b.)* $\Leftrightarrow$ S is a sort system, Ct is a concept-object system of the form $(\, X, V, tp, F \,)$ coordinated with the sort system S, Ql is a system of quantifiers and logical connectives coordinated with S and Ct, and the set $X \cup V$ doesn't include any of the symbols $/,/$ (comma), $/(/, /)/, / : /, / * /, /\langle /, /\rangle /, /\& /$.

We'll denote by $S(B)$, $Ct(B)$, and $Ql(B)$ the components of an arbitrary conceptual basis B of the form $(\, S, Ct, Ql \,)$. Each component with the name h of the mentioned systems of the forms

$$(\, St, P, Gen, Tol \,),$$

$$(\, X, V, tp, F \,),$$

$$(\, int_1, int_2, ref, eq, neg, binlog, ext \,)$$

will be denoted by $h(B)$.

For instance, the set of sorts, the distinguished sort "a meaning of proposition," and the primary informational universe of B will be denoted by $St(B)$, $P(B)$, $X(B)$ respectively.

We'll interpret the conceptual bases as formal enumerations of (a) primary informational units needed for building semantic representations of NL-texts and for describing knowledge about the world, (b) the information associated with these units and required for constructing the semantic representations of NL-texts and for forming knowledge fragments and representing goals of intelligent systems.

Example 2. Let S_1, Ct_1, Ql_1 be respectively the sort system, concept-object system, and the system of quantifiers and logical connectives determined above. Then, obviously, the triple $B_1 = (S_1, Ct_1, Ql_1)$ is a conceptual basis, and

$$St(B_1) = St_1, P(B_1) = prop,$$

$$X(B_1) = X_1, V(B_1) = V_0.$$

This example shows that the set of all conceptual bases is non void.

3.10 A Discussion of the Constructed Mathematical Model

3.10.1 Mathematical Peculiarities of the Model

The form of the constructed mathematical model for describing a system of primary units of conceptual level used by an applied intelligent system is original. Let's consider the distinguished features of this model seeming to be the most important both from the mathematical standpoint and from the standpoint of using the model in the design of linguistic processors.

1. The existence of the hierarchy of notions is constructively taken into account: with this aim, a partial order *Gen* on the set of sorts *St* is defined, it is called the generality relation.
2. Many entities considered in an application domain can be qualified from different points of view. For instance, people are, on one hand, intelligent systems, because they can read, solve tasks, compose music, poems, etc. But, on the other hand, people are physical objects being able to move in space. That is why we can metaphorically say that many entities have "the coordinates" on different "semantic axes."

   For taking into account this important phenomenon, we introduced a binary relation *Tol* on the set of sorts *St*, it is called the tolerance relation. The accumulated experience has shown that this original feature of the model is very important for elaborating the algorithms of semantic-syntactic analysis of natural language texts. The reason for this statement is that the use of the same word in several non similar contexts may be explained by the realization in these contexts of different "semantic coordinates" of the word.
3. The phrase "This notion is used both in physics and chemistry" (it applies, for instance, to the notion "a molecule") is very simple for a person having some command of English. However, the main popular approaches to the formalization of NL-semantics can't appropriately reflect the semantic structure of this phrase. The reason for this shortcoming is that such approaches don't offer a formal analogue of the conceptual (or informational, semantic) unit corresponding to the words "a notion," "a concept."

The situation is different for the model constructed above. First, the model introduces a special basic type $[\uparrow\ concept]$ interpreted as the type of the conceptual unit corresponding to the words "a notion," "a concept." Secondly, the model describes, in particular, the class of formal objects called concept-object systems. The component X of arbitrary concept-object system Ct of the form (X, V, tp, F) (this component is called the primary informational universe) can include an element (a symbol) interpreted as the conceptual unit corresponding to the words "a notion," "a concept" (see an example in Sect. 3.8).

This feature of the constructed mathematical model is important for the design of natural language processing systems dealing with scientific and scientific-technical texts and, besides, for the design of linguistic processors of applied intelligent systems extracting knowledge from encyclopedic dictionaries or updating electronic encyclopedic dictionaries by means of extracting knowledge from articles, monographs, textbooks, technical reports, etc.

4. One of the most important distinguishing features of the built model is an original definition of the set of types generated by arbitrary sort system, where the types are considered as formal characterizations of the entities belonging to considered thematic domains. In accordance with this definition, (a) the form of the types of objects from an application domain differs from the form of the notions qualifying these objects; (b) the form of the types of the objects differs from the form of the types of sets consisting of such objects; (c)) the form of the types of the notions (in other words, of the concepts) qualifying the objects differs from the form of the types of notions qualifying the sets consisting of such objects (for example, the type of the concept "a person" is different from the type of the concepts "an editorial board," "a student group").

5. The constructed model associates the types with the designations of the functions too. It may be noticed that the definition of the set of types generated by a sort system enables us to associate (in a reasonable way) the types with a number of rather non standard but practically important functions. In particular, this applies to functions with the values being (a) the set of concepts explained in an encyclopedic dictionary or in a Web-based ontology; (b) the set of concepts mentioned in the definition of the given concept in the given dictionary; (c) the set of semantic representations of the known definitions of the considered notion; (d) the number of elements of the considered set; (e) the set consisting of all suppliers of the considered firm; (f) the set consisting of all employees of the considered firm.

3.10.2 The Comparison of the Model with Related Approaches

Let's compare the constructed model with the approaches to describing primary units of conceptual level offered by first-order predicate logic, discourse representation theory, theory of conceptual graphs, and episodic logic.

In the standard first-order predicate logic, one considers the unstructured non-intersecting sets of constants, variables, functional symbols, and predicate symbols. More exactly, each functional symbol is associated with a natural number denoting the number of arguments of the corresponding function, and each predicate symbol is associated with a natural number denoting the number of attributes of the corresponding relation. In the multi-sorted first-order predicate logics, the set of constants consists of the non intersecting classes where each of them is characterized by a certain sort.

The mathematical model constructed above provides, in particular, the following additional opportunities in comparison with the multi-sorted first-order predicate logics:

- Due to the introduction of the tolerance relation as a component of a sort system, it is possible to associate with a primary unit of conceptual level not only one but, in many cases, several sorts being, metaphorically speaking, "the coordinates on orthogonal semantic axes" of the entities qualified or denoted by such a unit;
- The association of the types with the primary units of conceptual level means that the set of such units has a fine-grained structure; in particular, the types enable us to distinguish in a formal way the following: (a) the types of objects from thematic domains and the types of notions qualifying these objects; (b) the types of objects and the types of sets consisting of such objects; (c) the types of notions qualifying some objects and the types of notions qualifying the sets consisting of such objects;
- The constructed model allows for considering the primary units of conceptual level corresponding to the words and expressions "a certain," "definite," "any," "all," "several," "the majority," "the minority";
- The model provides the possibility to consider the primary unit of conceptual level corresponding to the words "a notion," "a concept."

It should be added that the proposed mathematical model enables us to consider the functions with the arguments and/or values being semantic representations (SRS) of the assertions (propositions) and narrative texts. For instance, one of such functions can associate each notion defined in an encyclopedic dictionary with a formula being an SR of this notion. But in the first-order predicate logics, the arguments and values of the functions can be only terms but not formulas, and terms are the designations of the objects from the application domains but not the designations of the meanings of assertions (propositions) and narrative texts.

The Discourse Representation Theory (DRT) can be interpreted as one of the variants of the first-order predicate logic combining the use of logical formulas and two-dimensional diagrams for visual representation of information. That is why the enumerated advantages of the constructed mathematical model for describing the systems of primary units of conceptual level used by computer intelligent systems apply to DRT too.

Both the Theory of Generalized Quantifiers (TGQ) and the mathematical model constructed above consider the units of conceptual level corresponding to the expressions "a certain," "definite," "all," "several." However, all other enumerated advantages of the model in comparison with the first-order predicate logic are simultaneously the advantages in comparison with the approach of TGQ.

Unlike the first-order predicate logic, the notation of the Theory of Conceptual Graphs (TCG) allows for distinguishing the designations of concrete objects (concrete cities, cars, firms, etc.) and the designations of notions qualifying these objects ("a city," "a car," "a firm," etc.). The other enumerated advantages of the proposed model are also the advantages in comparison with TCG.

All enumerated advantages of the constructed model can be interpreted as the advantages in comparison with the approach to structuring the collection of primary units of conceptual level provided by Episodic Logic.

The analysis carried out above allows for drawing the conclusion that the constructed model proposes a more "fine-grained" conceptual structure of application domains in comparison with the main popular approaches to the formalization of NL-semantics; the model considerably increases the resolution possibility of the spectrum of formal tools destined for investigating various application domains.

Since the end of the 1990s, the studies on the creation of ontologies for various application domains have been quickly progressing. In these studies, the term "ontology" is interpreted as a specification of conceptualization; practically this means that an ontology enumerates the notions used in a considered group of application domains and associates the notions with their definitions and knowledge pertaining to the classes of objects qualified by some notions. The first step in each computer project of the kind consists in selecting an initial (or basics) structure of the considered application domain or a group of application domains.

It can be conjectured that the constructed mathematical model for describing a system of primary units of conceptual level used by linguistic processors and for representing the information associated with such units can find applications in the projects aimed at the elaboration of more perfect ontologies in arbitrary application domains. The reason for this hypothesis is that the elaborated model proposes a formal tool with the highest "resolution possibility" in comparison with the other known approaches to the formalization of NL-semantics and, as a consequence, to the conceptual structuring of application domains.

Problems

1. What is an n-ary relation on an arbitrary non empty set Z, where $n \geq 1$?
2. What is a binary relation on an arbitrary non empty set Z?
3. What is a reflexive relation on an arbitrary non empty set Z?
4. What is an antireflexive relation on an arbitrary non empty set Z?
5. What is a symmetric relation on an arbitrary non empty set Z?
6. What is an antisymmetric relation on an arbitrary non empty set Z?
7. What is a transitive relation on an arbitrary non empty set Z?
8. What is a partial order on an arbitrary non empty set Z?
9. What is the interpretation of the generality relation being a component of a sort system?
10. What are the mathematical properties of a generality relation on the set of sorts?

11. What is the interpretation of the tolerance relation being a component of a sort system?
12. What are the mathematical properties of a tolerance relation on the set of sorts?
13. What is the difference between a generality relation and a tolerance relation?
14. What is the difference between the types of concrete things and the types of notions qualifying these things?
15. What is the difference between the types of concrete things and the types of sets consisting of these things?
16. What kind of information is given by the symbol ↑ being the beginning of a type?
17. What is the difference between the possible types of the notions "a student" and "an academic group" ("a student group")?
18. What is the difference between the possible types of a concrete ship and a concrete squadron?
19. Give an example of an elementary compound type.
20. Why is the set of elementary types broader than the set of elementary compound types?
21. What is the spectrum of an elementary type?
22. What is a refinement of a sort?
23. What are the name and interpretation of the component X of a concept-object system $Ct = (X, V, tp, F)$?
24. What is the interpretation of the referential quantifier?
25. Is it true or not that the logical connectives $\neg, \wedge, \vee$ are the elements of the primary informational universe being a component of arbitrary concept-object system?
26. What are the structure and interpretation of a conceptual basis?

Chapter 4
A Mathematical Model for Describing Structured Meanings of Natural Language Sentences and Discourses

Abstract The purpose of this chapter is to construct a mathematical model describing a system consisting of ten partial operations on the finite sequences with the elements being structured meanings of Natural Language (NL) expressions. Informally, the goal is to develop a mathematical tool being convenient for building semantic representations both of separate sentences in NL and of complex discourses of arbitrary big length pertaining to technology, medicine, economy, and other fields of professional activity. The starting point for developing this model is the definition of the class of conceptual bases introduced in the previous chapter. The constructed mathematical model includes the definition of a new class of formal systems, or calculuses – the class of K-calculuses (knowledge calculuses) and the definition of a new class of formal languages – the class of SK-languages (standard knowledge languages).

4.1 The Essence of a New Approach to Formalizing Semantic Structure of Natural Language Texts

The analysis shows that the task of modeling numerous expressive mechanisms interacting in NL goes far beyond the scope of first-order predicate logic and beyond the scope of popular approaches to the formalization of NL semantics. That is why it seems to be reasonable to develop an original formal approach to this problem, starting from a careful consideration of the fundamental presuppositions underlying a formalism to be elaborated.

4.1.1 Toward Expanding the Universe of Formal Study

Let's imagine that we want to investigate a problem with the help of formal means and, with this aim, to consider a set of entities (real and abstract) and, besides, some relations with the attributes being the elements of these sets and some functions

V.A. Fomichov, *Semantics-Oriented Natural Language Processing*, IFSR International Series on Systems Science and Engineering 27, DOI 10.1007/978-0-387-72926-8_4, © Springer Science+Business Media, LLC 2010

with the arguments and values from this set. Then let's call such a set of entities *a universe of formal study*.

Suppose, for instance, that we consider the problem of minimizing the cost of delivery of a certain set of goods from a factory to a certain set of shops. Then the universe of formal study includes a factory, the kinds of goods, the concrete goods, the sizes of the goods of each kind, the shops, the lengths of a number of roots, etc.

Under the framework of the first-order predicate logic (FOL), the universe of formal study and the set of formulas describing the properties of the entities from the universe of formal study and the relationships between these entities are two separate sets. It is forbidden, in particular, to construct formulas of the kind $p(d_1, \ldots, d_n)$, where $n \geq 1$, p is an n-ary predicate symbol, $d_1, \ldots, d_n$ are the attributes of p, and there exists such k, $1 \leq k \leq n$ that d_k is a formula (but not a term). Due to this restriction, FOL is not convenient, in particular, for expressing the conceptual structure of sentences with direct and indirect speech and with the subordinate clauses of purpose.

The analysis carried out by the author has shown that a broadly applicable or a universal approach to the formalization of NL-semantics is to proceed from *a new look at the universe of formal study*. We need to expand the universe of formal study by means of adding to the considered set of real and abstract entities (things, situations, numbers, colors, numerical values of various parameters, etc.) the sets consisting of the entities of the following kinds:

1. The simple and compound designations of the notions (concepts) qualifying the objects;
2. The simple and compound designations of the goals of intelligent systems and of the standard ways of using the things;
3. The simple and compound designations of the sets consisting of objects or notions or goals;
4. The semantic representations of the sentences and complicated discourses pertaining to the studied application domains;
5. The finite sequences of the elements of any of the mentioned kinds;
6. The mental representations of the NL-texts as informational items having both the content and the metadata (the list of the authors, the date and language of publication, the set of application domains, etc.).

A broadly applicable mathematical framework for the investigation of NL semantics is to allow for considering the relations with the attributes being the elements of an expanded universe of the kind and the functions with the arguments and values from such an expanded universe.

These are just the unique features possessed by the theory of SK-languages (see Fig. 4.1). In particular, this theory allows for including in the universe of formal study the following elements of new kinds:

- a compound designation of a notion

$$scholar * (Field\_of\_knowledge, biology)(Degree, Ph.D.);$$

- a compound designation of a goal of a young scholar

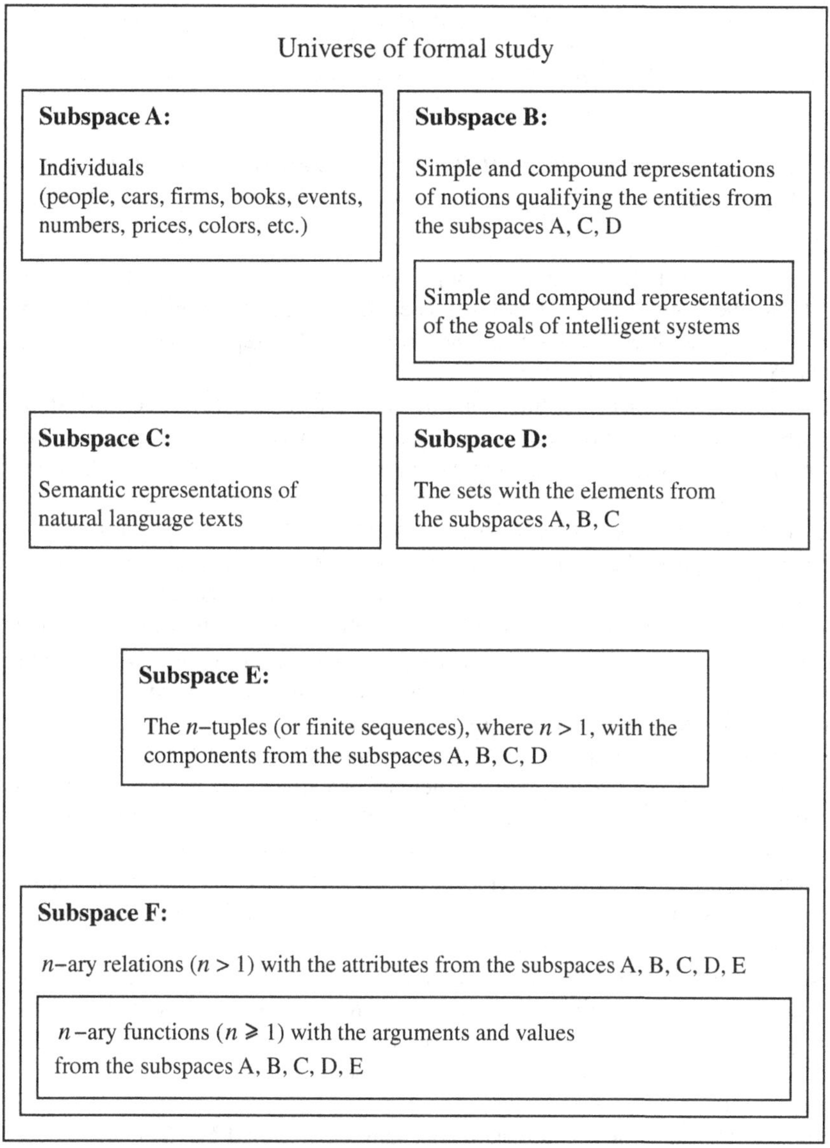

Fig. 4.1 The structure of the expanded universe of formal study

$$Defending2 * (Sci - institution, Stanford - University)$$

$$(Kind\_of\_dissertation, Ph.D.dissertation*$$

$$(Field\_of\_knowledge, computer\_science));$$

- a compound designation of a set

$$certain\ art - collection * (Quantity,\ 17)$$

$$(Qualitative\_composition,\ vase).$$

4.1.2 The Algebraic Essence of the Model Describing Conceptual Operations

During last decade, the most popular approaches to building formal representations of the meanings of NL-texts have been Discourse Representation Theory (DRT) [143, 144], Theory of Conceptual Graphs (TCG), represented, in particular, in [195, 196], and Episodic Logic (EL) [130–132, 182, 184]. In fact, DRT and TCG are oriented at describing the semantic structure of only sentences and short simple discourses. EL studies the structure of only a part of discourses, more exactly, of discourses where the time and causal relationships between the situations (called episodes) are realized.

The analysis shows that the frameworks of DRT, TCG, and EL don't allow for considering an expanded universe of formal study satisfying the requirements listed above. That is why the demand to consider an expanded universe of formal study led the author of this paper in the 1980s and 1990s to the creation of an original mathematical approach to describing conceptual (or semantic) structure of sentences and discourses in NL and operations on conceptual structures needed for building semantic representations of a broad spectrum of NL-texts.

The definition of the class of restricted standard knowledge languages (RSK-languages) [70, 76] became the first mathematically complete answer in English to the following question: how would it be possible to describe in a mathematical way a system of operations on conceptual structures allowing for building (after a finite number of steps) semantic representations (SRs) of arbitrarily complicated sentences and discourses from arbitrary application domains, starting from primary informational items.

In other words, an attempt was undertaken to elaborate a new theoretical approach enabling us to effectively describe structured meanings (or contents, or semantic structure, or conceptual structure) of real sentences and arbitrarily complicated discourses pertaining to technology, medicine, business, etc.

Expanding this approach to studying semantics of NL, let's consider the main ideas of determining a new class of formal languages called SK-languages. Our starting point will be the definition of the class of conceptual bases introduced in the preceding chapter.

Each conceptual basis B determines three classes of formulas, the first class $Ls(B)$ being considered as the principal one and being called *the SK-language (standard knowledge language) in the stationary conceptual basis B*. Its strings (they are called K-strings) are convenient for building SRs of NL-texts. We'll consider below only the formulas from the first class $Ls(B)$.

In order to determine for arbitrary conceptual basis B three classes of formulas, a collection of some rules $P[0], P[1], \ldots, P[10]$ for building well-formed expressions is defined. The rule $P[0]$ provides an initial stock of formulas from the first class. For example, there is such a conceptual basis B that, according to $P[0]$, $Ls(B)$ includes the elements

$$container1, blue, country, France, set, 12, all, arbitrary,$$

$$Height, Distance, Quantity, Authors, Friends, Suppliers, x1, x2, e3, P7.$$

For arbitrary conceptual basis B, let $Degr(B)$ be the union of all Cartesian m-degrees of $Ls(B)$, where $m \geq 1$. Then the meaning of the rules of constructing well-formed formulas $P[0], P[1], \ldots, P[10]$ can be explained as follows: for each k from 1 to 10, the rule $P[k]$ determines a partial unary operation $Op[k]$ on the set $Degr(B)$ with the value being an element of $Ls(B)$.

For instance, there is such a conceptual basis B that the value of the partial operation $Op[7]$ (it governs the use of logical connectives AND and OR) on the four-tuple

$$\langle \wedge, Belgium, The - Netherlands, Luxemburg \rangle$$

is the K-string

$$(Belgium \wedge The - Netherlands \wedge Luxemburg).$$

Thus, the essence of the basic model of the theory of SK-languages is as follows: this model determines a partial algebra of the form

$$(Degr(B), Operations(B)),$$

where $Degr(B)$ is the carrier of the partial algebra, $Operations(B)$ is the set consisting of the partial unary operations $Op[1], \ldots, Op[10]$ on $Degr(B)$.

4.1.3 Shortly About the Rules for Building Semantic Representations of Natural Language Texts

It was mentioned above that the goal of introducing the notion of conceptual basis is to get a starting point for constructing a mathematical model describing (a) the regularities of structured meanings both of separate sentences and of complex discourses in NL; (b) a collection of the rules allowing for building semantic representations both of sentences and complex discourses in NL, starting from the primary units of conceptual level and using a small number of special symbols.

Let's consider now the basic ideas underlying the definitions of the rules intended for building semantic representations of sentences and complex discourses in NL. These ideas are stated informally, with the help of examples. The exact mathematical definitions can be found in the next chapter.

Let's regard (ignoring many details) the structure of strings which can be obtained by applying any of the rules $P[1], \ldots, P[10]$ at the last step of inferring the formulas. The rule $P[1]$ enables us to build K-strings of the form *Quant Conc*, where *Quant* is a semantic item corresponding to the meanings of such words and expressions as "certain," "any," "arbitrary," "each," "all," "several," etc. (such semantic items will be called *intensional quantifiers*), and *Conc* is a designation (simple or compound) of a concept. The examples of K-strings for $P[1]$ as the last applied rule are as follows:

$$certn \, container1, \; all \, container1,$$

$$certn \, consignment,$$

$$certn \, container1 * (Content1, \, ceramics),$$

where the last expression is built with the help of both the rules $P[0]$, $P[1]$ and the rule with the number 4, the symbol *certn* is to be interpreted as the informational item corresponding to the expression "a certain."

The rule $P[2]$ allows for constructing the strings of the form $f(a_1, \ldots, a_n)$, where f is a designation of a function, $n \geq 1$, $a_1, \ldots, a_n$ are K-strings built with the help of any rules from the list $P[0], \ldots, P[10]$. The examples of K-strings built with the help of $P[2]$ are as follows:

$$Distance(Moscow, Paris),$$

$$Weight(certn \, container1 * (Color, \, blue)(Content1, \, ceramics)).$$

Using the rule $P[3]$, we can build the strings of the form $(a1 \equiv a2)$, where $a1$ and $a2$ are K-strings formed with the help of any rules from $P[0], \ldots, P[10]$, and $a1$ and $a2$ represent the entities being homogeneous in a certain sense. The following expressions are the examples of K-strings constructed as a result of employing the rule $P[3]$ at the last step of inference:

$$(Distance(Moscow, Paris) \equiv x1),$$

$$(y1 \equiv y3), \; (Height(certn \, container1) \equiv 2/m).$$

The rule $P[4]$ is intended, in particular, for constructing K-strings of the form $rel(a_1, \ldots, a_n)$, where *rel* is a designation of *n*-ary relation, $n \geq 1$, $a_1, \ldots, a_n$ are the K-strings formed with the aid of some rules from $P[0], \ldots, P[10]$. The examples of K-strings for $P[4]$:

$$Belong(Bonn, Cities(Germany)),$$

$$Subset(certn \, series1 * (Name - origin, \, tetracyclin), \, all \, antibiotic).$$

The rule $P[5]$ enables us to construct the K-strings of the form $Expr : v$, where *Expr* is a K-string not including v, v is a variable, and some other conditions are satisfied. Using $P[5]$, one can mark by variables in the semantic representation of any NL-text: (a) the descriptions of diverse entities mentioned in the text (physical objects, events, concepts, etc.), (b) the semantic representations (SRs) of sentences and of larger texts' fragments to which a reference is given in any part of a text. The examples of K-strings for $P[5]$ are as follows:

$$certn\,container1 : x3,$$

$$Higher(certn\,container1 : x3, certn\,container1 : x5) : P1.$$

The rule $P[5]$ provides the possibility to form SRs of texts in such a manner that these SRs reflect the referential structure of NL-texts. This means that an SR of an NL-text includes the variables being the unique marks of the various entities mentioned in the text; the set of such entities can include the structured meanings of some sentences and larger fragments of a discourse being referred to in this discourse.

The rule $P[6]$ provides the possibility to build the K-strings of the form $\neg Expr$, where $Expr$ is a K-string satisfying a number of conditions. The examples of K-strings for $P[6]$ are as follows:

$$\neg antibiotic,$$

$$\neg Belong(penicillin, certn\,series1 * (Name - origin, tetracyclin)).$$

Using the rule $P[7]$, one can build the K-strings of the form

$$(a_1 \wedge a_2 \wedge \ldots \wedge a_n)$$

or of the form

$$(a_1 \vee a_2 \vee \ldots \vee a_n),$$

where $n > 1$, $a_1, \ldots, a_n$ are the K-strings designating the entities which are homogeneous in some sense. In particular, $a_1, \ldots, a_n$ may be SRs of assertions (or propositions), descriptions of physical things, descriptions of sets consisting of things of the same kind, descriptions of concepts. The following strings are examples of K-strings for $P[7]$:

$$(streptococcus \vee staphylococcus),$$

$$(Belong((Bonn \wedge Hamburg \wedge Stuttgart), Cities(Germany))$$

$$\wedge \neg Belong(Bonn, Cities((Finland \vee Norway \vee Sweden)))).$$

The rule $P[8]$ allows us to build, in particular, the K-strings of the form

$$cpt * (rel_1, val_1), \ldots, (rel_n, val_n),$$

where cpt is an informational item from the primary informational universe $X(B)$ designating a concept (a notion), for $k = 1, \ldots, n$, rel_k is the name of a function with one argument or of a binary relation, val_k designates a possible value of rel_k for objects characterized by the concept cpt in case rel_k is the name of a function, and val_k designates the second attribute of rel_k in case rel_k is the name of a binary relation. The following expressions are the examples of K-strings obtained as a result of applying the rule $P[8]$ on the final step of inference:

$$container1 * (Content1, ceramics),$$

$$consignment * (Quantity, 12)(Compos1,$$

$$container1 * (Content1, ceramics)).$$

The rule $P[9]$ enables us to build, in particular, the K-strings of the forms $\exists v(conc)D$ and $\forall v(conc)D$, where $\forall$ is the universal quantifier, $\exists$ is the existential quantifier, $conc$ and D are K-strings, $conc$ is a designation of a primary notion ("person," "city," "integer," etc.) or of a compound notion ("an integer greater than 200," etc.). D may be interpreted as a semantic representation of an assertion with the variable v about any entity qualified by the concept $conc$. The examples of K-strings for $P[9]$ are as follows:

$$\forall n1 \, (integer)\exists n2 \, (integer)Less(n1, n2),$$

$$\exists y \, (country * (Location, Europe))Greater(Quantity(Cities(y)), 15).$$

The rule $P[10]$ is intended for constructing, in particular, the K-strings of the form $\langle a_1, \dots a_n \rangle$, where $n > 1$, $a_1, \dots, a_n$ are the K-strings. The strings obtained with the help of $P[10]$ at the last step of inference are interpreted as designations of n-tuples. The components of such n-tuples may be not only designations of numbers, things, but also SRs of assertions, designations of sets, concepts, etc.

4.1.4 The Scheme of Determining Three Classes of Formulas Generated by a Conceptual Basis

Let's consider in more detail the suggested original scheme of an approach to determining three classes of well-formed expressions called formulas.

The notions introduced above enable us to determine for every conceptual basis B a set of formulas $Forms(B)$ being convenient for describing structured meanings (SMs) of NL-texts and operations on SMs.

Definition 4.1. Let B be an arbitrary conceptual basis, $Specsymbols$ be the set consisting of the symbols $/,/$ (comma), $/(/, /)/, / : /, / * /, /\langle /, /\rangle /$. Then

$$D(B) = X(B) \cup V(B) \cup Specsymbols,$$

$$Ds(B) = D(B) \cup \{/\&/\},$$

$D^+(B)$ and $Ds^+(B)$ are the sets of all non empty finite sequences of the elements from $D(B)$ and $Ds(B)$ respectively.

The essence of the approach to determining conceptual formulas proposed in this book is as follows. As stated above, some assertions $P[0], P[1], \dots, P[10]$ will be defined; they are interpreted as the rules of building semantic representations of NL-texts from the elements of the primary universe $X(B)$, variables from $V(B)$, and several symbols being the elements of the set $Specsymbols$. The rule $P[0]$ provides an initial stock of formulas.

If $1 \leq i \leq 10$, then for an arbitrary conceptual basis B and for $k = 1, \ldots, i$, the assertions $P[0], P[1], \ldots, P[i]$ determine by conjoint induction some sets of formulas

$$Lnr_i(B) \subset D^+(B),$$

$$T^0(B), Tnr_i^1(B), \ldots, Tnr_i^i(B) \subset Ds^+(B),$$

$$Ynr_i^1(B), \ldots, Ynr_i^i(B) \subset Ds^+(B).$$

The set $Lnr_i(B)$ is considered as the main subclass of formulas generated by $P[0], \ldots, P[i]$. The formulas from this set are intended for describing structured meanings (or semantic content) of NL-texts.

If $1 \leq k \leq i$, the set $Tnr_i^k(B)$ consists of strings of the form $b \& t$, where $b \in Lnr_i(B)$, $t \in Tp(S(B))$, and b is obtained by means of applying the rule $P[k]$ to some simpler formulas at the final step of an inference. It should be added that for constructing b from the elements of $X(B)$ and $V(B)$, one may use any of the rules $P[0], \ldots, P[k], \ldots, P[i]$; these rules may be applied arbitrarily many times.

If a conceptual basis B is chosen to describe a certain application domain, then b can be interpreted as a semantic representation of a text or as a fragment of an SR of a text pertaining to the considered domain. In this case, t may be considered as the designation of the kind of the entity qualified by such SR or by a fragment of SR. Besides, t may qualify b as a semantic representation of a narrative text.

The number i is interpreted in these denotations as the maximal ordered number of such rules from the list $P[0], P[1], \ldots, P[10]$ that these rules may be employed for building semantic representations of NL-expressions or for constructing knowledge modules.

For instance, it will be shown below that the sets $Lnr_4(B_1), \ldots, Lnr_{10}(B_1)$ include the formulas

$$Elem(P.Somov, Friends(J.Price)),$$

$$Elem(Firm\_Ocean, Suppliers(Firm\_Rainbow)),$$

and the sets $Tnr_4^4(B_1), \ldots, Tnr_{10}^4(B_1)$ include the formulas

$$Elem(P.Somov, Friends(J.Price)) \& prop,$$

$$Elem(Firm\_Ocean, Suppliers(Firm\_Rainbow)) \& prop,$$

where $prop$ is the distinguished sort $P(B_1)$ ("a meaning of proposition").

Each string $c \in Ynr_i^k(B)$, where $1 \leq k \leq i$, can be represented in the form

$$c = a_1 \& \ldots \& a_m \& b,$$

where $a_1, \ldots, a_m, b \in Lnr_i(B)$. Besides, there is such type $t \in Tp(S(B))$ that the string $b \& t$ belongs to $Tnr_i^k(B)$.

The strings $a_1, \ldots, a_m$ are obtained by employing some rules from the list $P[0], \ldots, P[i]$, and b is constructed from "blocks" $a_1, \ldots, a_m$ (some of them could be a little bit changed) by applying just one time the rule $P[k]$. The possible quantity of "blocks" $a_1, \ldots, a_m$ depends on k. Thus, the set $Ynr_i^k(B)$ fixes the result of applying the rule $P[k]$ just one time.

For instance, we'll see below that the sets $Ynr_4^4(B_1), \ldots, Ynr_{10}^4(B_1)$ include the formulas

$$Elem \& P.Somov \& Friends(J.Price) \& Elem(P.Somov, Friends(J.Price)),$$

$$Elem \& Firm\_Ocean \& Suppliers(Firm\_Rainbow)$$

$$\& Elem(Firm\_Ocean, Suppliers(Firm\_Rainbow)).$$

Let for $i = 1, \ldots, 10$,

$$T_i(B) = T^0(B) \cup Tnr_i^1(B) \cup \ldots \cup Tnr_i^i(B);$$

$$Y_i(B) = {}^{\cdot}Ynr_i^1(B) \cup \ldots \cup Ynr_i^i(B);$$

$$Form_i(B) = Lnr_i(B) \cup T_i(B) \cup Y_i(B).$$

We'll interpret $Form_i(B)$ as the set of all formulas generated by the conceptual basis B with the help of the rules $P[0], \ldots, P[i]$. This set is the union of three classes of formulas, the principal class being $Lnr_i(B)$. The formulas from these three classes will be called l-formulas, t-formulas, and y-formulas respectively (see. Fig. 4.2).

The class of l-formulas is needed for assigning a type from $Tp(S(B))$ to each $b \in Lnr_i(B)$, where $i = 1, \ldots, 10$. For $q = 0, \ldots, 9$, $Lnr_q(B) \subseteq Lnr_{q+1}(B)$. The set $Lnr_{10}(B)$ is called *the standard knowledge language* (or *SK-language, standard K-language*) in the stationary conceptual basis B and is designated as $Ls(B)$. That's why l-formulas will be often called also *K-strings*.

The set $T_{10}(B)$ is designated as $Ts(B)$. For every conceptual basis B and arbitrary formula A from the set $Ts(B)$, there exist such type $t \in Tp(S(B))$ and such formula $C \in Ls(B)$ that $A = C \& t$. We'll employ only the formulas from the subclasses $Ls(B)$ and $Ts(B)$ (i.e., only l-formulas and t-formulas) for constructing semantic representations of NL-texts. y-formulas are considered as auxiliary ones, such formulas are needed for studying the properties of the sets $Ls(B)$ and $Ts(B)$.

The pairs of the form $(B, Rules)$, where B is a conceptual basis, $Rules$ is the set consisting of the rules $P[0], \ldots, P[10]$, will be called *the K-calculuses (knowledge calculuses)*.

4.2 The Use of Intensional Quantifiers in Formulas

The term "intensional quantifier" was introduced in Chap. 3 for denoting the conceptual items (in other words, semantic items, informational items) associated, in particular, with the words and word combinations "every," "a certain," "arbitrary," "any," "all," "almost all," "a few," "several," "many."

The set of intensional quantifiers consists of two subclasses Int_1 and Int_2. These subclasses are defined in the following way.

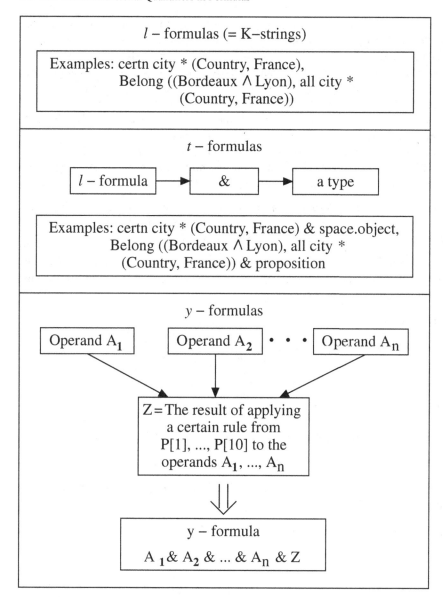

Fig. 4.2 Three classes of formulas determined by a conceptual basis

Each conceptual base B is a system of the form (S, Ct, Ql). The component Ql is a finite sequence of formal objects including, in particular, the distinguished sorts int_1 and int_2. This enables us to define the set $Int_m(B)$, where $m = 1, 2$, as the set of all such conceptual items qtr from the primary informational universe $X(B)$ that $tp(qtr) = int_m(B)$.

The elements of the set Int_1 designate, in particular, the meanings of the expressions "every," "a certain," "arbitrary," "any." The elements of the set Int_2 are interpreted as the denotations of the meanings, in particular, of the expressions "all," "almost all," "a few," "several," "many"; such words and word combinations are used for forming the designations of the sets. The minimal requirement to the set Int_1 is that this set includes the conceptual item associated with the word combination "a certain"; this conceptual item is called *the referential quantifier*. The minimal requirement to the set Int_2 is that Int_2 includes the conceptual item *all*.

The rule $P[1]$ allows us to join the intensional quantifiers to simple designations (a descriptor) or compound designations of the notions (concepts) . As a result of applying this rule one obtains, in particular

- the l-formulas of the form $qtr\,cpt$, where qtr is an intensional quantifier from the set $Int(B)$, cpt is a designation of a simple or compound notion;
- t-formulas of the form $qtr\,cpt\,\&\,t$, where t is a type from the set $Tp(S(B))$.

For instance, it is possible to define a conceptual basis B in such a way that, using the rule $P[0]$ during the first step of the construction process and the rule $P[1]$ during the final step, it will be possible to build:

- the l-formulas

$$certn\,city,\ certn\,city * (Name1,\ London),$$

$$every\,city,\ every\,person * (Profession,\ painter),$$

$$all\,city,\ all\,city * (Country,\ Russia),$$

- the t-formulas

$$certn\,city\,\&\,space.object,\ all\,city\,\&\,\{space.object\},$$

$$every\,person * (Profession,\ painter)\,\&\,ints * dyn.phys.ob,$$

where $ints$ is the sort "intelligent system," $dyn.phys.phys.ob$ is the sort "dynamic physical object."

Definition 4.2. If B is a conceptual basis, then for $m = 1, 2$,

$$Int_m(B) = \{qtr \in X(B) \mid tp(qtr) = int_m(B)\},$$

$$Int(B) = Int_1(B) \cup Int_2(B),$$

$$Tconc(B) = \{t \in Tp(S(B)) \mid t \text{ has the beginning } \uparrow\} \cup Spectp,$$

where

$$Spectp = \{[\uparrow entity],\ [\uparrow concept],\ [\uparrow object]\},$$

and the elements of the set $Spectp$ are interpreted as the types of informational units associated with the *words* "entity," "notion" ("concept"), "object," respectively.

Using the rules $P[0]$ and $P[l]$, we can build some strings of the form

$$quantifier\ concept\_descr,$$

where

$$quantifier \in Int(B),\ concept\_descr \in X(B),$$

$$tp(concept\_descr) \in Tconc(B).$$

For instance, if B_1 is the conceptual basis determined in Chap. 3, we can construct the l-formulas

$$certn\,person,\ certn\,tour.gr,\ certn\,concept,$$

$$all\,person,\ all\,tour.gr,\ all\,concept$$

generated by B_1.

It is possible also to build more complex strings of the form $qt\,cpt$, where qt is an intensional quantifier, with the help of the rule $P[1]$, using preliminary the rule $P[8]$ (see next sections) and, may be, some other rules (besides the rules $P[0]$ and $P[1]$) for constructing the string cpt denoting a concept. For instance, it will be possible to build the l-formulas

$$certain\,tour.group * (Numb,\ 12),$$

$$all\,tour.group * (Numb,\ 12)$$

in the conceptual basis B_1 constructed in Chap. 3. These formulas are interpreted as semantic representations of the expressions "a certain tourist group consisting of 12 persons" and "all tourist groups consisting of 12 persons."

The transition from an l-formula cpt designating a notion to a certain l-formula $qtr\,cpt$, where qtz is an intensional quantifier, is described with the help of a special function h.

Definition 4.3. Let B be a conceptual basis, $S = S(B)$. Then the mapping h : $\{1, 2\} \times Tp(S) \to Tp(S)$ is determined as follows: if $u \in Tp(S)$ and the string $\uparrow u$ belongs to Tp(S), then

$$h(1, \uparrow u) = u,\ h(2, \uparrow u) = \{u\};$$

$$h(1, [\uparrow entity]) = [entity],\ h(2, [\uparrow entity]) = \{[entity]\};$$

$$h(1, [\uparrow concept]) = [concept],\ h(2, [\uparrow concept]) = \{[concept]\};$$

$$h(1, [\uparrow object]) = [object],\ h(2, [\uparrow object]) = \{[object]\}.$$

From the standpoint of building semantic representations (SRs) of texts, the mapping h describes the transformations of the types in the course of the transition:

- from the notions "a person," "a tourist group" to the SRs of expressions "some person," "arbitrary person," "some tourist group," "arbitrary tourist group," etc. (in case the first argument of h is 1) and to the SRs of expressions "all people," "all tourist groups," etc. (if the first argument of h is 2);

- from the notions "an entity," "an object," "a concept" to the SRs of the expressions "some entity," "arbitrary entity," "some object," "arbitrary object," "some concept," "arbitrary concept," etc. (when the first argument of h is 1) and to the SRs of the expressions "all entities," "all objects," "all concepts," etc. (when the first argument of h is 2).

Definition 4.4. Denote by $P[1]$ the assertion "Let $cpt \in L(B) \setminus V(B)$, the type $u \in Tconc(B)$, $k \in \{0, 8\}$, and the string $cpt \& u$ belong to $T^k(B)$. Let $m \in \{1, 2\}$, $qtr \in Int_m$, $t = h(m, u)$, and b be the string of the form $qtr\, cpt$. Then $b \in L(B)$, the string of the form $b \& t$ belongs to $T^1(B)$, and the string of the form $qtr \& cpt \& b$ belongs to $Y^1(B)$."

Example 1. Let B be the conceptual basis B_1 constructed in Chap. 3;
$L(B)$, $T^0(B)$, $T^1(B)$, $Y^1(B)$ be the least sets of formulas jointly defined by the assertions $P[0]$ and $P[1]$. Then it is easy to verify that the following relationships take place:

$$person \in L(B) \setminus V(B),\ person \& \uparrow ints * dyn.phys.ob \in T^0(B),$$

$$certn \in Int_1(B),\ h(1, \uparrow ints * dyn.phys.ob) = ints * dyn.phys.ob \Rightarrow$$

$$certn\, person \in L(B),\ certn\, person \& ints * dyn.phys.ob \in T^1(B),$$

$$certn \& person \& certn\, person \in Y^1(B);$$

$$all \in Int_2(B),\ h(2, \uparrow ints * dyn.phys.ob) = \{ints * dyn.phys.ob\} \Rightarrow$$

$$all\, person \in L(B),\ all\, person \& \{ints * dyn.phys.ob\} \in T^1(B),$$

$$all \& person \& all\, person \in Y^1(B);$$

$$tour.group \in L(B) \setminus V(B),$$

$$tour.group \& \uparrow \{ints * dyn.phys.ob\} \in T^0(B),$$

$$h(1, \uparrow \{ints * dyn.phys.ob\}) = \{ints * dyn.phys.ob\},$$

$$h(2, \uparrow \{ints * dyn.phys.ob\}) = \{\{ints * dyn.phys.ob\}\} \Rightarrow$$

$$certn\, tour.group,\ all\, tour.group \in L(B),$$

$$certn\, tour.group \& \{ints * dyn.phys.ob\},$$

$$all\, tour.group \& \{\{ints * dyn.phys.ob\}\} \in T^1(B),$$

$$certn \& tour.group \& certn\, tour.gr \in Y^1(B),$$

$$all \& tour.group \& all\, tour.gr \in Y^1(B),$$

$$concept \in L(B) \setminus V(B),\ concept \& [\uparrow concept] \in T^0(B);$$

$$h(1, [\uparrow concept]) = [concept],$$

$$h(2, [\uparrow concept]) = \{[concept]\} \Rightarrow$$

$$certn\, concept,\, all\, concept \in L(B),$$

$$certn\, concept\,\&\,[concept],\ all\, concept\,\&\,\{[concept]\} \in T^1(B),$$

$$certn\,\&\,concept\,\&\,certn\, concept \in Y^1(B),$$

$$all\,\&\,concept\,\&\,all\, concept \in Y^1(B).$$

Comment to the rule P[1]. The fragment of the rule $P[1]$ "Let $cpt \in L(B) \setminus V(B)$, the type $u \in Tconc(B)$, $k \in \{0, 8\}$, and the string $cpt\,\&\,u$ belong to $T^k(B)$" means that u is the type of a notion, i.e. either u has the beginning $\uparrow$ or u is one of the symbols $[\uparrow entity]$, $[\uparrow concept]$, $[\uparrow object]$.

If $k = 0$, cpt designates a primitive (unstructured) concept. If $k = 1$, cpt is a compound denotation of a concept. Such compound denotations of the concepts will be constructed with the help of the rule $P[8]$; the examples of the kind are the expressions

$$person * (Activity.field,\, biology),\ city * (Country,\, Russia).$$

The rule $P[1]$ will be very often used below for constructing semantic representations of NL-texts, because it is necessary for building semantic images of the expressions formed by the nouns with dependent words. For example, let Qs1 be the question "Where has been the two-tonne green container delivered from?" Then, as a result of fulfilling the first step of constructing a SR of Qs1, it is possible to obtain the expression

$$certn\, container1 * (Weight,\, 2/tonne)(Color,\, green),$$

and after fulfilling the final step of constructing a SR of Qs1, one is able to obtain the expression

$$Question(x1,\, Situation(e1,\, delivery2 * (Goal - place,\, x1)$$

$$(Object1,\, certn\, container1 * (Weight,\, 2/tonne)(Color,\, green)))).$$

Often used notations. We'll define in what follows some rules $P[2], \ldots, P[10]$. For $k = 1, \ldots, 10$, the rule $P[k]$ states that a certain formula b belongs to $L(B)$, a certain formula $b\,\&\,t$ belongs to the set $T^k(B)$, where $t \in Tp(S(B))$, and a certain formula z belongs to the set $Y^k(B)$. If $1 \le s \le 10$, B is any conceptual basis, then the rules $P[0]$, $P[1], \ldots, P[s]$ determine by conjoint induction the sets of formulas

$$L(B),\, T^0(B),\, T^1(B), \ldots, T^s(B),\, Y^1(B), \ldots, Y^s(B).$$

Let's denote these sets by

$$Lnr_s(B),\, T^0(B),\, Tnr_s^1(B), \ldots, Tnr_s^s(B),\, Ynr_s^1(B), \ldots, Ynr_s^s(B)$$

and denote the family consisting of all these sets by $Globset_s(B)$.

Let $n > 1$, $Z_1, \ldots, Z_n \in Globset_s(B)$, $w_1 \in Z_1, \ldots, w_n \in Z_n$.

Then, if these relationships for the formulas $w_1, \ldots, w_n$ are the consequence of employing some rules $P[l_1], \ldots, P[l_m]$, where $m \geq 1$, we'll denote this fact by the expression of the form

$$B(l_1, \ldots, l_m) \Rightarrow w_1 \in Z_1, \ldots, w_n \in Z_n.$$

The sequence $l_1, \ldots, l_m$ may contain the repeated numbers.

In the expressions of the kind we'll often omit the symbol B in the designations of the sets $Z_1, \ldots, Z_n$; besides, we'll use the expressions $w_1, w_2 \in Z_1$, $w_3, w_4, w_5 \in Z_2$, and so on.

Example 2. It was shown above that

$$B_1(0, 1) \Rightarrow all\, person,\, all\, tour.group \in Lnr_1(B_1),$$

$$B_1(0, 1) \Rightarrow all\, person\, \&\, \{ints * dyn.phys.ob\},$$

$$all\, tour.group\, \&\, \{\{ints * dyn.phys.ob\}\} \in Tnr_1^1(B_1),$$

$$B_1(0,1) \Rightarrow all\, \&\, person\, \&\, all\, person \in Ynr_1^1(B_1),$$

$$all\, \&\, tour.group\, \&\, all\, tour.group \in Ynr_1^1(B_1).$$

where *tour.group* is the designation of the notion "a tourist group," *ints* is the sort "intelligent system."

The expression

$$B_1(0, 1) \Rightarrow all\, \&\, person\, \&\, all\, person \in Ynr_1^1(B_1)$$

is equivalent to the expression

$$B_1(0, 1) \Rightarrow all\, \&\, person\, \&\, all\, person \in Ynr_1^1.$$

4.3 The Use of Relational Symbols and the Marking-Up of Formulas

This section introduces and illustrates the application of the rules $P[2] - P[5]$ intended for building semantic representations of NL-texts.

4.3.1 The Rules for Employing Relational Symbols

The rule $P[2]$ enables us, in particular, to construct the K-strings of the form $f(a_1, \ldots, a_n)$, where f is a designation of a function with n arguments $a_1, \ldots, a_n$.

The rule $P[3]$ is intended for building the K-strings of the form $(a_1 \equiv a_2)$, where a_1 and a_2 denote the entities characterized by the types being comparable with respect to the relation $\vdash$.

Using consecutively the rules $P[2]$ and $P[3]$, we can build the K-strings of the form $(f(a_1, \ldots, a_n) \equiv b)$, where b is the value of the function f for $a_1, \ldots, a_n$.

Let's recall that, according to the definitions given in Chap. 3, for arbitrary sort system S, the set of main types

$$Mtp(S) = Tp(S) \setminus \{[\uparrow entity], [\uparrow concept], [\uparrow object]\}.$$

Definition 4.5. Let B be any conceptual basis, $S = S(B)$. Then

- $R_1(B)$ is the set of all such $d \in X(B)$ that for each d there is such $t \in Mtp(S)$ that (a) t has no beginning "(" and (b) $tp(d)$ is the string of the form $\{t\}$;
- for arbitrary $n > 1$, $R_n(B) = \{d \in X(B) \mid$ there are such $t_1, \ldots, t_n \in Mtp(S)$ that the type $tp(d)$ is the string of the form $\{(t_1, \ldots, t_n)\}\}$;
- for arbitrary $n > 1$, $F_n(B) = F(B) \cup R_{n+1}(B)$.

If $n > 1$, the elements of $R_n(B)$ will be called *n-ary relational symbols,* and the elements of $F_n(B)$ will be called additionally *n-ary functional symbols.*

It is easy to show that for arbitrary conceptual basis B and arbitrary $k, m > 1$, it follows from $k \neq m$ that $R_k(B) \cap R_m(B) = \emptyset$.

Definition 4.6. Denote by $P[2]$ the assertion "Let $n \geq 1$, $f \in F_n(B)$,

$$tp = tp(B), u_1, \ldots, u_n, t \in Mtp(S(B)),$$

$$tp(f) = \{(u_1, \ldots, u_n, t)\};$$

for $j = 1, \ldots, n$, $0 \leq k[j] \leq i$, $z_j \in Mtp(S(B))$, $a_j \in L(B)$; the string $a_j \& z_j$ belong to $T^{k[j]}(B)$; if a_j doesn't belong to the set of variables $V(B)$, then $u_j \vdash z_j$ (i.e. the type z_j is a concretization of the type u_j); if $a_j \in V(B)$, then u_j and z_j are comparable with respect to the concretization relation $\vdash$. Let b be the string of the form $f(a_1, \ldots, a_n)$. Then

$$b \in L(B), b \& t \in T^2(B),$$

$$f \& a_1 \& \ldots \& a_n \& b \in Y^2(B)."$$

It should be recalled before the formulation of the next definition that, according to the definition of conceptual basis, the symbol $\equiv$ is an element of the primary informational universe $X(B)$ for arbitrary conceptual basis B.

Definition 4.7. Denote by $P[3]$ the assertion "Let $a_1, a_2 \in L(B)$, the types u_1, u_2 belong to the set of main types $Mtp(S(B))$, u_1 and u_2 are comparable with respect to the concretization relation $\vdash$. Let for $m = 1, 2$, $0 \leq k[m] \leq i$, the string $a_m \& u_m \in T^{k[m]}(B)$; P be the sort "a meaning of proposition" of the conceptual basis B, and b be the string $(a_1 \equiv a_2)$. Then

$$b \in L(B), \ b \& P \in T^3(B),$$

and the string $a_1 \& \equiv \& a_2 \& b$ belongs to the set $Y^3(B)$."

In the rules $P[2]$ and $P[3]$, the symbol i designates an unknown integer, such that $2 \leq i \leq 10$. The interpretation of the symbol i is as follows:

The rules $P[0] - P[3]$ and further rules will be used together with a definition joining all these rules and having the initial phrase "Let B be an arbitrary conceptual basis, $1 \leq i \leq 10$." The number i will be interpreted as the maximal ordered number of a rule in the collection of the rules which will be used for constructing the formulas. For instance, if $i = 3$, we may use the rules with the numbers 0–3, but we can't employ the rules with the numbers 4–10. The parameter i in the rules for constructing the formulas enables us to define the language Lnr_{i+1} after introducing the rule $P[i+1]$ and to study the expressive possibilities of this language.

Example 1. Let B_1 be the conceptual basis defined in Chap. 2; $i = 3$;

$$b_1 = Suppliers(Firm\_Rainbow),$$

$$b_2 = Numb(Suppliers(Firm\_Rainbow)),$$

$$b_3 = (Numb(Suppliers(Firm\_Rainbow)) \equiv 12),$$

$$b_4 = Numb(all \, concept),$$

$$b_5 = Numb(all \, chemist),$$

$$b_6 = (all \, chemist \equiv S1),$$

$$b_7 = Numb(S1).$$

Then one can easily verify the validity of the following relationships (taking into account the notation introduced in the preceding section):

$$B_1(0) \Rightarrow Suppliers \& \{(org, \{org\})\} \in T^0,$$

$$Firm\_Rainbow \& org * space.ob * ints \in T^0,$$

$$Numb \& \{(\{[entity]\}, nat)\} \in T^0;$$

$$B_1(0, 1, 2) \Rightarrow b_1 \& \{org\} \in Tnr_3^2;$$

$$B_1(0, 2, 2) \Rightarrow b_2 \in Lnr_3, \ b_2 \& nat \in Tnr_3^2,$$

$$Numb \& b_1 \& b_2 \in Ynr_3^2;$$

$$B_1(0, 2, 2, 0, 3) \Rightarrow b_3 \in Lnr_3, \ b_3 \& prop \in Tnr_3^3;$$

$$B_1(0, 1, 2) \Rightarrow b_4, b_5 \in Lnr_3,$$

$$b_4 \& nat, \ b_5 \& nat \in Tnr_3^2;$$

$$B_1(0, 1, 3) \Rightarrow b_6 \in Lnr_3, \ b_6 \& prop \in Tnr_3^3;$$

$$S1 \in V(B_1), \, tp(S1) = \{[entity]\} \Longrightarrow$$

$$B_1(0, 2) \Rightarrow b_7 \in Lnr_3, \, b_7 \,\&\, nat \in Tnr_3^2.$$

Definition 4.8. Denote by $P[4]$ the assertion "Let

$$n \geq 1, \, r \in R_n(B) \setminus F(B), \, u_1, \ldots, u_n \in Mtp(S(B)), \, tp = tp(B),$$

$tp(r)$ be the string of the form $\{(u_1, \ldots, u_n)\}$ in case $n > 1$ or the string of the form $\{u_1\}$ in case $n = 1$; for $j = 1, \ldots, n$, $0 \leq k[j] \leq i$, $z_j \in Mtp(S(B))$, $a_j \in L(B)$;

the strings $a_j \,\&\, z_j$ belong to $T^{k[j]}(B)$; if a_j doesn't belong to the set of variables $V(B)$, then $u_j \vdash z_j$ (i.e. the type z_j is a concretization of the type u_j); if $a_j \in V(B)$, then u_j and z_j are comparable with respect to the concretization relation $\vdash$.

Let b be the string of the form $r(a_1, \ldots, a_n)$, $P = P(B)$ be the sort "a meaning of proposition" of the conceptual basis B. Then

$$b \in L(B), \, b \,\&\, P \in T^4(B),$$

$$r \,\&\, a_1 \,\&\, \ldots \,\&\, a_n \,\&\, b \in Y^4(B)."$$

Example 2. Let B_1 be the conceptual basis defined in Chapter 3; $i = 4$;

$$b_8 = Less(10000, Numb(all\ chemist)),$$

$$b_9 = Less(5000, Numb(all\ concept)),$$

$$b_{10} = Elem(P.Somov, all\ person),$$

$$b_{11} = Elem(Firm\_Ocean, Suppliers(Firm\_Rainbow)),$$

$$b_{12} = (R.Scott \equiv Director(Firm\_Ocean)),$$

$$b_{13} = Knows(N.Cope, (Numb(Suppliers(Firm\_Rainbow)) \equiv 12)),$$

$$b_{14} = Knows(P.Somov, (R.Scott \equiv Director(Firm\_Ocean))),$$

$$b_{15} = Less(10000, Numb(S1)),$$

$$b_{16} = Subset(all\ chemist, all\ person).$$

Taking into account the definition of the concept-object system introduced in the preceding chapter and employing the rules $P[0], \ldots, P[4]$, we have, obviously, the following relationships:

$$B_1(0, 1, 2, 3, 4) \Rightarrow b_8, \ldots, b_{16} \in Lnr_4,$$

$$b_8 \,\&\, prop, \ldots, b_{16} \,\&\, prop \in Tnr_4^4;$$

$$Subset \,\&\, all\ chemist \,\&\, all\ person \,\&\, Subset(all\ chemist, all\ person) \in Ynr_4^4.$$

Here the string *prop* is to be interpreted as the distinguished sort "a meaning of proposition," it belongs to the set of sorts $St(B_1)$.

4.3.2 The Rule for Marking Up the Formulas

The rule $P[5]$ is intended, in particular, for marking by variables in semantic representations (SRs) of NL-texts: (a) the descriptions of diverse entities mentioned in a text (physical objects, events, notions, etc.), (b) the fragments being SRs of sentences and of larger parts of texts to which a reference is given in any part of a text.

Definition 4.9. Denote by $P[5]$ the assertion "Let $a \in L(B) \setminus V(B)$, $0 \le k \le i$, $k \ne 5$, $t \in Mtp(S(B))$, $a \& t \in T^k(B)$; $v \in V(B)$, $u \in Mtp(S(B))$, $v \& u \in T^0(B)$, $u \vdash t$, v be not a substring of the string a. Let b be the string of the form $a : v$. Then the relationships

$$b \in L(B), \ b \& t \in T^5(B),$$

$$a \& v \& b \in Y^5(B)$$

take place."

    Example 3. Consider, as before, the conceptual basis B_1 constructed in the preceding chapter. Let $i = 5$, $a_1 = b_3 = (Numb(Suppliers(Firm\_Rainbow))) \equiv 12)$, $k_1 = 3$, $t_1 = prop = P(B_1)$, that is, the informational item $prop$ is the distinguished sort "a meaning of proposition." Then, obviously, $a_1 \& t_1 \in Tnr_3^3$.
    Suppose that $v_1 = P1$, $z_1 = prop = P(B_1)$. Then it follows from the definition of the conceptual basis B_1 and the rule $P[0]$ that

$$v_1 \& z_1 \in T^0(B_1).$$

Besides, $z_1 \vdash t_1$ (because $z_1 = t_1$), and v_1 is not a substring of the expression a_1.
    Let $b_{17} = (Numb(Suppliers(Firm\_Rainbow)) \equiv 12) : P1$. Then, according to the rule $P[5]$,

$$b_{17} \in Lnr_5(B_1), \ b_{17} \& prop \in Tnr_5^5(B_1),$$

$$a_1 \& v_1 \& b_{17} \in Ynr_5^5(B_1).$$

Let $b_{18} = Suppliers(Firm\_Rainbow) : S2$. Then it is easy to see that

$$B_1(0, 2, 5) \Rightarrow b_{18} \in Lnr_5, \ b_{18} \& \{org\} \in Tnr_5^5,$$

$$Suppliers(Firm\_Rainbow) \& S2 \& b_{18} \in Ynr_5^5.$$

    In the expression b_{17}, the variable $P1$ marks up the semantic representation of the phrase T1 = "The firm 'Rainbow' has 12 suppliers." That is why if the expression b_{17} is a fragment of a long formula, then it is possible to use the mark (the variable) $P1$ to the right from the occurrence of b_{17} for repeatedly representing (if necessary) the meaning of the phrase T1 instead of the much longer semantic representation of the phrase T1.
    In the expression b_{18}, the variable $S2$ marks up the set consisting of all suppliers of the firm "Rainbow."

It should be underlined that the rule $P[5]$ is very important for building semantic representations of discourses. It allows us to form SRs of discourses in such a manner that the SRs reflect the referential structure of discourses. The examples of the kind may be found in next sections.

4.4 The Use of Logical Connectives NOT, AND, OR

In comparison with the first-order predicate logic, the rules $P[6]$ and $P[7]$ in combination with the other rules allow for more complete modeling (on the level of semantic representations of NL-texts) of the manners of employing the logical connectives "not," "and," "or" in sentences and discourses in English, Russian, German, and many other languages. In particular, the existence of sentences of the kinds "This medicine was produced not in UK," "Professor Cope defended his Ph.D. dissertation not in the Stanford University," "This patent has been used in Austria, Hungary, The Netherlands, and France," has constructively been taken into account.

With this aim, first, it is permitted to join the connective $\neg$ not only to the expressions designating statements but also to the denotations of things, events, and notions. Second, it is permitted to use the connectives $\wedge$ (conjunction, logical "and") and $\vee$ (disjunction, logical "or") not only for joining the semantic representations of the statements but also for joining the denotations of things, events, and notions.

The rule $P[6]$ describes the use of the connective $\neg$ ("not").

Definition 4.10. Denote by $P[6]$ the assertion "Let $a \in L(B)$, $t \in Mtp(S(B))$, $0 \le k \le i$, k be not in the set $\{2, 5, 10\}$, $a \& t \in T^k(B)$, b be the string of the form $\neg a$. Then $b \in L(B)$, $b \& t \in T^6(B)$, the string of the form $\neg \& a \& b$ belongs to $Y^6(B)$."

In this assertion, the expression $\{2, 5, 10\}$ designates the set consisting of several forbidden values for k. This means the following: if an l-formula a is constructed step by step in any way with the help of some rules from the list $P[0], P[1], \ldots, P[i]$, then the rule $P[2]$ or $P[5]$ or $P[10]$ can't be applied at the last step of the inference.

Example 1. If B_1 is conceptual basis defined in the preceding chapter, $i = 6$, then one can easily verify that

$$B_1(0, 6) \Rightarrow \neg chemist \in Lnr_6,$$

$$chemist \& \uparrow ints * dyn.phys.ob \in Tnr_6^6;$$

$$B_1(0, 6, 4, 4) \Rightarrow Knows(P.Somov, Is1(N.Cope, \neg chemist)) \in Lnr_6,$$

$$B_1(0, 4, 4, 6) \Rightarrow \neg Knows(P.Somov, Is1(N.Cope, biologist)) \in Lnr_6.$$

We'll interpret the built l-formulas as possible semantic representations of the NL-expressions "not a chemist," "P. Somov knows that N. Cope is not a chemist," and "P. Somov doesn't know that N. Cope is a biologist."

The rule $P[7]$ describes the manners to use the connectives $\wedge$ (conjunction, logical "and") and $\vee$ (disjunction, logical "or") while constructing semantic representations of NL-texts. This rule allows for building, in particular, the l-formulas of the form $(a_1 \wedge a_2 \wedge \ldots \wedge a_n)$ and of the form $(a_1 \vee a_2 \vee \ldots \vee a_n)$.

For instance, the rule $P[7]$ jointly with some other rules enables us to construct the l-formulas

$$(chemist \vee biologist), (mathematician \wedge painter),$$

$$(First.name(x1, Pavel) \wedge Surname(x1, Somov)$$

$$\wedge Qualification(x1, chemist)).$$

Definition 4.11. Denote by $P[7]$ the assertion "Let

$$n > 1, t \in Mtp(S(B)), \text{ for } m = 1, \ldots, n,$$

$$0 \leq k[m] \leq i, a_m \in L(B), a_m \& t \in T^{k[m]}(B);$$

$s \in \{\wedge, \vee\}$, b be the string of the form

$$(a_1 \, s \, a_2 \, s \ldots s \, a_n).$$

Then the relationships

$$b \in L(B), b \& t \in T^7(B),$$

$$s \& a_1 \& \ldots \& a_n \& b \in Y^7(B)$$

take place."

Comment to the rule $P[7]$. According to this rule, all expressions joined during one step by a logical connective, are to be associated with the same type. Since this type t can be different from the distinguished sort "meaning of proposition," it is possible to employ the rule $P[7]$ for joining with the help of binary logical connectives not only semantic representations of statements but also the designations of various objects, simple and compound designations of the notions and of the goals of intelligent systems.

Example 2. Suppose that B_1 is the conceptual basis defined in the preceding chapter, $i = 6$. Let

$$b_1 = (R.Scott \wedge N.Cope),$$

$$b_2 = (chemist \vee biologist),$$

$$b_3 = ((Numb(Friends(J.Price)) \equiv 3) : P1$$

$$\wedge Knows(P.Somov, now, P1) \wedge$$

$$\neg Knows(P.Somov, now, Is1(J.Price, (chemist \vee biologist))))),$$

$$b_4 = Knows(P.Somov, now, Elem((R.Scott \wedge N.Cope), Friends(J.Price))),$$

$$b_5 = (Elem(R.Scott, Friends(N.Cope) : S3)$$

$$\wedge \neg Elem(P.Somov, S3)).$$

It is not difficult to show that

$$B_1(0, 7) \Rightarrow b_1, b_2 \in Lnr_7,$$

$$b_1 \& ints * dyn.phys.ob \in Tnr_7^7;$$

$$b_2 \& \uparrow ints * dyn.phys.ob \in Tnr_7^7;$$

$$B_1(0, 2, 2, 3, 5, 4, 7, 4, 4, 6, 7) \Rightarrow b_3 \in Lnr_7,$$

$$B_1(0, 7, 2, 4, 4) \Rightarrow b_4 \in Lnr_7, b_4 \& prop \in Tnr_7^4;$$

$$B_1(0, 2, 5, 4, 4, 6, 7) \Rightarrow b_5 \in Lnr_7,$$

$$b_5 \& prop \in Tnr_7^7.$$

4.5 Building Compound Designations of Notions and Objects

It will be shown below how to build compound representations of notions and, if necessary, to transform these representations of notions into the compound representations of objects (things, events, etc.), applying once the rule $P[1]$ and (it is optional) once the rule $P[5]$.

4.5.1 Compound Designations of Notions

Let's consider the rule $P[8]$ intended for constructing compound representations of notions (concepts) such as

$$text.book * (Field1, biology),$$

$$city * (Country, France),$$

$$concept * (Name.of.concept, \prime molecule\prime),$$

$$tourist.group * (Number.of.persons, 12)$$

$$(Qualitative\_composition, (chemist \vee biologist)).$$

Together with the rule $P[l]$ and other rules, it will enable us to build compound designations of things and sets of things in the form $qtr\ descr$, where qtr is an intensional quantifier, $descr$ is a compound designation of a notion formed with the help of the rule $P[8]$ at the last step of the inference.

For instance, in this way the following formulas can be built:

$$all\ person * (Age, 18/year),$$

$$certain\ person * (Age, 18/year),$$

$$certain\ tourist.group * (Number\_of\_persons, 12).$$

It should be recalled that for arbitrary conceptual basis B,

$$Tconc(B) = \{t \in Tp(S(B)) \mid t \text{ has the beginning } \uparrow\} \cup Spectp,$$

where

$$Spectp = \{[\uparrow entity], [\uparrow concept], [\uparrow object]\}.$$

Each element cpt of the primary universe $X(B)$ such that $tp(cpt) \in Tconc(B)$ is interpreted as a designation of a notion (a concept).

The set $R_2(B)$ consists of binary relational symbols (some of them may correspond to functions with one argument); $F(B)$ is the set of functional symbols. The element $ref = ref(B)$ from $X(B)$ is called the referential quantifier and is interpreted as a semantic item corresponding to the meaning of the expression "a certain" ("a certain book," etc.). $P(B)$ is the distinguished sort "a meaning of proposition" of the conceptual basis B.

Definition 4.12. Denote by $P[8]$ the assertion "Let

$$cpt \in X(B), tp = tp(B), t = tp(cpt), t \in Tconc(B), P = P(B), ref = ref(B).$$

Let $n \geq 1$, for $m = 1, \ldots, n$, $r_m \in R_2(B)$, c_m be the string of the form $ref\,cpt$, and $d_m, h_m \in L(B)$. If $r_m \in R_2(B) \cup F(B)$, let h_m be the string of the form $(r_m(c_m) \equiv d_m)$ and $h_m \& P \in T^3(B)$; if $r_m \in R_2(B) \setminus F(B)$, let h_m be the string of the form $r_m(c_m, d_m)$ and $h_m \& P \in T^4(B)$.

Let b be the string of the form

$$cpt * (r_1, d_1) \ldots (r_n, d_n).$$

Then the relationships

$$b \in L(B), \; b \& t \in T^8(B),$$

$$cpt \& h_1 \& \ldots \& h_n \& b \in Y^8(B).$$

take place."

Example 1. Suppose that B_1 is the conceptual basis defined in the final part of the preceding chapter, $i = 8$. Then consider a possible way of constructing the formula b_1 (defined below) corresponding to the notion "a tourist group consisting of 12 persons." Let

$$cpt = tour.group, \; t = tp(cpt) = \uparrow \{ints * dyn.phys.ob\},$$

$$P = prop, \; ref = certn, \; c_1 = 1, \; r_1 = Numb,$$

$$c_1 = ref\,cpt = certn\,tour.group, \; d_1 = 12,$$

$$h_1 = (r_1(c_1) \equiv d1) = (Numb(certn\,tour.group) \equiv 12).$$

Then

$$B_1(0, 1, 2, 3) \Rightarrow h_1 \in Lnr_3(B_1),$$

$$h_1 \,\&\, prop \in Tnr_8^3(B_1).$$

Let $b_1 = cpt * (r_1, d_1) = tour.group * (Numb, 12)$. Then it follows from the rule $P[8]$ that

$$b_1 \in Lnr_8(B_1),$$

$$b_1 \,\&\, \uparrow \{ints * dyn.phys.ob\} \in Tnr_8^8.$$

4.5.2 Compound Designations of Objects

Together with the rule $P[1]$ and other rules, the rule $P[8]$ allows for building compound designations of things, events, sets of things, and sets of events (all these entities are considered in this book as the particular kinds of objects). The compound designations of objects are constructed in the form

$$qtr\, cpt * (r_1, d_1) \ldots (r_n, d_n)$$

or in the form

$$qtr\, cpt * (r_1, d_1) \ldots (r_n, d_n) : v,$$

where qtr is an intensional quantifier, $cpt * (r_1, d_1) \ldots (r_n, d_n)$ is a compound designation of a notion formed with the help of the rule $P[8]$ at the last step of the inference, and v is a variable interpreted as an individual mark of the considered object.

For instance, in this way the following formulas can be constructed:

$$certain\, person * (Age, 18/year),$$

$$certain\, person * (Age, 18/year) : x5,$$

$$certain\, tourist.group * (Number.of.persons, 12),$$

$$certain\, tourist.group * (Number.of.persons, 12) : y3,$$

$$all\, person * (Age, 18/year),$$

$$all\, person * (Age, 18/year) : S1.$$

Example 2. Let's proceed from the same assumptions concerning the conceptual basis B_1 and the integer i as in Example 1. Then consider a possible way of building the formula b_2 (defined below) denoting a person characterized by the expression "a certain biologist from a certain tourist group consisting of 12 persons." Let

$$b_2 = certn\, biologist * (Elem, certn\, tour.group * (Numb, 12)).$$

Then

$$B_1(0, 1, 2, 3, 8, 1, 1, 4, 8, 1) \Rightarrow$$

$$b_2 \in Lnr_8(B_1), b_2 \,\&\, ints * dyn.phys.ob \in Tnr_8^1.$$

Example 3. Suppose that B_1 is the conceptual basis defined in the final part of the preceding chapter, $i = 8$. Then let's consider a way of constructing a possible semantic representation of the phrase "R. Scott has included N. Cope into a tourist group consisting of 12 persons." Let the variable $x1$ denote a moment of time, the variable $S3$ denote a concrete tourist group, and

$$b_3 = (Include1(R.Scott, N.Cope, x1, certntour.group*$$

$$(Numb, 12) : S3 \wedge Less(x1, \#now\#)).$$

Then

$$B_1(0, 1, 2, 3, 8, 1, 4, 5, 0, 4, 7) \Rightarrow$$

$$b_3 \in Lnr_8(B_1), b_3 \& prop \in Tnr_8^7.$$

4.6 Final Rules

This section introduces the rules $P[9]$ and $P[10]$ intended respectively for (a) the employment in the formulas of the existential and universal quantifiers, (b) constructing the representations of finite sequences.

4.6.1 The Use of Existential and Universal Quantifiers

The rule $P[9]$ describes how to join existential quantifier $\exists$ and universal quantifier $\forall$ to the semantic representations of statements (assertions, propositions). The distinctions from the manner of using these quantifiers in the first-order predicate logics are as follows: (a) the sphere of acting of quantifiers is explicitly restricted; (b) the variables used together with the quantifiers can denote not only the things, the numbers, etc., but also the sets of various entities.

For example, we'll be able to build a semantic representation of the sentence "For each country in Europe, there is a city with the number of inhabitants exceeding 3000" in the form

$$\forall x1 \, (country * (Location1, Europe)) \, \exists x2(city)$$

$$((Location1(x2, x1) \wedge Less(3000, Numb(Inhabitants(x2))))).$$

Here the expressions country $*(Location1, Europe)$ and city restrict the domain where the variables $x1$ and $x2$ can take values respectively.

Definition 4.13. Let's denote by $P[9]$ the assertion "Let $qex \in \{\exists, \forall\}$,

$$A \in L(B) \setminus V(B), P = P(B), k \in \{3, 4, 6, 7, 9\},$$

$$a \& P \in T^k(B), var \in V(B), tp = tp(B),$$

$tp(var) = [entity]$ is the basic type "entity," the string A includes the symbol var, $m \in \{0, 8\}$,

$$concept\_denot \in L(B) \setminus V(B),\ u \in Tc(B),$$

where $Tc(B)$ is the set of all types from the set $Tp(S(B))$ having the beginning $\uparrow$; the string $concept\_denot\ \&\ u$ belongs to the set $T^m(B)$.

Besides, let the string A don't include the substrings of the forms : var, $\exists var$, $\forall var$ and A don't have the ending of the form : z, where z is an arbitrary variable from $V(B)$, b be the string of the form

$$qex\ var\ (concept\_denot)\ A.$$

Then the relations

$$b \in L(B),\ b\ \&\ t \in T^9(B),$$

$$qex\ \&\ var\ \&\ concept\_denot\ \&\ A\ \&\ b \in Y^9(B)$$

take place."

Example 1. Let's construct a possible SR of the phrase T1 = "There are such a moment $x1$ and a tourist group $S3$ consisting of 12 persons that R. Scott included N.Cope into the group $S3$ at the moment $x1$." Let B_1 be the conceptual basis determined in the preceding chapter, $i = 9$, and

$$b_4\ =\ \exists x1\ (mom)\ \exists S3(tour.group * (Numb,\ 12))$$

$$(Include1(R.Scott,\ N.Cope,\ x1,\ S3) \wedge Before(x1,\ \#now\#)).$$

Then it is easy to show that

$$B_1(0,\ 4,\ 4,\ 7,\ 0,\ 1,\ 2,\ 3,\ 8,\ 9,\ 9) \Rightarrow$$

$$b_4 \in Lnr_9(B_1),\ b_4\ \&\ prop \in Tnr_9^9.$$

The formula b_4 is to be interpreted as a possible SR of T1.

Example 2. Let $i = 9$ and there is such conceptual basis B that the following relationships take place:

$$space.ob,\ nat.number,\ dyn.phys.ob,\ ints \in St(B),$$

$$city,\ country,\ Europe,\ Location1,\ Number\_of\_elem,$$

$$Inhabitants,\ Less \in X(B),$$

$$tp(city) = tp(country) = \uparrow space.ob,$$

$$tp(Europe) = space.ob,$$

$$tp(Location1) = \{(space.ob,\ space.ob)\},$$

$$tp(Number\_of\_elem) = \{(\{[entity]\},\ nat.number)\},$$

$$tp(Inhabitants) = \{(space.ob,\ \{dyn.phys.ob * ints\})\},$$

$$tp(Less) = \{(nat.number,\ nat.number)\},$$

$$Number\_of\_elem,\ Inhabitants \in F(B),$$

$$x1, x2 \in V(B), \, tp(x1) = tp(x2) = [entity],$$

$$3000 \in X(B), \, tp(3000) = nat.number.$$

$$prop = P(B).$$

The listed informational units are interpreted as follows:

$$space.ob, \, nat.number, \, dyn.phys.ob, \, ints$$

are the sorts "space object," "natural number," "dynamic physical object," "intelligent system"; *city, country* are the designations of the notions "a city" and "a country"; *Europe* is a designation of the world part Europe; *Location*1 is a designation of a binary relation between space objects; *Number_of_elem* is a designation of the function "Number of elements of a set"; *Inhabitants* is a designation of the function associating with every locality (a village, a city, a country, etc.) the set consisting of all inhabitants of this locality; *Less* is a designation of a binary relation on the set of natural numbers.

Let $qex_1 = \exists, \, var_1 = x2, \, concept\_denot_1 = city,$

$$A_1 = (Location1(x2, x1) \wedge Less(3000, Number\_of\_elem(Inhabitants(x2)))),$$

$$b_1 = qex_1 \, var_1 \, (concept\_denot_1) A_1.$$

Then it is not difficult to see that the following relationship takes place:

$$B(0, 2, 2, 3, 4, 4, 7, 9) \Rightarrow$$

$$b_1 \in Lnr_9(B), \, b_1 \, \& \, prop \in Tnr_9^9.$$

Let $qex_2 = \forall, \, var_2 = x1,$

$$concept\_denot_2 = country * (Location1, Europe),$$

$$A_2 = b_1, \, b_2 = qex_2 \, var_2 \, (concept\_denot_2) A_2.$$

Then it is easy to verify that

$$B(0, 2, 2, 3, 4, 4, 7, 9, 0, 1, 4, 8, 9) \Rightarrow$$

$$b_2 \in Lnr_9(B), \, b_2 \, \& \, prop \in Tnr_9^9.$$

4.6.2 The Representations of Finite Sequences

The rule $P[10]$ is intended for building the representations of finite sequences consisting of n elements, where $n > 1$, in the form $\langle a_1, a_2, \ldots, a_n \rangle$; such sequences are usually called in mathematics the n-tuples.

Definition 4.14. Denote by $P[10]$ the assertion "Let $n > 1$, for $m = 1, \ldots, n$, the following relationships take place:

$$a_m \in L(B), \ u_m \in Mtp(S(B)),$$

$$0 \le k[m] \le 10, \ a_m \& u_m \in T^{k[m]}.$$

Let t be the string of the form $(u_1, u_2, \ldots, u_n)$, and b be the string of the form

$$\langle a_1, a_2, \ldots, a_n \rangle.$$

Then the relationships

$$b \in L(B), \ b \& t \in T^{10}(B),$$

$$a_1 \& a_2 \& \ldots \& a_n \& b \in Y^{10}(B)$$

take place."

Example 3. Let B_1 be the conceptual basis defined in Chapter 3, $i = 10$, and b_3 be the string of the form

$$(Elem(x3, S1) \equiv '((x3 \equiv \langle certn \, real : x1, \, certn \, real : x2 \rangle) \wedge$$

$$(Less(x1, x2) \vee (x1 \equiv x2)))),$$

where the string b_3 is to be interpreted as a possible formal definition of the binary relation "Less or equal" on the set of real numbers. Then one can easily show that

$$B(0, 4, 1, 5, 1, 5, 4, 10, 3, 7, 7, 3) \Rightarrow$$

$$b_3 \in Lnr_{10}(B), \ b_3 \& prop \in Tnr_{10}^3.$$

4.6.3 A Summing-Up Information about the Rules P[0]–P[10]

The total volume of the definitions of the rules $P[0] - P[10]$ and of the examples illustrating these rules is rather big. For constructing semantic representations not only of discourses but also of a major part of separate sentences, one is to use a considerable part of these rules, besides, in numerous combinations.

Taking this into account, it seems to be reasonable to give a concise, non detailed characteristic of each rule from the list $P[0] - P[10]$. The information given below will make easier the employment of these rules in the course of constructing SRs of NL-texts and representing the pieces of knowledge about the world.

Very shortly, the principal results (the kinds of constructed formulas) of applying the rules $P[0] - P[10]$ are as follows:

- $P[0]$: An initial stock of l-formulas and t-formulas determined by (a) the primary informational universe $X(B)$, (b) the set of variables $V(B)$, and (c) the mapping tp giving the types of the elements from these sets.

- $P[1]$: l-formulas of the kind $qtr\,cpt$ or $qtr\,cpt * (r_1, d_1) \ldots (r_n, d_n)$, where qtr is an intensional quantifier, cpt is a simple (i.e., non structured) designation of a notion (a concept), $n \geq 1$, $r_1, \ldots, r_n$ are the designations of the characteristics of the entities (formally, these designations are unary functional symbols or the names of binary relations).
- $P[2]$: l-formulas of the kind $f(a_1, \ldots, a_n)$, where f is a functional symbol, $n \geq 1$, $a_1, \ldots, a_n$ are the designations of the arguments of the function with the name f; t-formulas of the kind $f(a_1 \ldots a_n) \,\&\, t$, where t is the type of the value of the function f for the arguments $a_1, \ldots, a_n$.
- $P[3]$: l-formulas of the kind $(a_1 \equiv a_2)$ and t-formulas of the kind $(a_1 \equiv a_2) \,\&\, P$, where P is the distinguished sort "a meaning of proposition."
- $P[4]$: l-formulas of the kind $rel\,(a_1, \ldots, a_n)$, where r is a relational symbol, $n \geq 1$, $a_1, \ldots, a_n$ are the designations of the attributes of the relation with the name rel; t-formulas of the kind $rel\,(a_1, \ldots, a_n) \,\&\, P$, where P is the distinguished sort "a meaning of proposition."
- $P[5]$: l-formulas of the kind $form : v$, where $form$ is l-formula, v is a variable from $V(B)$ being a mark of the formula $form$.
- $P[6]$: Proceeding from the l-formula $form$, in particular, the l-formula of the kind $\neg\, form$ is constructed.
- $P[7]$: Using, as the operands, a logical connective $s \in \{\vee, \wedge\}$ and some l-formulas $a_1, \ldots, a_n$, where $n > 1$, one obtains, in particular, the l-formula

$$(a_1 \, s \, a_2 \, s \ldots s \, a_n).$$

- $P[8]$: The operands of this rule are (a) an l-formula cpt from the primary informational universe $X(B)$ interpreted as a simple (non structured) designation of a notion (a concept), (b) the characteristics $r_1, \ldots, r_n$, where $n > 1$, of the entities qualified by the notion cpt, (c) the l-formulas $d_1, \ldots, d_n$; for $k = 1, \ldots, n$, if r_k is the name of a function with one argument, d_k is the value of this function for a certain entity qualified by the notion cpt; if r_k is the name of a binary relation, d_k designates the second attribute of this relation, where the first attribute is a certain entity qualified by the notion cpt. These operands are used for constructing (a) the l-formula

$$cpt * (r_1, d_1) \ldots (r_n, d_n)$$

  and (b) the t-formula

$$cpt * (r_1, d_1) \ldots (r_n, d_n) \,\&\, t,$$

  where t is a type from the set $Tp(S(B))$; such l-formula and t-formula are interpreted as the compound designations of the notions (concepts). For instance, there is such a conceptual basis B that the l-formula $city * (Country, France)$ and the t-formula $city * (Country, France) \,\&\, space.object$ can be constructed.
- $P[9]$: The l-formulas of the kind

$$qex\,var\,(concept\_denot)\,A$$

  and t-formulas of the kind

$$qex\, var\,(concept\_denot)\,A\,\&\,P$$

are constructed, where *qex* is either the existential quantifier ∃ or the universal quantifier ∀, *var* is a variable from $V(B)$, *concept_denot* is a simple (non structured) or compound designation of a notion (a concept), A is an l-formula reflecting the semantic content of a statement, P is the distinguished sort "a meaning of proposition."

- $P[10]$: The l-formulas of the kind $\langle a_1, \ldots, a_n \rangle$, where $n > 1$, are constructed; such formulas are interpreted as the designations of finite sequences containing n elements, or n-tuples.

4.7 SK-Languages: Mathematical Investigation of Their Properties

Let's remember the denotations introduced in the first section of this chapter. Suppose that B is an arbitrary conceptual basis, and *Specsymbols* is the set consisting of the symbols $/,/$ (comma), $/(/, /)/, / : /, / * /, /\langle/, /\rangle/$. Then

$$D(B) = X(B) \cup V(B) \cup Specsymbols,$$

$$Ds(B) = D(B) \cup \{/\&/\},$$

$D^+(B)$ and $Ds^+(B)$ are the sets of all non empty finite sequences of the elements from $D(B)$ and $Ds(B)$ respectively.

Definition 4.15. Let B be an arbitrary conceptual basis, $1 \leq i \leq 10$, and the sets of strings

$$L(B) \subset D^+(B),$$

$$T^0(B), T^1(B), \ldots, T^i(B) \subset Ds^+(B),$$

$$Y^1(B), \ldots, Y^i(B) \subset Ds^+(B)$$

are the least sets jointly determined by the rules $P[0], P[1], \ldots, P[i]$. Then denote these sets, respectively, by

$$Lnr_i(B), T^0(B), Tnr_i^1(B), \ldots, Tnr_i^i(B),$$

$$Ynr_i^1(B), \ldots, Ynr_i^i(B)$$

and denote the family (that is, the set) consisting of all listed sets of strings by $Globset_i(B)$.

Besides, let the following relationships take place:

$$T_i(B) = T^0(B) \cup Tnr_i^1(B) \cup \ldots \cup Tnr_i^i(B); \tag{4.1}$$

$$Y_i(B) = Ynr_i^1(B) \cup \ldots \cup Ynr_i^i(B); \tag{4.2}$$

$$Form_i(B) = Lnr_i(B) \cup T_i(B) \cup Y_i(B). \tag{4.3}$$

Definition 4.16. If B is an arbitrary conceptual basis, then:

$$Ls(B) = Lnr_{10}(B), \tag{4.4}$$

$$Ts(B) = T_{10}(B), \tag{4.5}$$

$$Ys(B) = Y_{10}(B), \tag{4.6}$$

$$Forms(B) = Form_{10}(B), \tag{4.7}$$

$$Ks(B) = (B, Rules), \tag{4.8}$$

where *Rules* is the set consisting of the rules $P[0], P[1], \ldots, P[10]$.

The ordered pair $Ks(B)$ is called *the K-calculus (knowledge calculus) in the conceptual basis B*; the elements of the set $Forms(B)$ are called *the formulas inferred in the stationary conceptual basis B*. The formulas from the sets $Ls(B), Ts(B)$, and $Ys(B)$ are called respectively *l-formulas, t-formulas,* and *y-formulas*. The set of *l*-formulas $Ls(B)$ is called *the standard knowledge language (or SK-language, standard K-language) in the conceptual basis B*.

The set $Ls(B)$ is considered as the main subclass of formulas generated by the collection of the rules $P[0], \ldots, P[10]$. The formulas from this set are intended for describing structured meanings (or semantic content) of NL-texts.

Theorem 4.1. *If B is an arbitrary conceptual basis, then:*
(a) the set $Lnr_0(B) \neq \emptyset$;
(b) for $m = 1, \ldots, 10, Lnr_{m-1}(B) \subseteq Lnr_m(B)$.

Proof. (a) For arbitrary conceptual basis B, $X(B)$ includes the non empty set $St(B)$; this follows from the definitions of a sort system and a conceptual basis. According to the rule $P[0]$, $St(B) \subset Lnr_0(B)$. Therefore, $Lnr_0(B) \neq \emptyset$. (b) The structure of the first definition in this section and the structure of the statements $P[0], P[l], \ldots, P[10]$ show that the addition of the rule $P[m]$, where $1 \leq m \leq 10$, to the list $P[0], \ldots, P[m-1]$ either expands the set of *l*-formulas or doesn't change it (if the set of functional symbols $F(B)$ is empty, $Lnr_1(B) = Lnr_2(B)$). The proof is complete.

Taking into account this theorem, it is easy to see that previous sections of this chapter provide numerous examples of *l*-formulas, *t*-formulas, and *y*-formulas inferred in a stationary conceptual basis and, as a consequence, numerous examples of the expressions of SK-languages.

Theorem 4.2. *If B is any conceptual basis, then:*

$$Ls(B), Ts(B), Ys(B) \neq \emptyset.$$

Proof. Let B be a conceptual basis. Then $Ls(B) \neq \emptyset$ according to the Theorem 4.1 and the relationship (4.4). It follows from the definition of a concept-object system that the set of variables $V(B)$ includes a countable subset of such variables *var* that

$tp(var) = [entity]$. Let v_1 and v_2 be two variables from this subset, b be the string of the form $(v_1 \equiv v_2)$, c be the string of the form

$$v_1 \,\&\, \equiv \,\&\, v_2 \,\&\, (v_1 \equiv v_2),$$

$P = P(B)$ be the sort "a meaning of proposition" of B. Then, according to the rules $P[0]$ and $P[3]$,

$$b \,\&\, P \in Tnr_i^3(B), \; c \in Ynr_i^3(B)$$

for each $i = 3, \ldots, 10$. Therefore, with respect to the relationships (4.1), (4.2), (4.5), and (4.6), the sets of formulas $Ts(B)$ and $Ys(B)$ are non empty.

Theorem 4.3. *If B is an arbitrary conceptual basis, then:*
(a) If $w \in Ts(B)$, then w is a string of the form $a \,\&\, t$, where $a \in Ls(B)$, $t \in Tp(S(B))$, and such a representation depending on w is unique for each w;
(b) if $y \in Ys(B)$, then there are such $n > 1$ and $a_1, \ldots, a_n, b \in Ls(B)$ that y is the string of the form

$$a_1 \,\&\, \ldots \,\&\, a_n \,\&\, b;$$

besides, such representation depending on y is unique for each y.

Proof. The structure of the rules $P[0], \ldots, P[10]$ and the definition of a conceptual basis immediately imply the truth of this theorem.

Theorem 4.4. *If B is a conceptual basis, $d \in X(B) \cup V(B)$, then there are no such integer k, where $1 \leq k \leq 10$, $n > 1$, and the strings $a_1, \ldots, a_n \in Ls(B)$ that*

$$a_1 \,\&\, \ldots \,\&\, a_n \,\&\, d \in Ynr_{10}^k(B).$$

Interpretation. If d is an element of the primary informational universe $X(B)$ or a variable from $V(B)$, then it is impossible to obtain d with the help of any operations determined by the rules $P[1] - P[10]$.

Proof. Assume that there are such $k \in \{1, \ldots, 10\}$, $n > 1$, $a_1, \ldots, a_n \in Ls(B)$ that the required relationship takes place. For arbitrary $m \in \{1, \ldots, 10\}$, the set $Ynr_{10}^m(B)$ may include a string $a_1 \,\&\, \ldots \,\&\, a_n \,\&\, d$, where d contains no occurrences of the symbol $\&$, only in case d is obtained from the elements $a_1, \ldots, a_n$ by means of applying one time the rule $P[m]$. Then it follows from the structure of the rules $P[l], \ldots, P[10]$ that d must contain at least two symbols. But we consider the elements of the set $X(B) \cup V(B)$ as symbols. Therefore, we get a contradiction, since we assume that $d \in X(B) \cup V(B)$. The proof is complete.

Theorem 4.5. *Let B be an arbitrary conceptual basis,*

$$z \in Ls(B) \setminus (X(B) \cup V(B)).$$

Then there is one and only one such $n + 3$-tuple $(k, n, y_0, y_1, \ldots, y_n)$, where $1 \leq k \leq 10$, $n \geq 1$, $y_0, y_1, \ldots, y_n \in Ls(B)$ that

$$y_0 \,\&\, y_1 \,\&\, \ldots \,\&\, y_n \,\&\, z \in Ynr_{10}^k(B).$$

Interpretation. If an l-formula z doesn't belong to the union of the sets $X(B)$ and $V(B)$, then there exists the unique rule $P[k]$, where $1 \le k \le 10$, and the unique finite sequence of l-formulas $y_0, y_1, \ldots, y_n$ that the string z is constructed from the "blocks" $y_0, y_1, \ldots, y_n$ by means of applying just one time the rule $P[k]$.

The truth of this theorem can be proved with the help of two lemmas. In order to formulate these lemmas, we need

Definition 4.17. Let B be an arbitrary conceptual basis, $n \ge 1$, for $i = 1, \ldots, n$, $c_i \in D(B)$, $s = c_1 \ldots c_n$, $1 \le k \le n$. Then let the expressions $lt_1(s, k)$ and $lt_2(s, k)$ denote the number of the occurrences of the symbol $'('$ and symbol $'\langle'$ respectively in the substring $c_1 \ldots c_k$ of the string $s = c_1 \ldots c_n$.

Let the expressions $rt_1(s, k)$ and $rt_2(s, k)$ designate the number of the occurrences of the symbol $')'$ and the symbol $'\rangle'$ in the substring $c_1 \ldots, c_k$ of the string s. If the substring $c_1, \ldots, c_k$ doesn't include the symbol $'('$ or the symbol $'\langle'$, then let respectively

$$lt_1(s, k) = 0, \ lt_2(s, k) = 0,$$

$$rt_1(s, k) = 0, \ rt_2(s, k) = 0.$$

Lemma 1. Let B be an arbitrary conceptual basis, $y \in Ls(B)$, $n \ge 1$, for $i = 1, \ldots, n$, $c_i \in D(B)$, $y = c_1 \ldots c_n$. Then
  (a) if $n > 1$, then for every $k = 1, \ldots, n - 1$ and every $m = 1, 2$,

$$(a) \ \ lt_m(y, k) \ge rt_m(y, k);$$

$$(b) \ \ lt_m(y, n) = rt_m(y, n).$$

Lemma 2. Let B be an arbitrary conceptual basis, $y \in Ls(B)$, $n > 1$, $y = c_1 \ldots c_n$, where for $i = 1, \ldots, n$, $c_i \in D(B)$, the string y include the comma or any of the symbols $\equiv, \wedge, \vee$, and k be such arbitrary natural number that $1 < k < n$. Then
  (a) if c_k is one of the symbols $\equiv, \wedge, \vee$, then

$$lt_1(y, k) > rt_1(y, k) \ge 0;$$

(b) if c_k is the comma , then at least one of the following relationships takes place:

$$lt_1(y, k) > rt_1(y, k) \ge 0,$$

$$lt_2(y, k) > rt_2(y, k) \ge 0.$$

The proofs of Lemma 1, Lemma 2, and Theorem 4.5 can be found in the Appendix to this monograph.

Definition 4.18. Let B be an arbitrary conceptual basis, $z \in Ls(B) \setminus (X(B) \cup V(B))$, and there is such $n+3$-tuple $(k, n, y_0, y_1, \ldots, y_n)$, where $1 \le k \le 10$, $n \ge 1$, $y_0, y_1, \ldots, y_n \in Ls(B)$ that

$$y_0 \& y_1 \& \ldots \& y_n \& z \in Ynr_{10}^k(B).$$

Then the $n+3$-tuple $(k, n, y_0, y_1, \ldots, y_n)$ will be called *a form-establishing sequence of the string z.*

With respect to this definition, the Theorem 4.5 states that for arbitrary conceptual basis B, each string from $Ls(B) \setminus (X(B) \cup V(B))$ has the unique form-establishing sequence.

Theorem 4.6. *Let B be an arbitrary conceptual basis, $z \in Ls(B)$. Then there is one and only one such type $t \in Tp(S(B))$ that $z \& t \in Ts(B)$.*

Proof. Let's consider two possible cases.
Case 1.
Let B be an arbitrary conceptual basis, $z \in X(B) \cup V(B), t \in Tp(S(B)), tp(z) = t$. Then the rule $P[0]$ implies that $z \& t \in Ts(B)$.

Assume that w is such type from $Tp(S(B))$ that $z \& w \in Ts(B)$. The analysis of the rules $P[0] - P[10]$ shows that this relationship can follow only from the rule $P[0]$. But in this case, w is unambiguously determined by this rule, therefore, w coincides with t.
Case 2.
Let B be an arbitrary conceptual basis, $z \in Ls(B) \setminus (X(B) \cup V(B))$. In accordance with Theorem 4.5, there are such integer k, n where $1 \leq k \leq 10$, $n \geq 1$, and such $y_0, y_1, \ldots, y_n \in Ls(B)$ that the string z is constructed from these elements by applying one time the rule $P[k]$. That is why there is such type $t \in Tp(S(B))$ that $z \& t \in Tnr_{10}^k(B)$ and, as a consequence, $z \& t \in Ts(B)$.

It follows from Theorem 4.5 that z unambiguously determines k, n,
$y_0, y_1, \ldots, y_n$ of the kind. But in this situation, the $n+3$-tuple $(k, n, y_0, y_1, \ldots, y_n)$ unambiguously determines such type u that $z \& u \in Tnr_{10}^k(B)$. Therefore, the type u coincides with the type t.

Taking this into account, Theorem 4.5 states that every string of the SK-language $Ls(B)$, where B is an arbitrary conceptual basis, can be associated with the only type t from $Tp(S(B))$.

Definition 4.19. Let B be an arbitrary conceptual basis, $z \in Ls(B)$. Then *the type* of l-formula z is such element $t \in Tp(S(B))$, denoted $tp(z)$, that $z \& t \in Ts(B)$.

Problems

1. What is the interpretation of the notion "a universe of formal study?"
2. What is the difference between l-formulas and t-formulas?
3. What is the difference between l-formulas and y-formulas?
4. What are K-strings?
5. What is the scheme of determining step by step three classes of formulas generated by a conceptual basis?
6. What is a K-calculus?
7. Describe the classes of formulas obtained due to employing only the rule $P[0]$.
8. What new manners of using the logical connectives $\wedge$ and $\vee$ (in comparison with the first-order predicate logic) are determined by the rule $P[7]$?
9. Why only one rule but not two rules govern the employment of the logical connectives $\wedge$ and $\vee$?

10. What rule enables us to join the referential quantifier to a simple or compound designation of a notion?

11. Describe the structure of compound designations of notions obtained as a result of employing the rule $P[8]$ at the last step of inference.

12. What rule is to be employed immediately after the rule $P[8]$ in order to obtain a compound designation of a concrete thing, event or a concrete set of things, events?

13. Infer (describe the steps of constructing) a K-string designating the set of all students of the Stanford University born in France. For this, introduce the assumptions about the considered conceptual basis.

14. Formulate the necessary assumptions about the considered conceptual basis and infer (describe the steps of constructing) the K-string:

$$(a) \;\; Subset(all\, car,\, all\, transport\_means);$$

$$(b) \;\; (Manufacturer(certain\, car : x7) \equiv$$

$$certain\, company1 * (Name1, \text{``}Volvo\text{''}));$$

$$(c) \;\; Belong((Barcelona \wedge Madrid \wedge Tarragona),$$

$$all\, city * (Country,\, Spain));$$

$$(d) \;\; Belong(certain\, scholar * (First\_name, \text{``}James''\text{)}(Surname, \text{``}Hendler\text{''}) : x1,$$

$$Authors(certain\, inf\_object * (Kind1,\, sci\_article)$$

$$(Name1, \text{``}The\, Semantic\, Web\text{''})(Date\_of\_publ,\, 2001))).$$

Chapter 5
A Study of the Expressive Possibilities of SK-Languages

Abstract In this chapter we will continue the analysis of the expressive possibilities of SK-languages. The collection of examples considered above doesn't demonstrate the real power of the constructed mathematical model. That is why let's consider a number of additional examples in order to illustrate some important possibilities of SK-languages concerning the construction of semantic representations of sentences and discourses and describing the pieces of knowledge about the world. The advantages of the theory of SK-languages in comparison, in particular, with Discourse Representation Theory, Episodic Logic, Theory of Conceptual Graphs, and Database Semantics of Natural Language are set forth. If the string *Expr* of a certain SK-language is a semantic representation of a natural language expression *T*, the string *Expr* will be called *a possible K-representation (KR) of the expression T*.

5.1 A Convenient Method of Describing Events

The key role in the formation of sentences is played by the verbs and lexical units which are the derivatives of verbs – the participles, gerunds, and verbal nouns, because they express the various relations between the objects in the considered application domain.

In computer linguistics, a conceptual relation between a meaning of a verbal form and a meaning of a word combination (or a separate word) depending on this verbal form is called *a thematic role (or conceptual case, deep case, semantic case, semantic role)*.

In such different languages as Russian, English, German, and French, it is possible to observe the following regularity: in the sentences with the same verb mentioning an event, the different quantity of the thematic roles connected with a meaning of this verb is explicitly realized.

For example, let T1 = "Professor Novikov arrived yesterday" and T2 = "Professor Novikov arrived yesterday from Prague." Then in the sentence T1 two thematic roles are explicitly realized, these roles can be called *Agent1 (Agent of action)* and *Time*.

V.A. Fomichov, *Semantics-Oriented Natural Language Processing*, IFSR International Series on Systems Science and Engineering 27, DOI 10.1007/978-0-387-72926-8_5, © Springer Science+Business Media, LLC 2010

Meanwhile, in the sentence T2 three thematic roles *Agent1, Time, and Place 1* (the latter designates the relation connecting an event of moving in space and an initial spatial object) are explicitly realized.

Let's consider a flexible way of constructing semantic representations of the event descriptions taking into account this phenomenon of natural language. For this purpose it is required to formulate a certain assumption about the properties of the considered conceptual basis B.

Assumption 1.

The set of sorts $St(B)$ includes the distinguished sort *sit* ("situation"); the set of variables $V(B)$ includes such countable subset $Vsit = \{el, e2, e3, ...\}$ that for every $v \in Vsit$, $tp(v) = sit$; the primary informational universe $X(B)$ includes such binary relational symbol *Situation* that its type $tp(Situation)$ is the string of the form $\{(sit, \uparrow sit)\}$.

The meaning of the expression $\uparrow sit$ in the right part of the relationship

$$tp(Situation) = \{(sit, \uparrow sit)\}$$

is as follows: we will be able to build the expressions of the form

$$Situation(e_k, concept\_descr),$$

where e_k is a variable denoting an event (a sale, a purchase, a flight, etc.) and *concept_descr* is a simple or compound denotation of a notion being a semantic characteristic of the event.

Let's agree that a connection between a mark of a situation and a semantic description of this situation will be given by means of the formulas of the kind

$$Situation(var, cpt * (rel_1, d_1) \ldots (rel_n, d_n)),$$

where *var* is a variable of type *sit*, the element *cpt* belongs to $X(B)$ and is interpreted as a notion qualifying a situation, $n \geq 1$, for $k = 1, ..., n$, rel_k is a characteristic of the considered situation, d_k is the value of the characteristic rel_k.

Example. Let the expressions *Expr1* and *Expr2* be defined as follows:

$$Expr1 = \exists e1\,(sit)\,(Situation(e1, arrival * (Time, x1)$$

$$(Agent1, person * (Qualif, professor)(Surname,$$

$$'Novikov') : x2)) \wedge Before(x1, \#now\#)),$$

$$Expr2 = \exists el\,(sit)\,(Situation(e1, arrival * (Time, x1)$$

$$(Agent1, person * (Qualif, professor)(Surname, 'Novikov') : x2)$$

$$(Place1, city * (Name1, 'Prague') : x3)) \wedge Before(x1, \#now\#)).$$

Then it is easy to see that it is possible to construct such conceptual basis B that Assumption 1 will be true, and the following relationships will take place:

$$B(0, 1, 2, 3, 4, 5, 7, 8, 9) \Rightarrow Exprl, Expr2 \in Ls(B),$$

$$Exprl \& prop \in Ts(B), Expr2 \& prop \in Ts(B),$$

where $prop = P(B)$ is the distinguished sort "a meaning of proposition" of the conceptual basis B.

This method of describing the events will be used many times below. Most often, for the sake of compactness, the existential quantifier followed by a variable marking-up an event will be omitted. For example, instead of the formula $Exprl$, we will consider the following formula $Expr3$ of the form

$$(Situation(e1, arrival * (Time, x1)(Agent1, person*$$

$$(Qualif, professor)(Surname, 'Novikov') : x2)) \wedge Before(xl, \#now\#)).$$

5.2 Formalization of Assumptions About the Structure of Semantic Representations of Sets

The statements, questions, commands can include the designations of sets. In order to have a unified approach to constructing semantic representations of sets' descriptions, it is reasonable to formulate a number of additional assumptions about the considered conceptual bases.

With respect to the fact that the designations of sets in texts often include the designations of natural numbers ("five three-tonne containers", etc.), we will suppose that the following requirements are satisfied for the considered conceptual basis B:

Assumption 2.

The set of sorts $St(B)$ includes the distinguished sort nat ("natural number"), the primary informational universe $X(B)$ includes a subset of strings Nt consisting of all strings of the form $d_1 \ldots d_k$, where $k \geq 1$, for $m = 1, \ldots, k$, d_m is a symbol from the set

$$\{'0', 1, '2', '3', '4', '5', '6', 7, '8', '9'\},$$

and it follows from $d_1 = '0'$ that $k = 1$. Besides, for each z from the set Nt, $tp(z) = nat$.

Let's also demand that the primary information universe $X(B)$ includes the distinguished elements

$$set, Numb, Qual - compos, Object - compos$$

interpreted as follows: set is the designation of the concept "a finite set," $Numb$ is the name of an one-argument function "The quantity of elements of a set," $Qual - compos$ is the name of the binary relation "Qualitative composition of a set," $Object - compos$ is the name of the binary relation "Object composition of a set."

Assumption 3.

The primary informational universe $X(B)$ includes such elements set, $Numb$, $Qual - compos$, $Object - compos$ that

$$tp(set) = \uparrow \{[entity]\},$$

$$t(Numb) = \{(\{[entity]\}, nat)\},$$

$$tp(Qual - compos) = \{(\{[entity]\}, [concept])\},$$

$$tp(Object - compos) = \{(\{[entity]\}, [entity])\},$$

where $[entity]$, $[concept]$, $[object]$ are the basic types "entity," "concept," "object" (see Chap. 3).

Let's consider the interpretation of the distinguished elements of the primary informational universe $X(B)$ mentioned in the Assumption 3.

Using the elements set, $Numb$, and any string $number$ from Nt, we can construct a semantic representation of the expression "a certain set containing $number$ elements" in the form

$$certn \, set * (Numb, number),$$

where $certn = ref(B)$ is the referential quantifier of the considered conceptual basis B.

The purpose of considering the binary relational symbol $Qual - compos$ is as follows. Let v be a variable designating a certain set, and $conc$ be a simple or compound designation of a notion. Then the expression

$$Qual - compos(v, conc)$$

designates the meaning of the statement "Each element of the set v is qualified by the notion $conc$," and this statement can be true or false. The examples of such expressions are the formulas

$$Qual - compos(S1, container1),$$

$$Qual - compos(S2, paper1),$$

$$Qual - compos(S3, container1 * (Material, aluminum)),$$

$$Qual - compos(S4, paper1 * (Area1, biology)).$$

On the other hand, the symbol $Qual - compos$ will be also employed in constructing the compound designations of the sets in the form

$$ref \, set * (Qual - compos, conc) : v,$$

where ref is the the referential quantifier of the considered conceptual basis B; $conc$ is a simple or compound designation of a notion (a concept), v is a variable.

In particular, the conceptual basis B can be chosen so that the language Ls (B) includes the expressions

$$certn\,set * (Qual - compos, container1) : S1,$$

$$certn\,set * (Qual - compos, paper1) : S2,$$

$$certn\,set * (Qual - compos, container1 * (Material, aluminum)) : S3,$$

$$certn\,set * (Qual - compos, paper1 * (Area1, biology)) : S4.$$

A fragment of a text designating a set can be an explicit enumeration of the elements of this set. For instance, the text "Two customers, the joint-stock company 'Rainbow' and the Open Company 'Zenith,' have not paid September deliveries" contains a fragment of the kind.

The binary relational symbol $Object - compos$ is intended, in particular, for constructing the expressions of the form

$$Object - compos(v, (x_1 \wedge x_2 \ldots \wedge x_n)),$$

where $v, x_1, x_2, \ldots, x_n$ are the variables, v designates a set, and $x_1, x_2, \ldots, x_n$ are the designations of all elements belonging to the set with the designation v.

For example, we'll consider the expression

$$(Object - compos(y1, (x1 \wedge x2)) \wedge$$

$$Is1(x1, joint - stock - comp) \wedge Is1(x2, open - comp) \wedge$$

$$Name(x1, \text{"Rainbow"}) \wedge Name(x2, \text{"Zenith"}))$$

as a semantic representation of the statement "The set $y1$ consists of the joint-stock company 'Rainbow' and the Open Company 'Zenith'," where the first company is designated by $x1$, and the second company has the designation $x2$.

At the same time we should have the opportunity (if necessary) to build the formulas with the structure

$$ref\,set * (Object - compos, (x_1 \wedge x_2 \wedge \ldots, x_n)) : y1,$$

where ref is the referential quantifier, and $x_1 \ldots, x_n, y_1$ are the variables of type $[entity]$ (the basic type "entity").

For instance, we should have the opportunity of constructing the expression

$$certn\,set * (Object - compos, (x1 \wedge x2)) : y1.$$

Example. Let $Setdescr1$ and $Setdescr2$ be the expressions "3 containers with ceramics from India" and "a party consisting of the boxes with the numbers 3217, 3218, 3219" respectively.

Then it is possible to construct such conceptual basis B that the Assumption 2 will be true, and $Ls(B)$ will include the formula

$$certn\,set*(Numb,\,3)(Qual-compos,\,container1*$$

$$(Content1,\,certn\,set*(Qual-compos,\,manufact\_product*$$

$$(Kind,\,ceramics)(Country,\,India))))$$

and the formula

$$certn\,set*(Numb,\,3)(Object-compos,\,(certn\,box1*$$

$$(Number1,\,3217):x1\wedge certn\,box1*(Number1,\,3218):x2$$

$$\wedge\,certn\,box1*(Number1,\,3219):x3)):S1.$$

The constructed formulas will be interpreted as possible K-representations of the expressions $Setdescr1$ and $Setdescr2$; here xl, $x2$, $x3$ are the labels of boxes, $S1$ is the label of a set.

5.3 Semantic Representations of Questions with the Role Interrogative Words

Let $Interrog\_expr$ be the designation of the union of the set of all interrogative adverbs and the set of all word groups composed by interrogative pronouns and by interrogative pronouns with associated prepositions. Then it is possible to distinguish in the set $Interrog\_expr$ a subset including, in particular, the words and word groups "who," "to whom," "from whom," "when," "where." In order to formulate the property of each element of this subset, we shall introduce the designation nil for an empty preposition. If an interrogative pronoun $qswd$ is used in any question without preposition, let's agree to say that this pronoun $qswd$ is associated in this question with the empty preposition nil.

Therefore, for each pronoun $qswd$ from the considered subset, there is such preposition $prep$ that the pair $(prep, qswd)$ corresponds to a certain thematic role. Rather often, there are several prepositions $prep$ satisfying this condition.

For example, the pairs

$$(to,\,whom),\,(for,\,whom),\,(from,\,whom)$$

correspond to the thematic roles $Addressee$, $Addressee$, $Source1$, respectively. We see that the different pairs (to, who), (for, who) correspond to one thematic role $Addressee$, and it is confirmed by the analysis of the phrases "Who is the book sent to?" and "Who is the book sent for?" The thematic role "Source1" is realized, in particular, in the sentence "This book was sent by Yves."

The pronouns and adverbs belonging to the specified subset will be called below *the role interrogative words*.

Assumption 4.

The primary informational universe $X(B)$ of the conceptual basis B includes the symbol *Question*, and

$$tp(Question) = \{([entity], P)\},$$

where $tp = tp(B)$ is a mapping associating a type with an informational unit, $[entity]$ is the basic type "entity," $P = P(B)$ is the distinguished sort "a meaning of proposition."

Let Assumption 4 be true for the conceptual basis B. Then the semantic representation of a question with n role interrogative words can be presented in the form

$$Question(v_1, A)$$

in case $n = 1$ and in the form

$$Question((v_1 \wedge \ldots \wedge v_n), A)$$

in case $n > 1$, where $v_1 \ldots v_n$ are the variables, and A is an l-formula depending on the variables $v_1 \ldots v_n$ and displaying the content of a statement (i.e. being a semantic representation of a statement).

Example 1. Let Qs1 be the question "Where has the three-ton aluminum container arrived from?" and *Expr1* be the string

$$Question(x1, Situation(el, receipt1 * (Time, certn\,moment*$$

$$(Before, \#now\#) : t1)(Location1, x1)(Object1,$$

$$certn\,container1 * (Weight, 3/ton)(Material, aluminum) : x2))).$$

Then it is easy to construct such conceptual basis B that Assumption 1 and Assumption 4 are true for B, $P(B) = prop$, and

$$B(0, 1, 2, 3, 4, 5, 8) \Rightarrow Expr1 \in Ls(B),$$

$$Expr1 \,\&\, prop \in Ts(B).$$

The expression *Expr1* is a possible K-representation of the question Qs1. In this expression, the symbols $xl, x2, el, t1$ are the variables, $receipt1$ is the informational unit (in other words, semantic unit) corresponding to the noun "receipt" and expressing the meaning "transporting a physical object to a spatial object" (unlike the meaning "receipt of a Ph.D. degree").

Example 2. Let Qs2 = "When and where has the three-ton aluminum container arrived from?" Then a K-representation of the question Qs2 can be the expression

$$Question((t1 \wedge x1), Situation(el, receipt1 * (Time, certn\,moment*$$

$$(Before, \#now\#) : t1)(Place1, x1)(Object1,$$

$$certn\,container1 * (Weight, 3/ton)(Material, aluminum) : x2))).$$

5.4 Semantic Representations of Questions About the Quantity of Objects and Events

Assumption 5.

Let $X(B)$ be a conceptual basis of the form (S, Ct, Ql), where S is a sort system, Ct is a concept-object system, and Ql is a system of quantifiers and logical connectives of the form

$$(ref, int1, int2, eq, neg, binlog, ext),$$

and let $X(B)$ include such elements *arbitrary, all, Elem* that

$$tp(arbitrary) = int1, \, tp(all) = int2,$$

$$tp(Elem) = \{([entity], \{[entity]\})\}.$$

The elements *arbitrary, all, Elem* are interpreted as informational units "any" ("arbitrary"), "all," and "An element of a set" (the name of the relation "To be an element of a set").

It is necessary to notice that $int1$ and $int2$ are the distinguished elements of the set of sorts $St(B)$. By definition (see Sect. 3.8), the elements $int1$ and $int2$ are the types of intensional quantifiers from the first and second considered classes respectively.

Example 1. Let Qs1 = "How many copies of the books by P.N. Somov are available in the library?" Then it is possible to define such conceptual basis B that Assumptions 4 and 5 are true for B, and the expression

$$Question(x1, (x1 \equiv Numb(all\,copy1 * (Inform - object,$$

$$arbitrary\,book * (Authors, certn\,person * (Initials, 'P.N.')$$

$$(Surname,'Somov') : x2) : x3)(Storage - place, certn\,library : x4))))$$

belongs to the SK-language $Ls(B)$. Therefore, this expression is a possible K-representation of the question Qs1.

Example 2. Let Qs2 = "How many people participated in the creation of the textbook on statistics?", then a possible K-representation of the question Qs2 can have the form

$$Question(x1, ((x1 \equiv Numb(all\,person * (Elem, S1)))) \wedge$$

$$Description(arbitrary\,person * (Elem, S1) : y1, (Situation(e1,$$

$$participation1 * (Agent1, y1)(Time, x2)(Kind - of - activity,$$

$$creation1 * (Product1, certn\,textbook * (Area1, statistics)))) \wedge$$

$$Before(x2, \#Now\#))))).$$

Example 3. Let Qs3 = "How many books did arrive in January of this year to the library No. 18?" Then the formula

$$Question(x1, ((x1 \equiv Numb(all\ book * (Elem, S1)))) \wedge$$

$$Description(arbitrary\ book * (Elem, S1) : y1,$$

$$Situation(e1, receipt2 * (Object1, y1)$$

$$(Time, \langle 01, \#current\_year\# \rangle)(Place2,$$

$$certn\ library * (Number1, 18) : x2)))))$$

is a possible K-representation of the question Qs3.

Example 4. The question Qs4 = "How many times did Mr. Stepan Semyonov fly to Mexico?" can have the following possible K-representation:

$$Question(x1, (x1 \equiv Numb(all\ flight * (Agent1,$$

$$certn\ person * (Name, 'Stepan')(Surname, 'Semyonov') : x2)$$

$$(Place2, certn\ country * (Name1, 'Mexico') : x3)$$

$$(Time, arbitrary\ moment * (Before, \#now\#))))).$$

5.5 Semantic Representations of Questions with an Interrogative Pronoun Attached to a Noun

The method proposed above for constructing K-representations of the questions with the role interrogative words can also be used for building K-representations of the questions with the pronoun "what" attached to a noun in singular or plural.

Example 1. Let Qs1 = "What publishing house has released the novel 'The Winds of Africa'?" Then the string

$$Question(x1, (Situation(e1, releasing1 * (Time, x2)$$

$$(Agent2, certn\ publish\_house : x1)(Object, certn\ novel1 *$$

$$(Name1, 'The\ Winds\ of\ Africa') : x3)) \wedge Before(x2, \#now\#)))$$

can be interpreted as a K-representation of the question Qs1.

Example 2. Let Qs2 = "What foreign publishing houses the writer Igor Nosov is collaborating with?" Then the formula

$$Question(S1, (Qual - compos(S1, publish\_house *$$

$$(Kind - geogr, foreign)) \wedge Description(arbitrary\ publish\_house *$$

$$(Elem, S1) : y1, Situation(e1, cooperation1 * (Agent,$$

$$certn\ person * (Profession, writer)(First\_name, 'Igor')$$

$(Surname, 'Nosov') : x1)(Organization1, y1)(Time, \#now\#)))))$

is a possible K-representation of the question Qs2.

5.6 Semantic Representations of General Questions

The questions with the answer "Yes/No" are called in linguistics *general questions*. The form of representing the meaning of questions with the interrogative words proposed above can also be used for constructing semantic representations of general questions. With this aim, each question of the kind will be interpreted as a request to specify the truth value of a certain statement. For example, it is possible to interpret the question Qs1 = "Is Gent a city of Belgium?" as a request to find the truth value of the statement "Gent is one of the cities of Belgium." For realizing this idea, we introduce

Assumption 6.

The set of sorts $St(B)$ includes the distinguished sort *boolean* called "logical value"; the primary informational universe $X(B)$ includes the different elements *truth, false*, and $tp(truth) = tp(false) = boolean$; the set of functional symbols $F(B)$ includes such unary functional symbol $Truth - value$ that

$$tp(Truth - value) = \{(P, boolean)\},$$

where $P = P(B)$ is the distinguished sort "a meaning of proposition."

Example 1. Suppose that for the considered conceptual basis B the Assumption 6 is true. Then a K-representation of the question Qs1 = "Is Gent a city of Belgium?" can have the form

$$Question(x1, (x1 \equiv Truth - value(Elem(certn\,city*$$

$$(Name, "Gent") : x2, all\,city*(Part1,$$

$$certn\,country*(Name, "Belgium") : x3)))))$$.

In this formula, the symbol *certn* is the referential quantifier $ref(B)$ of the conceptual basis B; $xl, x2, x3$ are the variables from $V(B)$.

Example 2. Let Qs2 = "Did the international scientific conference 'COLING' take place in Asia ?" Then the formula

$$Question(xl, (x1 \equiv Truth - value((Situation(el, holding1*$$

$$(Event1, certn\,conf*(Type1, intern)(Type2, scient)$$

$$(Name, 'COLING') : x2)(Place, certn\,continent*$$

$$(Name, "Asia") : x3)(Time, x4)) \wedge Before(x4, \#now\#)))))$$

can be interpreted a K-representation of the question Qs2.

5.7 Describing Semantic Structure of Commands

Let's proceed from two basic ideas. First, when we speak about a command or about an order, we always mean that there is one intelligent system forming the command (it is designated by the expression *#Operator#*) and another intelligent system (or a finite set of intelligent systems) which should execute the command or an order (it is designated by expression *#Executor#*). Secondly, a verb in an imperative mood or the infinitive form of a verb will be replaced by the corresponding verbal noun.

Assumption 7.

The set of sorts $St(B)$ includes the distinguished elements *ints* (the sort "intelligent system"), *mom* (the sort "a moment of time"); the primary informational universe $X(B)$ includes such elements *Command, #Operator#, #Executor#, #now#* that

$$tp = (Command) = \{(ints, ints, mom, \uparrow sit)\},$$

$$tp(\#Operator\#) = tp(\#Executor\#) = ints,$$

$$tp(\#now\#) = mom.$$

Example. Let Comm1 = "Deliver a box with the details to the warehouse No. 3," where Comm1 is the command transmitted by the operator of a flexible industrial system to an intelligent transport robot. Let *Semrepr*1 be the K-string

$$Command(\#Operator\#, \#Executor\#, \#now\#,$$

$$delivery1 * (Object1, certn\,box * (Content1,$$

$$certn\,set * (Qual - compos, detail1)) : x1)$$

$$(Place2, certn\,warehouse * (Number1, 3) : x2)).$$

Then the basis B can be defined in such a way that Assumptions 1 and 7 will be true, and the following relationship will take place:

$$B(0, 1, 2, 3, 4, 5, 8) \Rightarrow Semrepr1 \in Ls(B).$$

The K-string *Semrepr*1 can be interpreted as a K-representation of the command Comm1.

5.8 Representation of Set-Theoretical Relationships and Operations on Sets

Example 1. Let T1a be the sentence "Namur is one of the cities of Belgium." Then we shall proceed from the semantically equivalent text T1b = "Namur belongs to set of all cities of Belgium" in order to construct a K-representation of the sentence T1a as the K-string *Expr*1 of the form

$$Elem(certn\,city * (Name1, \text{``}Namur\text{''}) : x1,$$

$$all\,city * (Part1, certn\,country * (Name1, \text{``}Belgium\text{''}) : x2)) \,.$$

Then there is such conceptual basis B that

$$B(0,1,2,3,8,1,5,0,1,2,3,8,1,5,2,4) \Rightarrow Expr1 \in Ls(B).$$

While constructing the K-string $Expr1$, the following assumptions were used:

$$Elem, country, city, certn, Part1 \in X(B),$$

$$tp(Elem) = \{([entity], \{[entity]\})\},$$

$$tp(country) = tp(city) = \uparrow space.object;$$

$$tp(Part1) = \{(space.object, space.object)\},$$

and $certn = ref(B)$ is the referential quantifier of the conceptual basis B.

Example 2. Let T2a be the command = "Include the container No. 4318 into the party to be sent to Burgas." We shall transform the command T2a into the statement T2b = "An operator has ordered to include the container No. 4318 into a certain party sent to the city Burgas."

Then it is possible to construct a K-representation of the texts T2a and T2b in the form

$$Command(\#Operator\#, \#Executor\#, \#Now\#, inclusion1*$$

$$(Object1, certn\,container1 * (Number1, 4318) : x1)(Target - set,$$

$$certn\,party2 * (Place - destin, certn\,city*$$

$$(Name, \text{``}Burgas''\text{)} : x2) : S1)),$$

where $S1$ is the label of a party of production. It is possible to present in a similar way the orders about the division of set of objects into certain parts and about the assembly of several sets into one, as in case of overloading the details from several boxes into one box.

5.9 Semantic Representations of Phrases with Subordinate Clauses of Purpose and Indirect Speech

Example 1. Let 1 = "Alexander entered the State University – Higher School of Economics (HSE) in order to acquire the qualification 'Business Informatics,' " and let $Semrepr1$ be the K-string

$$(Situation(e1, entering2 * (Agent1, certn\,person*$$

$$(Name, \text{``}Alexander\text{''}) : x1)(Lern\_institution, certn\,university*$$

$$(Name1, \text{``HSE''}) : x2)(Time, t1)(Purpose,$$

$$acquisition1 * (New\_property, certn\, qualification1 * (Name1,$$

$$business\_informatics) : x3))) \wedge Before(t1, \#now\#)).$$

Then there is such conceptual basis B that

$$B(0, 1, 2, 3, 4, 5, 7, 8) \Rightarrow Semrepr1 \in Ls(B) ,$$

$$Semrepr1 \& prop \in Ts(B) .$$

Example 2. Let 2 = "The director said that the reorganization of the firm is planned on February," and let $Semrepr2$ be the K-string

$$(Situation(e1, oral\_communication * (Agent1,$$

$$Director(certn\, org : xl))(Time, t1)(Content2, Planned(t1,$$

$$certn\, reorganization * (Object\_org, certn\, firm1 : x1),$$

$$Nearest(February, t1)))) \wedge Before(t1, \#now\#)).$$

Then it is easy to construct such conceptual basis B that the following relationships take place:

$$B(0, 2) \Rightarrow Nearest(February, t1) \in Ls(B),$$

$$Nearest(February, t1) \& time\_interval \in Ts(B);$$

$$B(0, 1, 2, 4, 5, 7, 8) \Rightarrow Semrepr2 \in Ls(B) ,$$

$$Semrepr2 \& prop \in Ts(B) .$$

5.10 Explicit Representation of Causal and Time Relations in Discourses

As it was mentioned above, a discourse, or a coherent text, is a sequence of the phrases (complete or incomplete, elliptical) with interconnected meanings. The *referential structure of an NL-text* is a correspondence between the groups of words from this text and the things, events, processes, meanings designated by these groups of words. The SK-languages provide broad possibilities of describing semantic structure of discourses, in particular, their referential structure.

Example. Let T1 = "The first-year student Peter Gromov didn't notice that the schedule had changed. As a result, he missed the first lecture on linear algebra." In this text, we can observe the following feature of discourses: the personal pronoun "he" is used instead of the longer combination "The first-year student Peter

Gromov." One says that the last expression and the pronoun "he" have the same referent being a certain person, a student of a college or a university.

Obviously, in order to specify the referent structure of a text, it is necessary to connect the labels with the entities designated by the groups of words from this text or implicitly mentioned in the text. We shall make this in the following way:

- the implicitly mentioned educational institution – the label $x1$;
- "The first-year student Peter Somov," "he" – the label $x2$;
- "the schedule" – the label $x3$;
- "the first lecture on linear algebra " – the label $x4$;
- "didn't notice" – the label $e1$ (the event 1);
- "has changed" – the label $e2$;
- $e3$ – the label of the situation described by the first sentence of T1;
- "missed" – the label $e4$ (event).

We shall suppose that a semantic representation of the text T1 should include the fragment $Cause\ (e3, e4)$.

Let $Semrepr1$ be the K-string

$$((Situation(e1, \neg\,noticing1 * (Agent1, certn\,person*$$

$$(Name, ``Peter'')(Surname, ``Gromov'')(Qualif, student*$$

$$(Year1, 1)(Learn\_institution, x1)) : x2)(Time, t1)$$

$$(Object\_of\_attention, e2)) \wedge Before(e1, \#now\#) \wedge$$

$$Situation(e2, change1 * (Object1, certn\,schedule : x3)(Time, t2))$$

$$\wedge Before(t2, t1)) : P1 \wedge Characterizes(P1, e3)).$$

Then $Semrepr1$ is a possible K-representation of the first sentence $S1$ of the discourse T1.

Let $Semrepr2$ be the K-string

$$(Situation(e4, missing1 * (Agent1, x2)(Event1,$$

$$certn\,lecture1 * (Discipline, linear-algebra)$$

$$(Learn\_institution, x1) : x4)(Time, t3))$$

$$\wedge Before(t3, \#now\#)).$$

Then $Semrepr2$ can be interpreted as a possible K-representation of the second sentence $S2$ from the discourse T1. Let

$$Semdisc1 = (Semrepr1 \wedge Semrepr2 \wedge Cause(e3, e4)).$$

Then the formal expression $Semdisc1$ is a possible semantic representation of the discourse T1 being its K-representation.

5.11 Semantic Representations of Discourses with the References to the Meanings of Phrases and Larger Parts of the Text

Example. Let T1 = "The join-stock company 'Rainbow' will sign the contract till December 15th. The deputy director Igor Panov has told this." Here the pronoun "this" designates the reference to the meaning of the first sentence of the discourse T1. Let *Semrepr*1 be the expression

$$(Situation(e1, signing1 * (Agent2, certn\,organization*$$

$$(Type, joint\_stock\_comp)(Name1, ``Rainbow") : x1)(Time, t1)$$

$$(Inf\_object, certn\,contract1 : x2))\wedge$$

$$Before(t1, Date(12, 15, \#current\_year\#)))),$$

and *Semrepr*2 be the expression of the form

$$(Semrepr1 : P1 \wedge Situation(e2, oral\_message * (Agent1,$$

$$certn\,person * (Name, ``Igor")(Surname, ``Panov") : x3)(Time, t2)$$

$$(Content2, P1)) \wedge Before(t2, \#now\#)\wedge$$

$$Deputy\_Director(x3, certn\,organization : x4)).$$

Then there is such conceptual basis *B* that

$$B(0, 1, 2, 3, 4, 5, 7, 8) \Rightarrow Semrepr2 \in Ls(B),$$

$$Semrepr2 \,\&\, prop \in Ts(B),$$

where $prop = P(B)$ is the distinguished sort "a meaning of proposition."

The rule $P[5]$ allows for attaching a variable v to a semantic representation *Semrepr* of arbitrary narrative text and for obtaining the formula

$$Semrepr : v,$$

where v is a variable of the sort $P(B)$ ("a meaning of proposition").

Therefore, the variables of the sort $P(B)$ will be the images of the expressions "about this," "this method," "this question," etc., in the complete semantic representation of the considered discourse (in the same way as in the last example).

5.12 Representing the Pieces of Knowledge About the World

Example 1. Let T1 = "The notion 'a molecule' is used in physics, chemistry and biology." It is possible to define such conceptual basis *B* that the set of sorts $St(B)$

includes element *activity_field*, the primary informational universe $X(B)$ includes the elements

$$activity\_field, string, notion, \text{``molecule''}, Use, Notion - name,$$

$$physics, chemistry, biology,$$

and the types of these elements are set by the relationships

$$tp(notion) = [\uparrow concept], tp(\text{``molecule''}) = string,$$

$$tp(physics) = tp(chemistry) = tp(biology) = activity\_field,$$

$$tp(Use) = \{([concept], activity\_field)\},$$

$$tp(Notion - name) = \{([concept], string)\}.$$

Let *certn* be the referential quantifier, *Use* and *Notion − name* be the binary relational symbols not being the names of the functions, and

$$s1 = Notion - name(certn\,notion, \text{``molecule''}),$$

$$s2 = notion * (Notion - name, \text{``molecule''}),$$

$$s3 = Use(certn\,notion * (Notion - name, \text{``molecule''}),$$

$$(physics \wedge chemistry \wedge biology)).$$

Then $B(0, 1, 4) \Rightarrow s1 \in Ls(B)$; $B(0, 1, 4, 8) \Rightarrow s2 \in Ls(B)$;

$$B(0, 1, 4, 8, 1, 0, 7, 4) \Rightarrow s3 \in Ls(B).$$

The built formula $s3$ is a possible semantic representation of the sentence T1.

Example 2. Let T2 be the definition "Teenager is a person at the age from 12 to 19 years"; *semdef* be the K-string

$$((teenager \equiv person * (Age, x1)) \wedge$$

$$\neg Less1(x1, 12/year) \wedge \neg Greater1(x1, 19/year)).$$

Then *semdef* is a possible K-representation of T2.

5.13 Object-Oriented Representations of Knowledge Pieces

We can build complex designations of objects and sets of objects, using SK-languages.

Example 1. We can build the following K-representation of a description of the international scientific journal "Informatica":

$$certn \, int.sc. journal * (Title, \text{``}Informatica\text{''})$$

$$(Country, \, Slovenia)(City, \, Ljubljana)(Fields,$$

$$(artif.intel \wedge cogn.science \wedge databases)) : k225 \, ,$$

where $k225$ is the mark of the knowledge module with the data about "Informatica."

The definition of the class of SK-languages allows for building the formal conceptual representations of texts as informational objects reflecting not only the meaning but also the external characteristics (or metadata) of the text: the authors, the date, the application fields of described methods and models, etc.

Example 2. In a way similar to the way used in the previous example, we can construct a knowledge module stating the famous Pythagorean Theorem and also indicating its author and field of science. For instance, such a module may be the following expression of a certain SK-language:

$$certn \, textual\_object * (Kind1, theorem)(Fields1,$$

$$geometry)(Authors, Pythagoras)(Content\_inf\_ob,$$

$$\exists x_1 \, (geom) \, \exists x_2 \, (geom) \, \exists x_3 \, (geom) \, \exists x_4 \, (geom)$$

$$If-then((Is1(x_1, right-triangle) \wedge$$

$$Hypotenuse(x_2, x_1) \wedge Leg1((x_3 \wedge x_4), x_1)),$$

$$(Square(Length(x_2)) \equiv Sum(Square(Length(x_3)),$$

$$Square(Length(x_4)))))) : k81.$$

5.14 The Marked-Up Conceptual Bases

The analysis shows that it is possible and reasonable to select a compact collection of primary informational units to be used for constructing semantic representations of NL-texts independently on application domains, in particular, for building SRs of questions, commands, and descriptions of sets.

The Assumptions 1–7 formulated above indicate such primary informational units. The purpose of introducing the definitions below is to determine the notion of a marked-up conceptual basis, i.e., a conceptual basis satisfying the Assumptions 1–7.

Definition 5.1. Let B be an arbitrary conceptual basis, $St(B)$ be the set of sorts of the basis B, $P(B)$ be the sort "a meaning of proposition," $X(B)$ be the primary informational universe of the basis B. Then a system Qmk of the form

$$(sit, \, Vsit, \, Situation, \, Question, \, boolean, \, true, \, false, \, Truth-value) \qquad (5.1)$$

is called *a marking-up of questions of the conceptual basis B* $\Leftrightarrow$ when

- the elements *sit*, *boolean* belong to the set-theoretical difference of the set of sorts $St(B)$ and the set $Concr(P)$, consisting of all sorts being the concretizations of the sort $P(B)$ with respect to the generality relation $Gen(B)$,
- $X(B)$ includes the different elements *Situation*, *Question*, *boolean*, *true*, *false*, *Truth − value*,
- Assumptions 1, 4, and 6 are true for the components of this system.

Definition 5.2. Let B be an arbitrary conceptual basis. Then a system *Setmk* of the form

$$(nat, Nt, set, Numb, Qual - compos, Object - compos, arbitrary, all, Elem)$$
(5.2)

is called *a set-theoretical marking-up of the conceptual basis B* $\Leftrightarrow$ when

- the element *nat* belongs to the set-theoretical difference of the set of sorts $St(B)$ and the set $Concr(P)$, consisting of all sorts being the concretizations of the sort $P(B)$ with respect to the generality relation $Gen(B)$,
- Nt is a subset of the primary informational universe $X(B)$,
- the elements *set*, *Numb*, *Qual − compos*, *Object − compos*, *arbitrary*, *all*, *Elem* are different elements of the set $X(B)$,
- Assumptions 2, 3, and 5 are true for the components of this system.

Definition 5.3. Let B be an arbitrary conceptual basis, Qmk be a marking-up of questions of the form (5.1) of the basis B. Then a system Cmk of the form

$$(ints, mom, \#now\#, \#Operator\#, \#Executor\#, Command)$$
(5.3)

is called *a marking-up of commands of the basis B coordinated with the marking-up of questions Qmk* $\Leftrightarrow$ when

- *ints*, *mom*, $\#now\#$, $\#Operator\#$, $\#Executor\#$, *Command* are different elements of the set $X(B)$,
- *ints*, *mom* $\in St(B) \setminus (Concr(P) \cup Concr(sit) \cup \{boolean\})$, where $Concr(P)$ and $Concr(sit)$ are the sets of all sorts being respectively the concretizations of the sort $P(B)$ and of the sort *sit* with respect to the generality relation $Gen(B)$,
- Assumption 7 is true for the components of the system Cmk.

The formal notions introduced above enable us to make the final step and to join these notions in the definition of the class of marked-up conceptual bases.

Definition 5.4. A *marked-up conceptual basis (m.c.b.)* is an arbitrary four-tuple Cb of the form

$$(B, Qmk, Setmk, Cmk),$$
(5.4)

where B is an arbitrary conceptual basis, Qmk is a marking-up of questions of the form (5.1) for the basis B, Qmk is a set-theoretical marking-up of the conceptual basis B, Cmk is a marking-up of commands of the basis B coordinated with the marking-up of questions Qmk, and the following conditions are satisfied:

- all components of the systems *Qmk*, *Setmk*, *Cmk*, except for the component *Nt* of the system *Setmk*, are the different elements of the primary informational universe $X(B)$;
- if *Stadd* = {*sit*, *boolean*, *ints*, *mom*, *nat*} and *Concr*(*P*) is the set of all sorts being the concretizations of the sort $P(B)$ ("a meaning of proposition") with respect to the generality relation *Gen*(*B*), then *Stadd* is a subset of the set $St(B) \setminus Concr(P)$, and every two different elements of the set *Stadd* are incomparable both for the generality relation *Gen*(*B*) and for the tolerance relation *Tol*;
- if *s* is an arbitrary element of the set *Stadd*, and *u* is an arbitrary element of the set *Concr*(*P*), then the sorts *s* and *u* are incomparable for the tolerance relation *Tol*.

Definition 5.5. Let's agree to say that a marked-up conceptual basis *Cb* is a marked-up basis of the standard form ⇔ *Cb* is a system of the form (5.4), *Qmk* is a system of the form (5.1), *Setmk* is a system of the form (5.2), and *Cmk* is a system of the form (5.3).

We'll consider below the marked-up conceptual bases of only standard form.
The class of formal languages

$$\{ Ls(B) \mid B \text{ is the first component of arbitrary m.c.b. } Cb \}$$

will be used as the class of semantic languages while describing the correspondences between NL-texts and their semantic representations.

5.15 Related Approaches to Representing Semantic Structure of NL-Texts

Proceeding from the ideas stated in this chapter, one is able to easily simulate the expressive mechanisms provided by Discourse Representation Theory (DRT), Episodic Logic, and Theory of Conceptual Graphs (TCG).

For instance, the example in Sect. 5.10 shows how it would be possible to describe causal and time relationships between the events mentioned in a discourse without the episodic operator.

The manner of building short compound (though rather simple) designations of the notions and sets of objects proposed by TCG can be replaced by much more general methods of the theory of K-representations.

Example 1. Let T = "Sue sent the gift to Bob," and a possible TCG-representation of T1 is

$$[Person : Sue] \leftarrow (Agent) \leftarrow [Send] \rightarrow$$

$$(Theme) \rightarrow [gift] : \#]$$

$$(Recipient) \rightarrow [Person : Bob].$$

Then a possible K-representation of T1 is

$$Situation(e1, sending1 * (Time, certn\,mom*$$

$$(Before, \#now\#) : t1)Agent, certn\,person*$$

$$(Name, \text{``Sue''}) : x1)(Theme, certn\,gift : x2)$$

$$(Recipient, certn\,person * (Name, \text{``Bob''}) : x3)).$$

Example 2. A certain group of vehicles can be denoted in TCG by the expression

$$[vehicle : \{*\}]$$

and by the K-string

$$certn\,set * (Qual - compos, vehicle).$$

Example 3. Let's agree that the string *at* designates a special symbol used in e-mail addresses. Then a certain set consisting of 70 books can be denoted in TCG by the expression

$$[book : \{*\}\,at\,70]$$

and by the K-string

$$certn\,set * (Numb, 70)(Qual - compos, book).$$

Example 4. A set consisting of three concrete persons with the names Bill, Mary, and Sue can have a TCG-representation

$$[person : \{Bill, Mary, Sue\}]$$

and a K-representation

$$certn\,set * (Object - compos, (certn\,person*$$

$$(Name, \text{``Bill''}) \land certn\,person * (Name, \text{``Mary''})$$

$$\land certn\,person * (Name, \text{``Sue''}))).$$

Example 5. The set of all farmers located in the state of Maine can have a TCG-representation

$$[[Farmer : \lambda] \rightarrow (Loc) \rightarrow [State : Maine] : \{*\}\forall]$$

and a K-representation

$$all\,farmer * (Loc, certn\,state2*$$

$$(Name1, \text{``Maine''})).$$

The material of this chapter helps also to understand how it would be possible to simulate other expressive mechanisms of TCG, in particular, the manners to represent the finite sequences and the semantic structure of the sentences with direct or indirect speech.

The principal advantage of the theory of SK-languages is that the convenient ways of simulating numerous expressive mechanisms of NL are the specific combinations of discovered general operations on conceptual structures.

Due to this feature, there is no necessity, as in TCG, to invent special combinations of symbols (for instance, $\{*\}$ or $\{*\}$ *at* 70 or $\{*\}\forall$) as the indicators of special constructions of semantic level.

The concrete advantages of the theory of SK-languages in comparison with first-order logic, Discourse Representation Theory (DRT), and Episodic Logic (EL) are, in particular, the possibilities to

1. distinguish in a formal way objects (physical things, events, etc.) and notions qualifying these objects;
2. build compound representations of notions;
3. distinguish in a formal manner the objects and the sets of objects, the notions and the sets of notions;
4. build complicated representations of sets, sets of sets, etc.;
5. describe set-theoretical relationships;
6. effectively describe structured meanings (SMs) of discourses with the references to the meanings of phrases and larger parts of discourses;
7. describe SMs of sentences with the words "concept," "notion";
8. describe SMs of sentences where the logical connective "and" or "or" joins not the expressions – assertions but the designations of things, sets, or notions;
9. build complex designations of objects and sets;
10. consider non traditional functions with arguments and/or values being the sets of objects, of notions, of texts' semantic representations, etc.;
11. construct formal analogues of the meanings of infinitives with dependent words and, as a consequence, to represent proposals, goals, obligations, and commitments.

It should be added that the model has at least three global distinctive features as regards its structure and destination in comparison with EL.

The first feature is as follows: In fact, the purpose of this monograph is to represent in a mathematical form a hypothesis about the general mental mechanisms (or operations) underlying the formation of complicated conceptual structures (or semantic structures, or knowledge structures) from basic conceptual items.

EL doesn't undertake an attempt of the kind, and 21 Backus-Naur forms used in [129] for defining the basic logical syntax rather disguise such mechanisms (operations) in comparison with the more general 10 rules described in Chap. 4.

The second global distinctive feature is that this book formulates a hypothesis about *a complete collection of operations of conceptual level* providing the possibility to build effectively the conceptual structures corresponding to arbitrarily complicated real sentences and discourses pertaining to science, technology, business, medicine, law, etc.

The third global distinction is that the form of describing in EL the basic collection of informational units is not a strictly mathematical one.

For example, the collection of Backus-Naur forms used with this purpose in [129] contains the expressions

$$\langle l - place - pred - const \rangle ::= happy \mid person \mid certain \mid probable \mid \ldots,$$

$$\langle 1 - fold - pred - modifier - const > ::=$$

$$plur \mid very \mid \mid former \mid almost \mid in - manner \mid \ldots.$$

The only way to escape the use of three dots in productions is to define an analogue of the notion of a conceptual basis introduced in this monograph.

The items (4)–(8), (10), (11) in the list above indicate the principal advantages of the theory of SK-languages in comparison with the Theory of Conceptual Graphs(TCG). Besides, the expressive possibilities of the new theory are much higher than the possibilities of TCG as regards the items (1), (2), (9).

There are numerous technical distinctions between the theory of SK-languages and Database Semantics of Natural language [119, 120]. The principal global distinction is that the theory of SK-languages puts forward (in a mathematical form) a hypothesis about the organization of structured meanings associated not only with separate sentences in NL but also with arbitrary complex discourses.

On the contrary, the Database Semantics of NL doesn't propose the formal tools being convenient for studying the problem of formalizing semantic structure of complex long discourses.

It is the same principal distinction and principal advantage as in the case of the comparison with the approach to representing semantic structure of NL-sentences and short discourses with the help of the language UNL (Universal Networking Language). A number of concrete advantages of SK-languages in comparison with UNL, in particular, concerning the representation of complex concepts (called scopes) is analyzed in [84, 85, 90, 93, 94].

Problems

1. Describe the components of a marking-up of questions of a conceptual basis.
2. Explain the assumptions about the types of distinguished informational units

$$set, Numb, Qual - compos, Object - compos, arbitrary, all, Elem$$

being the components of a set-theoretical marking-up of a conceptual basis.
3. What is the difference between a conceptual basis and a marked-up conceptual basis?
4. How is it proposed to build semantic representations of events?
5. What are the main ideas of constructing compound semantic descriptions of the sets?
6. How is it possible to build the K-representations of commands?
7. Construct the K-representations of the following NL-texts:

(a) How many parcels from Reading have been received?

(b) Is it possible for a pensioner to get a credit?

(c) What journal published for the first time an article about BMW 330?

(d) Which scholars from Belgium did attend the conference?

(e) BMW 750 was put for sale in the year 1985.

(f) It is planned to inaugurate next year the offices of this bank in Omsk and Tomsk.

(g) Dr. William Jones, the president of the university, visited in March the University of Heidelberg (Germany). This was written in the newsletter of the University of Heidelberg.

Chapter 6
The Significance of a New Mathematical Model for Web Science, E-science, and E-commerce

Abstract The significance of the theory of K-representations for e-science, e-commerce, and Web science is shown. The following possibilities provided by SK-languages are analyzed: building semantic annotations of Web-sources and Web-services, constructing high-level conceptual descriptions of visual images, semantic data integration, and the elaboration of formal languages intended for representing the contents of messages sent by computer intelligent agents (CIAs). It is also shown that the theory of SK-languages opens new prospects of building formal representations of contracts and records of commercial negotiations carried out by CIAs. A theoretically possible strategy of transforming the existing Web into Semantic Web of a new generation is proposed.

6.1 The Problem of Semantic Data Integration

6.1.1 The Purpose of Semantic Data Integration in E-Science and Other E-Fields

In the modern world, the objective demands of science and technology often urge research groups in different countries to start the projects aimed at solving the same or similar tasks. A considerable part of obtained results is available via Web. Since the obtained results of the studies can be expressed in different formats, it is important to elaborate the software being able to semantically integrate the data stored in Web-documents, that is, to present the meaning of available documents in a unified format.

As a consequence, a researcher or research group starting to investigate a problem will be able to quickly get adequate information about the state of affairs in the field of interest.

On the other hand, the problem of semantic data integration emerges in connection with numerous practical tasks. Imagine, for instance, that it is required to

V.A. Fomichov, *Semantics-Oriented Natural Language Processing*, IFSR International Series on Systems Science and Engineering 27, DOI 10.1007/978-0-387-72926-8_6,
© Springer Science+Business Media, LLC 2010

elaborate an itinerary for a ship transporting goods across an ocean and several seas. In this case, it is necessary to take into account the geophysical, economical, and political data about many areas of the world, in particular, about many areas of the ocean and several seas.

One of the fields of professional activity where the necessity of semantic data integration is most acute is *e-science*. This term (it emerged only in this decade) unifies the studies in different fields of science based on the extensive use of large volumes of obtained data stored on the Web-servers. This applies, in particular, to the studies on bioinformatics, physics, ecology, life sciences [19, 35, 116, 117, 202, 209, 212].

In e-science, a considerable part of obtained results is described in natural language texts being available via Web: in scientific articles, technical reports, encyclopedic dictionaries, etc. That is why an important demand of e-science consists in developing the formal means allowing for representing in a unified format both the meanings of NL-texts and the pieces of knowledge about the application domains.

It is one of the principal goals of *semantic e-science* – a subfield of e-science aimed at creating the semantic foundations of e-science [25, 115].

6.1.2 Ontologies in Modern Information Society

The notion of ontology is one of the most important notions for the studies on semantic data integration.

An ontology can be defined as a specification of a conceptualization [110]. The term "conceptualization" is used for indicating a way an intelligent system structures its perceptions about the world. A specification of a conceptualization gives a meaning to the vocabulary used by an intelligent system for processing knowledge and interacting with other intelligent systems.

In the last decade, one has been able to observe a permanent growth of interest in building and studying ontologies. The reason is that the researchers and systems developers have become more interested in reusing or sharing knowledge across systems. Different computer systems use different concepts and terms for describing application domains. These differences make it difficult to take knowledge out of one system and use it in another. Imagine that we are able to construct ontologies that can be used as the basis for multiple systems. In this case different systems can share a common terminology, and this will facilitate sharing and reuse of knowledge.

In a similar way, if we are able to create the tools that support merging ontologies and translating between them, then sharing knowledge is possible even between systems based on different ontologies.

The main source for automatically building ontologies is a great amount of available texts in natural language (NL). Taking this into account, we need the powerful formal means for building semantic representations (SRs) of (a) NL-definitions of concepts and (b) sentences and discourses in NL expressing knowledge of other kinds about an application domain.

The analysis of formal approaches to representing knowledge provided by the Theory of Conceptual Graphs, Episodic Logic, and Description Logic shows that these approaches give formal means with rather restricted expressive possibilities as concerns building SRs of definitions of notions and SRs of sentences and discourses representing the pieces of knowledge about the world.

That is why we need to have much more powerful and convenient formal means (in comparison with the broadly used ones) for describing structured meanings of natural language (NL) texts and, as a consequence, for building ontologies.

6.1.3 The Language UNL and the Problem of Sharing Knowledge

Since the second half of the 1990s, one has been able to observe the progress of two parallel approaches to adding to the existing Web the ability to understand the meanings of electronic documents.

On one hand, it has been the activity of the research laboratories of the World Wide Web Consortium (W3C) and a number of other research centers in the world aimed at developing a Semantic Web. Though officially the task of creating Semantic Web was announced in the beginning of 2001, the possibility to pose this task was created by a number of preliminary studies resulted, in particular, in the development of Resource Description Framework (RDF) – a language for describing the metadata about informational sources and RDF Schema Specification Language (RDFS). In the first decade of this century, RDF and RDFS became the basis for the development of DAML + OIL and its successor OWL – two languages intended for constructing ontologies [123, 136, 186, 204].

On the other hand, the following fundamental problem has emerged in the mid-1990s: how to eliminate the language barrier between the end users of the Internet in different countries. For solving this problem, H. Uchida et al. [199] proposed a new language-intermediary, using the words of English language for designating informational units and several special symbols. This language, called the Universal Networking Language (UNL), is based on the idea of representing the meanings of separate sentences by means of binary relations.

The second motive for the elaboration of UNL was an attempt to create the language means allowing for representing in one format the various pieces of knowledge accumulated by mankind and, as a consequence, to create objective preconditions for sharing these pieces of knowledge by various computer systems throughout the world [211].

Since 1996, UNO has been funding a large-scale project aimed at the design of a family of natural language processing systems (NLPSs) transforming the sentences in various natural languages into the expressions of UNL and also transforming the UNL-expressions into sentences in various natural languages. For several years the coordinator of this project was the UNO Institute for Advanced Studies by the Tokyo University. At the moment, under the framework of this project, the NLPSs for six official UNO languages are being elaborated (English, Arabic, Spanish, Chinese, Russian, and French), and also for nine other languages, including Japanese, Italian, and German. Since the beginning of the 2000s, the studies in this direction have been coordinated by the Universal Networking Digital Language Foundation.

The initially scheduled duration of the UNL project started in 1996 is 10 years. That is why it is just the time to analyze the achieved results and to take the right decisions concerning further studies in this direction. It is shown in the papers [84, 90, 93, 94] with respect to the online monographs [200, 201] that the expressive possibilities of UNL are rather restricted.

First of all, the language UNL is oriented at representing the contents of only separate sentences but not arbitrary discourses. Even the UNL specifications published in 2006 don't contain a theory of representing the meanings of discourses.

That is why in the papers [84, 90, 93, 94] it is proposed to interpret the language UNL (despite the linguistic meaning of its title) as a semantic networking language of the first generation.

With respect to the fact that the expressive power of UNL is rather restricted, it seems reasonable to look for another, more powerful formal approach to describing structured meanings of natural language texts with the aim to find (if possible) a model for constructing a universal or broadly applicable semantic networking language for adding a meaning-understanding ability to the existing Web and for contributing to semantic integration of Web data.

6.2 Building Semantic Annotations of Web Data

The analysis of a number of publications studying the problem of transforming the existing Web into Semantic Web allows for drawing the following conclusion: an ideal configuration of Semantic Web would be a collection of interrelated resources, where each of them has both an annotation in natural language (NL) and a formal annotation reflecting the meaning or generalized meaning of this resource, i.e. *a semantic annotation*. NL-annotations would be very convenient for the end users, and semantic annotations would be used by question-answering systems and search engines.

Most likely, the first idea concerning the formation of semantic annotations of Web data would be to use the formal means for building semantic representations of NL-texts provided by mathematical and computational linguistics.

However, the analysis shows that the expressive power of the main popular approaches to building SRs of NL-texts, in particular, of Discourse Representation Theory, Theory of Conceptual Graphs, and Episodic Logic is insufficient for effectively representing contents of arbitrary Web data, in particular, arbitrary biological, medical, or business documents.

First of all, the restrictions concern describing semantic structure of (a) infinitives with dependent words (e.g., representing the intended manner of using things and procedures); (b) constructions formed from the infinitives with dependent words by means of the logical connectives "and," "or," "not"; (c) the complicated designations of sets; (d) the fragments where the logical connectives "and," "or" join not the designations of assertions but the designations of objects ("the product A is distributed by the firms B1, B2, ..., BN"); (e) the explanations of the terms being unknown to an applied intelligent system; (f) the fragments containing the references to the meanings of phrases or larger fragments of a discourse ("this method," etc.); (g) the designations of the functions whose arguments and/or values may be the sets of objects ("the staff of the firm A," "the number of the suppliers of the firm A," etc.).

Taking into account this situation and the fact that the semantic annotations of Web-sources are to be compatible with the format of representing the pieces of knowledge in ontologies, a number of researchers undertook the efforts of constructing computer intelligent systems, using the languages RDF, RDFS, or OWL for building semantic annotations of Web-sources [161, 176].

However, the expressive power of RDF, RDFS, or OWL is insufficient for being an adequate formal tool of building semantic annotations of scientific papers, technical reports, etc.

Meanwhile, the formulated idea of where to get the formal means for building semantic annotations from is correct. The main purpose of this section is to set forth the principal ideas of employing the SK-languages for building semantic annotations of informational sources, in particular, Web-based sources.

Example. Let's consider a possible way of employing SK-languages for building a semantic annotation of the famous paper "The Semantic Web" by T. Berners-Lee, J. Hendler, and O. Lassila published in "Scientific American" in May 2001 [18].

Suppose that there is a Web-source associating the following NL-annotation with this paper:

It is proposed to create such a net of Web-based computer intelligent agents (CIAs) being able to understand the content of almost every Web-page that a considerable part of this net will be composed by CIAs being able to understand natural language. The authors consider the elaboration of ontologies as a precondition of sharing knowledge by CIAs from this net and believe that it is reasonable to use the languages RDF and RDFS as primary formal tools for the development of ontologies of the kind.

A semantic annotation corresponding to this NL-annotation can be the K-string (or l-formula) of the form

$$certn\,inf.ob * (Kind1, sci\_article)(Source1,$$

$$certn\,journal1 * (Name1, \text{``}Scientific\_American\text{''}) : x1)$$

$$(Year, 2001)(Month, May)(Authors,$$

$$certn\,group1 * (Numb, 3)(Elements1,$$

$$(\langle 1, certn\,scholar * (First\_name, \text{``}Tim\text{''})(Surname,$$

$$\text{``}Berners-Lee\text{''}) : x2\rangle \wedge \langle 2, \, certn\,scholar*$$

$$(First\_name, \text{``}James\text{''})(Surname, \text{``}Hendler\text{''}) : x3\rangle$$

$$\wedge \langle 3, \, certn\,scholar * (First\_name, \text{``}Ora''\text{''})(Surname,$$

$$\text{``}Lassila''\text{''}) : x4\rangle)) : S1)(Central\_ideas,$$

$$(\langle 1, \, Semrepr1\rangle \wedge \langle 2, \, Semrepr2\rangle \wedge$$

$$\langle 3, \, Semrepr3\rangle \wedge \langle 4, \, Semrepr4\rangle)) : v,$$

where the variable $S1$ designates the group consisting of all authors of this article, v is a variable being a mark of the constructed semantic annotation as an informational object, and $Semrepr1 - Semrepr4$ are the K-strings defined by the following relationships:

$$Semrepr1 = Proposed(S1, creation1 * (Product1,$$

$$certn\,family1 * (Qual - compos, \, intel\_comp\_agent*$$

$$(Property, web - based)(Ability, understanding1 * (Inf\_object,$$

$$Content(almost\_every\,web\_page)))) : S2)$$

$$(Time, certn\,time\_interval * (Part1,$$

$$Nearest\_future(decade1, \#now\#)))),$$

$$Semrepr2 = Proposed(S1, achieving\_situation*$$

$$(Description1, \exists S3(set)(Subset(S3, S2) \wedge$$

$$Estimation1(Numb(S3)/Numb(S2), considerable1) \wedge$$

$$Qual - compos(S3, intel\_comp\_agent * (Property,$$

$$web - based)(Ability, understanding1 * (Inf\_object,$$

$$almost\_every\,text * (Language1, certn\,language*$$

$$(Belong, NL\_family)))))))),$$

$$Semrepr3 = Believe(S1, Precondition(elaboration*$$

$$(Product1, certn\,family1 * (Qual - compos,$$

$$ontology) : S4), knowledge\_sharing*$$

$$(Group\_of\_intel\_systems, S2)),$$

$$Semrepr4 = Believe(S1, Reasonable(\#now\#,$$

$$using * (Object1, (RDF \wedge RDFS))$$

$$(Role1, primary\ formal\_tool)(Purpose,$$

$$elaboration1 * (Product1, S4)))).$$

To sum up, a comprehensive formal tool for building semantic annotations of Web data is elaborated. This tool is the theory of SK-languages. A very important additional expressive mechanism of SK-languages in comparison with the mechanisms illustrated in the example above is the convenience of building semantic representations of discourses with references to the meanings of phrases and larger parts of a discourse.

The analysis of expressive power of the class of SK-languages allows for conjecturing that it is both possible and convenient to construct semantic annotations of arbitrary Web data by means of SK-languages. That is why the theory of SK-languages can be interpreted as a powerful and flexible (likely, universal) formal metagrammar of semantic annotations of Web data.

6.3 Conceptual Descriptions of Visual Images

It is interesting that SK-languages, initially developed for representing structured meanings of NL-texts, open also new prospects for building high-level conceptual descriptions of visual images. This fact can be explained very easily: we are able to use NL for constructing high-level conceptual descriptions of visual images. Since the expressive possibilities of SK-languages are very rich, it is possible to use SK-languages for building conceptual descriptions of arbitrary visual images.

Example. We are able to describe Fig. 6.1 as follows:

The scene contains two groups of objects. The quadrant of the scene including the upper-left corner contains a figure being similar to a ellipse; this ellipse is formed by eight squares, the side of each square is 1 cm.

The quadrant of the scene including the bottom-right corner contains a figure being similar to a rectangle formed by 10 circles, the diameter of each circle is 1 cm. The longer sides of this rectangle are horizontally oriented, each of them consists of 4 circles.

This meaning can be expressed by the following K-string:

$$Number - of - groups(certn\ scene1 : x1, 2) \wedge$$

$$Groups(x1, (Gr1 \wedge Gr2)) \wedge Isolated(Gr1, Gr2)$$

$$\wedge Loc(Gr1, top - left - quadrant(x1)) \wedge$$

$$Similar\_shape(Gr1, certn\ ellipse : z1) \wedge$$

$$(Horiz\_diameter(z1) \equiv Multip\ (0.5, Length(x1))) \wedge$$

$$(Vertic\_diameter(z1) \equiv Multip\ (0.25, Length(x1))) \wedge$$

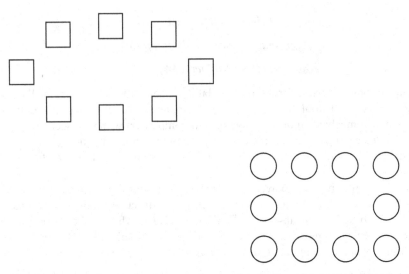

Fig. 6.1 An example of a visual scene to be associated with a high-level conceptual description being a K-string

$$Loc(Gr2, bottom - right - quadrant(x1)) \land$$

$$Similar\_shape(Gr2, certn\,rectangle : z2) \land$$

$$(Height(z2) \equiv Multip\ (0.5, Height(x1))) \land$$

$$(Length(z2) \equiv Multip\ (0.4, Length(x1))) \land$$

$$Indiv - composition(Gr1, (x1 \land x2 \land ... \land x8))$$

$$\land Is1((x1 \land x2 \land ... \land x8),$$

$$square1 * (Side1, Multip\ (0.06, Length(x1))) \land$$

$$Indiv - composition(Gr2, (y1 \land y2 \land ... \land y10))$$

$$\land Is1((y1 \land y2 \land ... \land y10),$$

$$circle * (Diameter, Multip\ (0.055, Length(x1))) \land$$

$$(Object\_number(every\,side1 * (Part1, z2)(Orient, horiz)) \equiv 4)).$$

Though the image presented on Fig. 6.1 is rather simple, the used method of building a high-level conceptual description of this image in the form of a K-string is rather general. This method is as follows:

1. Distinguish the principal groups of objects presented on the image and their number.
2. Calculate the positions of these groups.
3. Describe a shape (if possible) of each group.
4. Describe the number and a shape of objects being the elements of each group.
5. Describe the connections of distinguished groups of objects.

Thus, the theory of K-representations can be used in the design of multi media databases (with Web access too) for building in the same form, as the expressions of SK-languages, both semantic annotations of textual informational sources and high-level conceptual descriptions of visual images, in particular, being the components of textual informational sources.

6.4 Representation of Knowledge in Biology and Ecology

Let's consider a number of new important possibilities of building formal definitions of concepts provided by standard K-languages. If T is an expression in NL and a string E from an SK-language can be interpreted as a semantic representation (SR) of T, then E is called a K-representation (KR) of the expression T.

Example 1. Let Def1 = "A flock is a large number of birds or mammals (e.g. sheep or goats), usually gathered together for a definite purpose, such as feeding, migration, or defence." Def1 may have the K-representation $Expr1$ of the form

$$Definition1(flock, dynamic-group*(Compos1,$$

$$(bird \vee mammal*(Examples, (sheep \wedge goal)))), S1,$$

$$(Estimation1(Quantity(S1), high) \wedge$$

$$Goal-of-forming(S1,$$

$$certn\,purpose*(Examples,$$

$$(feeding \vee migration \vee defence))))).$$

Example 2. The definition Def1 is taken from a certain book published in a certain year by a certain publishing house. The SK-languages allow for building SRs of definitions in an object-oriented form reflecting their external connections. For instance, object-oriented SR of the definition Def1 can be the expression

$$certn\,inform-object*(Kind, definition)$$

$$(Content1, Expr1)(Source1, certn\,dictionary*$$

$$(Title, Longman\_Dictionary\_of\_Scientifc\_Usage)$$

$$(Publishing\_house, (Longman\_Group\_Limited/Harlow$$

$$\wedge Russky\_Yazyk\_Publishers/Moscow))$$

$$(City, Moscow)(Year, 1989)).$$

Example 3. Let T1 = "All granulocytes are polymorphonuclear; that is, they have multilobed nuclei." Then T1 may have the following K-representation:

$$(Property(arbitr\,granulocyte : x1,$$

$$polymor phonuclear) : P1 \wedge Explanation(P1,$$

$$If - then(Have1(x1, certn\,nucleus : x2),$$

$$Property(x2, multilobed)))).$$

Here $x1$ is the variable marking an arbitrary granulocyte, $x2$ is the variable marking the nucleous of the granulocyte $x1$, and $P1$ is the variable marking the meaning of the first phrase of T1.

Example 4. Consider the text D1 = "An adenine base on one DNA strand links only with a thymine base of the opposing DNA strand. Similarly, a cytosine base links only with a guanine base of the opposite DNA strand."

For constructing a KR of D1, the following remark may be helpful. A molecule of deoxyribonucleic acid (a DNA molecule) is composed of thousands of nucleotides (combinations of three basic elements: deoxyribose, phosphate, and a base). There are four kinds of bases: adenine, guanine, cytosine, and thymine. The nucleotides of a DNA molecule form a chain, and this chain is arranged in two long strands twisted around each other.

Taking into account this remark, one can associate with the first sentence of D1 a KR *Semrepr*1 of the form

$$\forall x1 \, (dna - molecule)(Link(arbitr\,base1 * (Is1, adenine)$$

$$(Part, arbitr\,strand1 * (Part, x1) : y1) : z1, certn\,base1*$$

$$(Is1, thymine)(Part, certn\,strand1 * (Part, x1)$$

$$(Opposite, y1) : y2) : z2) \wedge$$

$$\neg \exists z3 \, (base1)(Is1(z3, \neg thymine) \wedge$$

$$Part(z3, y2) \wedge Link(z1, z3) : P1.$$

In the string *Semrepr*1, the variables $y1$ and $y2$ are used to mark the descriptions of two strands of arbitrary DNA molecule $x1$; the variables $z1, z2, z3$ mark the bases.

The variable $P1$ (with it the sort "meaning of proposition" is associated) is used to mark the semantic representation of the first sentence of the discourse D1. This allows for building a compact semantic representation of the second sentence of D1, because the occurrence of the word "similarly" in the second sentence of D1 indicates the reference to the meaning of the first sentence.

In particular, the second sentence of D1 in the context of the first sentence may have a K-representation *Semrepr*2 of the form.

$$(Similarly(P1, P2) \wedge (P2 \equiv \forall x1 \, (dna - molecule)$$

$$(Link(arbitr\,base1 * (Is1, cytosine)$$

$$(Part, arbitr\,strand1 * (Part, x1) : y3) : z4,$$

$$certn\,base1 * (Is1,\,guanine)(Part,$$

$$certn\,strand1 * (Part, x1)(Opposite, y3)\,:\,y4)\,:\,z5) \wedge$$

$$\neg\exists z6(base1)(Is(z6, \neg guanine)$$

$$\wedge Part(z6, y4) \wedge Link(z4, z6))))).$$

Then we can associate with the text D1 the K-string *Semrepr3* of the form

$$(Semrepr1 \wedge Semrepr2),$$

where *Semrepr1* and *Semrepr2* are the K-strings defined above. Such string can be interpreted as a possible K-representation of the discourse D1.

The K-string *Semrepr3* illustrates an important opportunity afforded by SK-languages: to mark by variables the fragments of K-strings being semantic representations of narrative texts, infinitive groups, or questions. This opportunity allows us to effectively describe structured meanings of discourses with references to the meanings of fragments being statements, infinitive groups, or questions.

The presence of such references in discourses is often indicated by the following words and word combinations: "this recommendation," "for instance," "e.g.," "that is," "i.e.," "the idea discussed above," "in other words," etc.

The constructed KR *Semrepr3* of the discourse D1 illustrates several additional original features of K-strings (besides features discussed above). First, the symbol $\equiv$ connects a variable and a semantic representation of a sentence. Second, the symbol of negation $\neg$ can be connected with designations of notions. In such a way the substrings $\neg thymine$ and $\neg guanine$ are built.

Some additional useful properties of SK-languages are analyzed below.

6.5 Representation of Knowledge in Medicine

Example 1. Let T1 be the definition "The Eustachian tube is a canal leading from the middle ear to the pharynx." One can associate with T1, in particular, the following K-string interpreted as a semantic representation of T1:

$$Definition1(Eustachian - tube, canal1, x1,$$

$$\exists z(person)Lead1(x1, certn\,middle - ear *$$

$$(Part, z), certn\,pharynx * (Part, z))).$$

Example 2. If T2 = "Sphygmomanometer is instrument intended to measure blood pressure," then T2 may have a K-representation

$$(sphygmo - manometer \equiv instrument * (Purpose1, measuring1 *$$

$$(Param, blood - pressure)(Subject, any\,person))).$$

Here the semantic item *Purpose*1 is to be interpreted as the name of a binary relation. If a pair (A, B) belongs to this relation, A must be a physical object, and B must be a formal semantic analogue of an infinitive group expressing the intended manner of using this physical object.

Example 3. Let T3 be the definition "Thrombin is an enzyme helping to convert fibrinogen to fibrin during coagulation." Then the K-string

$$(thrombin \equiv enzyme * (Purpose1, helping * (Action,$$

$$converting1 * (Object1, certn\, fibrinogen)(Result1, certn\, fibrin)$$

$$(Process, any\, coagulation))))$$

can be interpreted as a possible KR of T3.

6.6 Representation of Semantic Content in Business

Let's demonstrate the possibility to represent semantic content of business sentences and discourses with references to the meanings of the fragments being phrases or larger parts of the discourse.

Example 1. If T1 = "Freight forward is a freight to be paid in the port of destination" then T1 may have a KR of the form

$$(freight - forward \equiv freight * (Description,$$

$$\langle x1, Payment - at(x1, certn\, port1 * (Destination - of, x1))))).$$

The element *certn* is interpreted as the informational item corresponding to the word combination "a certain."

Example 2. Let T2 be the following definition: "A small or medium enterprise is a company with at most 50 employees". This definition may have a K-representation

$$Definition(small\_med\_enterpr, \forall x1\, (company1)\, (Is1(x1, small\_med\_enterpr) \equiv$$

$$\neg\, Greater(Number(Employees(x1)), 50))))$$

or a K-representation

$$((small\_med\_enterpr \equiv company1 * (Description, P1)) \wedge$$

$$(P1 \equiv \forall x1\, (company1)(Is1(x1, small\_med\_enterpr) \equiv$$

$$\neg\, Greater(Number(Employees(x1)), 50)))).$$

Example 3. If T3 = "Mr. Green had asked to send him three containers with ceramics. That request was fulfilled on March 10," a possible KR of T3 may be constructed as follows:

$$((Situation(e1, request * (Agent1,$$

$$certn\,man * (Name, 'Green') : x1)(Goal,$$

$$sending1 * (Object1, certn\,set * (Number, 3)$$

$$(Qualitative - composition, container1*$$

$$(Contain1, ceramics)) : x2)(Time, t1)))$$

$$\wedge Before(t1, \#now\#)) : P1 \wedge$$

$$Fulfilled(certn\,request * (Description, P1) : x3, t2) \wedge$$

$$(t2 \equiv \langle 10, 03, current - year \rangle) \wedge$$

$$Before(t2, \#now\#) \wedge Before(t1, t2)).$$

6.7 SK-Languages as a Tool for Building E-Contracts

6.7.1 Formal Languages for E-Contracting

During the last several years in E-commerce two interrelated fields of researches have emerged called e-negotiations and electronic contracting. The birth of these fields was formally denoted by means of the organization at the beginning of the 2000s of several international conferences and workshops.

The collection of central problems faced by the researchers in these fields includes the creation of formal languages for representing contents of the records of negotiations conducted by computer intelligent agents (CIAs) and for forming contracts concluded in the course of such negotiations. These tasks can be considered as important particular cases of the problem of constructing general-purpose formal languages for business communication [118, 148].

Hasselberg and Weigand underline in [118] that if the messages in the field of E-commerce are to be processed automatically, the meaning must be formalized. This idea coincides with the opinion of Kimbrough and Moore [148] about the necessity of developing logical-semantic foundations of constructing formal languages for business communication (FLBC).

It is suggested in [145–148] to use first-order logic insofar as possible and reasonable for expressions in any FLBC. However, the expressive possibilities of the class of first-order logic languages are very restricted as concerns describing semantic structure of arbitrary business documents.

The analysis shows that the records of commercial negotiations and contracts can be formed with the help of expressive means of natural language (NL) used for the construction of arbitrary NL-texts pertaining to medicine, technology, law, etc. In particular, the texts from such documents may include

(a) the infinitives with dependent words expressing goals, offers ("to sell 30 boxes with apples"), promises, commitments, or intended manners of using things;
(b) the constructions formed from the infinitives with dependent words by means of the logical connectives "and," "or," "not," and expressing compound designations of goals, offers, promises, commitments or destinations of things;
(c) complex designations of sets ("a consignment consisting of 50 boxes with apples");
(d) the fragments where the logical connectives "and," "or" join not the designations of assertions but the designations of objects;
(e) the fragments containing the references to the meanings of phrases or larger fragments of a discourse ("this proposal," "that order," "this promise," etc.);
(f) the designations of the functions whose arguments and/or values may be the sets of objects ("the staff of the firm A," "the suppliers of the firm A," "the number of the suppliers of the firm A");
(g) the questions with the answer "Yes" or "No";
(h) the questions with interrogative words.

Meanwhile, the first-order predicate logic provides no possibility to build the formal analogues (on the semantic level) of the texts from business documents where the NL phenomena listed in items (a)–(g) are manifested.

That is why the problem of developing formal languages allowing for representing contents of the records of commercial e-negotiations carried out by CIAs and for forming contracts concluded in the course of such negotiations is very complicated. Hence it seems to be reasonable to use for solving this problem the most broadly applicable theories (ideally, universal) of representing meanings of NL-texts provided by mathematical linguistics and mathematical computer science.

6.7.2 The Possibilities of Forming Contracts and Records of E-Negotiations by Means of SK-Languages

The analysis shows that the SK-languages possess the expressive possibilities being necessary and sufficient for representing in a formal way the contents of contracts and of the records of commercial negotiations.

For illustrating an important part of such possibilities, let's consider a multi-partner scenario of the interaction of business partners in the course of handling a car damage claim by an insurance company (called AGFIL). The names of the involved parties are Europe Assist, Lee Consulting Services (Lee C.S.), Garages, and Assessors. Europe Assist offers a 24-h emergency call answering service to the policyholders. Lee C.S. coordinates and manages the operation of the emergency service on a day-to-day level on behalf of AGFIL. Garages are responsible for car repair. Assessors conduct the physical inspections of damaged vehicles and agree upon repair figures with the garages.

The process of a car insurance case can be described as follows: The policyholder phones Europe Assist using a free phone number to notify a new claim. Europe Assist will register the information, suggest an appropriate garage, and notify AGFIL which will check whether the policy is valid and covers this claim. After AGFIL receives this claim, AGFIL sends the claim details to Lee C.S. AGFIL will send a letter to the policyholder for a completed claim form. Lee C.S. will agree upon repair costs if an assessor is not required for small damages, otherwise an assessor will be assigned. The assessor will check the damaged vehicle and agree upon repair costs with the garage.

After receiving an agreement of repairing car from Lee C.S., the garage will then commence repairs. After finishing repairs, the garage will issue an invoice to Lee C.S., which will check the invoice against the original estimate. Lee C.S. returns all invoices to AGFIL. This firm processes the payment. In the whole process, if the claim is found invalid, all contractual parties will be contacted and the process will be stopped [31, 210].

This scenario provides the possibility to illustrate some properties of SK-languages making them a convenient tool for formally describing contracts.

Property 1. The possibility to build compound designations of goals.

Example 1. Let T1 = "The policyholder phones Europe Assist to inform about a car damage." Then T1 may have the following K-representation (KR), i.e., a semantic representation being an expression of a certain SK-language:

$$Situation(e1, phone - communic * (Agent1, \; certn \, person*$$

$$(Hold1, \; certn \, policy1 \; : \; x1) \; : \; x2)(Object2, \; certn \, firm1*$$

$$(Name1, \; "EuropeAssist") \; : \; x3)(Purpose,$$

$$inform - transfer * (Theme1,$$

$$certn \, damage1 * (Object1, \; certn \, car1) \; : \; x4))).$$

Property 2. The existence of the means allowing for representing in a compact way the time and causal relations between the situations.

Property 3. The possibility to construct compact semantic representations of such fragments of sentences which are obtained by means of joining the designations of things, events, concepts, or goals with the help of logical connectives AND, OR.

Example 2. Let T2 = "After receiving a repair invoice from the firm 'Lee C.S.' and a claim from the policyholder, the company 'AGFIL' pays the car repair to the garage." Then a KR of T2 can be the expression

$$(Situation(e1, \; (receiving1 * (Agent2, \; certn \, firm1*$$

$$(Name1, \; "AFGIL") \; : \; x1)(Object1, \; certn \, invoice*$$

$$(Theme, \; certn \, repair \; : \; e2) \; : \; x2)(Sender1,$$

$$certn\, firm1 * (Name1,\ "LeeC.S.")\ :\ x3) \wedge receiving1*$$

$$(Agent2,\ x1)(Object1,\ certnclaim1\ :\ x4)(Sender1,$$

$$certn\, person * (Hold1,\ certn\, policy1\ :\ x5)\ :\ x6))) \wedge$$

$$Situation(e2,\ payment1 * (Agent2,\ x1)(Addressee1,$$

$$certn\, garage\ :\ x7)(Sum,\ Cost(e2))) \wedge Before(e1,\ e2)).$$

The analysis of additional precious properties of SK-languages (as concerns the applications of the kind) can be found in [87].

6.8 Simulation of the Expressive Mechanisms of RDF, RDFS, and OWL

This section shows how it is possible to simulate the expressive mechanisms of the language systems RDF (Resource Description Framework) [171–173], RDF Schema (RDFS) [174, 175], and OWL (Ontology Web Language) being the basic languages of the Semantic Web project [162–164].

6.8.1 Simulation of the Expressive Mechanisms of RDF and RDFs

Example 1. According to [171], the sentence T1 = "The students in the course 6.001 are Amy, Tim, John, Mary, and Sue" is translated (in some pragmatic context) into the RDF structure

$$\langle rdf\ :\ RDF \rangle \langle rdf\ :\ Description$$

$$about\ =\ "U1/courses/6.001" \rangle$$

$$\langle s\ :\ students \rangle \langle rdf\ :\ Bag \rangle$$

$$\langle rdf\ :\ Rliresource\ =\ "U1/stud/Amy"/\rangle$$

$$\langle rdf\ :\ Rliresource\ =\ "U1/stud/Tim"/\rangle$$

$$\langle rdf\ :\ Rliresource\ =\ "U1/stud/Jolm"/\rangle$$

$$\langle rdf\ :\ Rliresource\ =\ "U1/stud/Mary"/\rangle$$

$$\langle rdf\ :\ Rliresource\ =\ "U1/stud/Sue"/\rangle$$

$$\langle /rdf\ :\ RBag \rangle \langle /s\ :\ students \rangle$$

$$\langle /rdf\ :\ RDescription \rangle \langle /rdf\ :\ RRDF \rangle, where\, U1\, is\, an\, URL$$

In this expression, the item *Bag* is the indicator of a bag container object. It is possible to construct the following similar expression of a certain SK-language:

$$certn\,course1 * (W3ad, \text{``}U1/courses/6.001\text{''})$$

$$(Students,\ certn\,bag * (Compos2,$$

$$(certn\,stud * (W3ad, \text{``}U1/stud/Amy\text{''}) \wedge$$

$$certn\,stud * (W3ad, \text{``}U1/stud/Tim\text{''}) \wedge$$

$$certn\,stud * (W3ad, \text{``}U1/slud/John\text{''}) \wedge$$

$$certn\,stud * (W3ad, \text{``}U1/stud/Mary\text{''}) \wedge$$

$$certn\,stud * (W3ad, \text{``}U1/stud/Sue\text{''})))).$$

Here the symbol *certn* is interpreted as the referential quantifier, i.e., as the informational item corresponding to the word combination "a certain" in cases when it is used for building the word combinations in singular ("a certain book," "a certain personal computer," etc.).

Example 2. Following [171], the model for the sentence T2 = "The source code for $X11$ may be found at $U3$, $U4$, or $U5$" (where $U3$, $U4$, $U5$ are some URLs) may be written in RDF (with respect to a certain pragmatic context) as

$$\langle rdf\ :\ RDF \rangle$$

$$\langle rdf\ :\ Descriptionabout = \text{``}U2/packages/X11\text{''} \rangle$$

$$\langle s\ :\ DistributionSite \rangle \langle rdf\ :\ Alt \rangle$$

$$\langle rdf\ :\ liresource = \text{``}U3\text{''}/ \rangle$$

$$\langle rdf\ :\ liresource = \text{``}U4\text{''}/ \rangle$$

$$\langle rdf\ :\ liresource = \text{``}U5\text{''}/ \rangle$$

$$\langle /rdf\ :\ Alt \rangle \langle /s : DistributionSite \rangle$$

$$\langle /rdf\ :\ Description \rangle \langle /rdf\ :\ RDF \rangle.$$

Here the informational item *Alt* is the indicator of an alternative container object. The theory of K-representations suggests the following similar expression:

$$certn\,resource * (W3ad, \text{``}U2/packages/X11\text{''})$$

$$(DistributionSite,\ (certn\,resource * (W3ad, \text{``}U3\text{''}) \vee$$

$$certn\,resource * (W3ad, \text{``}U4\text{''}) \vee certn\,resource * (W3ad, \text{``}U5\text{''}))).$$

Example 3. Consider the sentence T3 = "Ora Lassila is the creator of the resource U6" and the corresponding RDF-structure

$$\langle rdf\ :\ RDF \rangle \langle rdf\ :\ Descriptionabout = \text{``}U6\text{''}$$

$$s\ :\ Creator = \text{``}OraLassila\text{''}/ \rangle \langle /rdf\ :\ RDF \rangle.$$

Using a certain SK-language, we can build the following description of the mentioned resource:

$$certn\ resource * (W3ad,\ ``U6")$$

$$(Creator,\ ``OraLassila").$$

Example 4. The theory of K-representations enables us to build reified conceptual representations of statements, i.e. the representations in the form of named objects having some external ties: with the set of the authors, the date, etc. For instance, we can associate the sentence T3 = "Ora Lassila is the creator of the resource U6" with the expression of an SK-language

$$certn\ info - piece * (RDF - type,\ Statement)$$

$$(Predicate,\ Creator)(Subject,\ ``U6")$$

$$(Object,\ OraLassila)\ :\ il024,$$

where *il*024 is the name of an information piece.

This form is very close to the RDF-expression [171]

$$\{type,\ [X],\ [RDF\ :\ statement]\}$$

$$\{predicate,\ [X],\ Creator]\}$$

$$\{subject,\ [X],\ [U6]\}$$

$$\{object,\ [X],\ ``OraLassila"\}.$$

Proceeding from the ideas considered in the examples above and in the previous sections of this chapter, we would be able to approximate all RDF-structures by the similar expressions of RSK-languages.

Example 5. The RDF Schema (RDFS) description of the class "Marital status" from [174]

$$\langle rdfs\ :\ Classrdf\ :\ ID\ =\ ``MarStatus"/\rangle$$

$$\langle MarStatusrdf\ :\ ID\ =\ ``Married"/\rangle$$

$$\langle MarStatusrdf\ :\ ID\ =\ ``Divorced"/\rangle$$

$$\langle MarStatusrdf\ :\ ID\ =\ ``Single"/\rangle$$

$$\langle /rdfs\ :\ Class\rangle$$

can be represented by the following K-string:

$$(any\#MarStatus\ =\ (Married \lor Divorced \lor Single)).$$

The theory of K-representations provides the possibility to approximate all RDF-structures by the similar expressions of SK-languages. The same applies to the RDF Schema Specification Language (RDFS) [174, 175]. The analysis of RDF and RDFS expressive means supports such basic ideas of the theory of K-representations as: building compound formal designations of sets; joining by logical connectives not only the designations of assertions but also the designations of things, events, and concepts; considering assertions as objects having some external ties: with a date, the set of the authors, a language, etc.

6.8.2 Simulation of OWL Expressive Mechanisms

OWL (Ontology Web Language) is the principal language for constructing ontologies under the framework of the Semantic Web project [162–164].

It is easy to show that the expressive mechanisms of SK-languages demonstrated above in this and previous chapters allow for simulating all manners of describing the classes of objects in OWL and allow for defining such algebraic characteristics of the properties as reflexivity, symmetricity, and transitivity. Consider only several examples of the kind.

Example 5 above may be interpreted as an illustration of the manner to define the classes in OWL by means of explicit enumeration of all elements of this class. Another illustration is the following K-string being the definition of the notion "a working day of the week":

$$(arbitrary\, working\_day \equiv (Monday \vee$$

$$Tuesday \vee Wednesday \vee Thursday \vee Friday)).$$

For defining the classes as the unions or the intersections of other classes, the theory of K-representations enables us to use the relational symbols $Union$ and $Intersection$ with the type

$$\{(\{[entity]\}, \{[entity]\})\},$$

where $[entity]$ is the basic type "entity," for constructing the K-strings (or l-formulas) of the form

$$Union(Y, Z),\ Intersection(Y, Z),$$

where Y and Z are the designations of the classes.

An alternative way of defining the unions of classes illustrates the K-string

$$(air\_transp\_means \equiv (glider \vee airplane \vee$$

$$helicopter \vee deltaplane \vee dirigible))$$

interpreted as a semantic representation of the definition "Air transport means is a glider, airplane, helicopter, deltaplane, or dirigible."

The restrictions on the cardinality of the sets can be expressed with the help of the function $Numb$ (the number of elements of a set) with the type

$$tp(Numb) = \{(\{[entity]\}, nat)\},$$

where *nat* is the sort "natural number."

Example 6. The knowledge piece "All people have two parents" can have the following K-representation:

$$\forall x1 \,(person)\, Implies((S1 \equiv all\; person*$$

$$(Parent, x1)), (Numb(S1) \equiv 2)).$$

The SK-languages allow for expressing such algebraic characteristics of the properties (or binary relations) as reflexivity, symmetricity, and transitivity.

Example 7. The property of transitivity of a binary relation on the set of all space objects (the concretizations are physical objects and geometric figures) can be expressed by the following K-string:

$$\forall r1 \,(property1)\, (Transitive(r1) \equiv$$

$$\forall x1 \,(space.ob)\, \forall x2 \,(space.ob)\, \forall x3 \,(space.ob)$$

$$Implies((Associated(r1, x1, x2) \wedge Associated(r1, x2, x3)),$$

$$Associated(r1, x1, x3))).$$

The construction of this formula is based on the following assumptions:

- $r1, x1, x2, x3$ are the variables from the set $V(B)$, where B is the considered conceptual basis;
- $tp(r1) = tp(x1) = tp(x2) = tp(x3) = [entity]$;
- $tp(property1) = \uparrow \{(space.ob, space.ob)\}$;
- *Implies*, *Associated* $\in X(B)$, where $X(B)$ is the primary informational universe of the conceptual basis B;
- $tp(Implies) = \{(P, P)\}$, where $P = P(B)$ is the distinguished sort "a meaning of proposition";
- $tp(Associated) = \{([entity], [entity], [entity])\}$.

The principal advantage of the theory of K-representations in comparison with the language systems RDF, RDFS, and OWL is that it indicates a small collection of operations enabling us to build semantic representations of arbitrary NL-texts and, as a consequence, to express in a formal way arbitrarily complicated goals and plans of actions, to represent the content of arbitrary protocols of negotiations and to construct formal contracts concluded in the course of e-negotiations.

6.9 A Metaphor of a Kitchen Combine for the Designers of Semantic Technologies

It seems that a metaphor can help to better grasp the significance of the theory of SK-languages for the designers of semantic informational technologies, first of all, for the designers of Natural Language Processing Systems (NLPSs). It establishes a connection between the problems of house keeping and the problems associated with the development of semantic informational technologies.

When a woman having a full-time job enters the kitchen, she has a lot of things to do in a short time. The kitchen combines are constructed in order to make the work in the kitchen easier and to diminish the time needed for the work of the kind. For this, the kitchen combines can chop, slice, stir, grate, blend, squeeze, grind, and beat.

The designers of NLPSs have a lot of things to do in a very restricted time. That is why they need effective formal tools for this work. Like a kitchen combine for housekeeping, the theory of SK-languages can help the designers of semantic informational technologies to do many things. In particular, the theory of SK-languages is convenient for

- constructing formal definitions of concepts,
- representing knowledge associated with concepts,
- building knowledge modules (in particular, definitions of concepts) as the units having both the content (e.g., a definition of a concept) and the external characteristics (e.g., the authors, the date of publishing, the application fields),
- representing the goals of intelligent systems,
- building semantic representations of various algorithms given in a natural language form,
- representing the intermediate results of semantic-syntactic processing of NL-texts (in other words, building underspecified semantic representations of the texts),
- forming final semantic representations of NL- texts,
- representing the conceptual macro-structure of an NL discourse,
- representing the speech acts,
- building high-level conceptual descriptions of the figures occurring in the scientific papers, textbooks, technical patents, etc.

No other theory in the field of formal semantics of NL can be considered as a useful tool for all enumerated tasks. In particular, it applies to Montague Grammar and its extensions and to Discourse Representation Theory, Theory of Conceptual Graphs, and Episodic Logic.

In case of technical systems, a highly precious feature is the simplicity of construction. Very often, this feature contributes to the reliability of the system and the easiness of its exploitation.

The theory of SK-languages satisfies this criterion too, because it makes the following discovery in both non mathematical and mathematical linguistics: a system of such 10 operations on structured meanings (SMs) of NL-texts is found that, using primitive conceptual items as "blocks," we are able to build SMs of arbitrary NL-texts (including articles, textbooks, etc.) and arbitrary pieces of knowledge about the world. Such operations will be called quasilinguistic conceptual operations. Hence the theory of K-representations suggests a complete collection of quasilinguistic conceptual operations (it is a hypothesis supported by many weighty arguments).

The useful properties of SK-languages stated above allow for the conclusion that the theory of SK-languages can be at least not less useful for the designers of NLPSs and for a number of other semantic informational technologies as a kitchen combine is of use for making easier the work in the kitchen

6.10 The Significance of the Theory of K-Representations for Semantic Web and Web Science

6.10.1 Theory of K-Representations as a Universal Resources and Agents Framework

It appears that RDF, RDFS, and OWL are only the first steps of the World Wide Web Consortium along the way of developing semantically structured (or conceptual) formalisms, and hence the next steps will be made in the future. The emergence of the term "Web X.0" supports this conclusion. That is why let's try to imagine what may be the result of the evolution of consequent Web conceptual formalisms, for instance, one decade later.

In order to formulate a reasonable assumption, let's consider such important applications of Web as Digital Libraries and Multi media Databases. If the resources are articles, books, pictures, or films, then, obviously, important metadata of such resources are semantic representations of summaries (for textual resources and films) and high-level conceptual descriptions of pictures. As for e-commerce, the conceptual (or semantic) representations of the summaries of business documents are important metadata of resources. That is why it seems that the Web conceptual formalisms will evolve during the nearest decade to a Broadly Applicable Conceptual Metagrammar.

Hence the following fundamental problem emerges: how to construct a Universal Conceptual Metagrammar (UCM) enabling us to build semantic representations (in other words, conceptual representations) of arbitrary sentences and discourses in NL? Having a UCM, we will be able, obviously, to build high-level conceptual descriptions of visual images too.

The first answer to this question was proposed in [70]: the hypothesis is put forward that the theory of restricted K-calculuses and K-languages (the RKCL-theory) may be interpreted as a possible variant of a UCM. With respect to this hypothesis and the fact that the RKCL-theory enables us to effectively approximate the expressive means of RDF and RDFS, we may suppose that the more general theory of K-representations can be used as an effective tool and as a reference-point for developing comparable, more and more powerful and flexible conceptual formalisms for the advanced Web.

The analysis (in particular, carried out above) shows that the theory of K-representations is a convenient tool for constructing formal representations of the contents of arbitrary messages sent by intelligent agents, for describing communicative acts and metadata about the resources, and for building high-level conceptual representations of visual images. That is why the theory of K-representations together with the recommendations concerning its application in the mentioned directions may be called a Universal Resources and Agents Framework (URAF); this term was introduced for the first time in [78].

6.10.2 The Need for the Incentives for Semantic Web

During last several years, it has been possible to observe that the achieved state of Semantic Web and a state to be relatively soon achieved are considerably different from the state of affairs outlined as the goal in the starting publication on Semantic Web by Berners-Lee et al. [18].

The principal reason for this conclusion is the lack of large-scale applications implemented under the framework of Semantic Web project. This situation is implied by the lack of a sufficiently big amount (of "a critical mass") of formally represented content conveyed by numerous informational sources in many fields. This means the lack of a sufficiently big amount of Web-sources and Web-services with semantic annotations, of the visual images stored in multimedia databases and linked with the high-level conceptual descriptions, rich ontologies, etc.

This situation is characterized in the Call for Papers of the First International Symposium on Incentives for Semantic Web (Germany, Karlsruhe, October 2008) as the lack of *a critical mass of semantic content*.

That is why it has been possible to observe the permanent expansion in the scientific literature of the following opinion: a Semantic Web satisfying the initial goal of this project will be created in an evolutionary way as a result of the efforts of *many research groups in various fields*. In particular, this opinion is expressed in [10, 153].

It is important to underline that this point of view is also expressed in the article "Semantic Web Revisited" written by the pioneers of Web: N. Shadbolt et al. [191]. In this chapter, the e-science international community is indicated as a community playing now one of the most important roles in quick generation of semantic content in a number of fields. The activity of this community seems to give a sign of future success of Semantic Web project.

One of the brightest manifestations of the need for new, strong impulses to developing Semantic Web is the organization of the First International Symposium on Incentives for Semantic Web under the framework of the Semantic Web International Conference – 2008.

The content of this section is to be considered in the context of the broadly recognized need for the incentives for Semantic Web, in particular, for the incentives on the models stimulating the development of Semantic Web.

6.10.3 Toward a New Language Platform for Semantic Web

In [191], the authors ground the use of RDF as the basic language of the Semantic Web project with the help of *the principle of least power*: "the less expressive the language, the more reusable the data."

However, it seems that the stormy progress of e-science, first of all, urges us to find a new interpretation of this principle in the context of the challenges faced nowadays by the Semantic Web project. E-science needs to store on the Web the semantic content of the definitions of numerous notions, the content of scientific articles, technical reports, etc. The similar requirements are associated with semantics-oriented computer processing of the documents pertaining to economy, law, and politics. In particular, it is necessary to store the semantic content of the articles from newspapers, of TV-presentations, etc.

That is why it can be conjectured that, in the context of the Semantic Web project, the following new interpretation of the principle of least power is reasonable: an adequate language platform for Semantic Web is to allow for reflecting the results of applying ten partial operations on conceptual structures explicated by the mathematical model constructed in Chaps. 3 and 4 of this monograph.

The reason for this conclusion is the hypothesis set forth in the final part of the previous chapter: there are weighty grounds to believe that, combining ten partial operations determined, in essence, by the rules $P[1]-P[10]$, we are able to construct (and it is convenient to do) a semantic representation of arbitrarily complex NL-text pertaining to arbitrary field of professional activity.

6.10.4 A Possible Strategy of Developing Semantic Web of a New Generation

Let's consider the principal ideas of a new, theoretically possible strategy aimed at transforming the existing Web into a Semantic Web (Fig. 6.2).

The proposed strategy is based on (a) the mathematical model constructed in Chaps. 3 and 4 and describing a system of ten partial operations on conceptual structures and (b) the analysis of the expressive mechanisms of SK-languages carried out in this and previous chapters. The new strategy can be very shortly formulated as follows:

1. An XML-based format for representing the expressions of SK-languages (standard knowledge languages) will be elaborated. Let's agree that the term "a K-representation of an NL-text T" means in this chapter a semantic representation of T built in this format and that the term "a semantic K-annotation" will be interpreted below as a K-representation of an NL-annotation of an informational source. The similar interpretations will have the terms "a K-representation of a knowledge piece" and "a high-level conceptual K-description of a visual image."

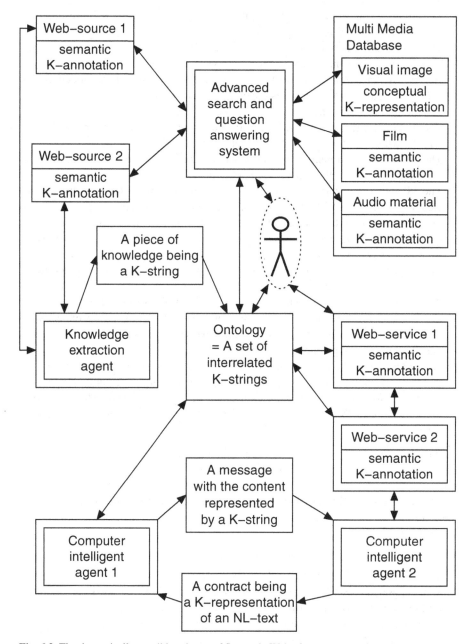

Fig. 6.2 The theoretically possible scheme of Semantic Web of a new generation

2. The NL-interfaces for different sublanguages of NL (English, Russian, German, Chinese, Japan, etc.) helping the end users to build semantic K-annotations of Web-sources and Web-services are being designed.

3. The advanced ontologies being compatible with OWL and using K-representations of knowledge pieces are being elaborated.

4. The new content languages using K-representations of the content of messages sent by computer intelligent agents (CIAs) in multi agent systems are being worked up. In particular, this class of languages is to include a subclass convenient for building the contracts concluded by the CIAs as a result of successful commercial negotiations.

5. The visual images of the data stored in multimedia databases are being linked with high-level conceptual K-descriptions of these images.

6. The NL-interfaces transforming the NL-requests of the end users of Web into the K-representations of the requests are being designed.

7. The advanced Web-based search and question-answering systems are being created that are able (a) to transform (depending on the input request) the fragments of a discourse into the K-representations, (b) to analyze these K-representations of the discourse fragments, and (c) to analyze semantic K-annotations of Web-sources and Web-services.

8. The NL processing systems being able to automatically extract knowledge from NL-texts, to build the K-representations of knowledge pieces, and to inscribe these K-representations into the existing ontologies are being elaborated.

9. The generators of NL-texts (the recommendations for the users of expert systems or of recommender systems, the summaries of Web-documents, etc.) using the SK-languages for representing the meaning of an NL-text to be synthesized are being constructed. Besides, a reasonable direction of research seems to be the design of applied intelligent systems able to present the semantic content of a message for the end user as an expression of a non standard K-language being similar to an NL-expression but containing, maybe, a number of brackets, variables, and markers.

Fulfilling these steps, the international scientific community will create in a reasonable time a digital conceptual space unified by a general-purpose language platform.

The realization of this strategy will depend on the results of its discussion by the international scientific community.

Problems

1 Describe the main ideas of building semantic annotations of informational sources with the help of SK-languages.

2 What new expressive mechanisms of SK-languages can be used for building high-level conceptual descriptions of visual images?

3 What new expressive mechanisms of SK-languages are useful for building compound denotations of notions?

4 Describe the proposed new interpretation of the principle of the language of least power for Semantic Web project.

Part II
Formal Methods and Algorithms for the Design of Semantics-Oriented Linguistic Processors

Chapter 7
A Mathematical Model of a Linguistic Database

Abstract In this chapter a broadly applicable mathematical model of linguistic database is constructed, that is, a model of a collection of semantic-syntactic data associated with primary lexical units and used by the algorithms of semantic–syntactic analysis for building semantic representations of natural language texts.

7.1 The Principles of Designing Semantics-Oriented Linguistic Processors

Most often, semantics-oriented natural language processing systems, or *linguistic processors (LPs)*, are complex computer systems, their design requires a considerable time, and their cost is rather high. Usually, it is necessary to construct a series of LPs, step by step expanding the input sublanguage of NL and satisfying the requirements of the end users.

On the other hand, the same regularities of NL are manifested in the texts pertaining to various thematic domains.

That is why, in order to diminish the total expenses of designing a family of LPs by one research center or group during a certain several-year time interval and in order to minimize the duration of designing each particular system from this family of LPs, it seems reasonable to pay more attention to (a) the search for best typical design solutions concerning the key subsystems of LPs with the aim to use these solutions in different domains of employing LPs; (b) the elaboration of formal means for describing the main data structures and principal procedures of algorithms implemented in semantic-syntactic analyzers of NL-texts or in the synthesizers of NL-texts.

That is why it appears that the adherence to the following two principles in the design of semantics-oriented LPs by one research center or a group will contribute, in the long-term perspective, to reducing the total cost of designing a family of LPs and to minimizing the duration of constructing each particular system from this family:

V.A. Fomichov, *Semantics-Oriented Natural Language Processing*, IFSR International Series on Systems Science and Engineering 27, DOI 10.1007/978-0-387-72926-8_7,

- the *Principle of Stability* of the used language of semantic representations (LSR) in the context of various tasks, various domains, and various software environments (stability is understood as the employment of a unified collection of rules for building the structures of LSR as well as domain- and task-specific variable set of primitive informational units);
- the *Principle of Succession* of the algorithms of LP based on using one or more compatible formal models of a linguistic database and unified formal means for representing the intermediate and final results of semantic-syntactic analysis of natural-language texts in the context of various tasks, various domains and various software environments (the succession means that the algorithms implemented in basic subsystems of LP are repeatedly used by different linguistic processors).

The theoretical results stated in Chaps. 2, 3, and 4 of this monograph provide a basis for following up the principle of stability of the used language of semantic representations. Chapter 4 defines a class of SK-languages that enable us to build semantic representations of natural language texts in arbitrary application domains.

This chapter is based on the results stated in previous chapters and is aimed at creating the necessary preconditions for implementing the succession principle in the design of LP algorithms.

In this and next chapters, we introduce a new method of transforming natural language texts into their semantic representations for the sublanguages of English, Russian, and German languages being of practical interest. This involves solving the following problems:

- Formalizing the structure of a linguistic database allowing for finding various conceptual relations, e.g., in the combinations "Verb + Preposition + Noun," "Verb + Noun," "Noun1 + Preposition + Noun2," "Numeral + Noun," "Adjective + Noun," "Noun1 + Noun2," "Participle + Noun," "Participle + Preposition + Noun," "Interrogative pronoun + Verb," "Preposition + Interrogative pronoun + Verb," "Interrogative Adverb + Verb," "Verb + Numerical Value Representation" (a number representation + a unit of measurement representation).
- Formalizing the structure of data used as an intermediate pattern of the input natural language text semantic structure to provide a basis for building later a semantic representation of the input text.
- Using the solutions to Problems 1 and 2 for developing a domain-independent method of transforming an input NL-text (question/command/statement) from the sublanguages of English, Russian, and German languages into its semantic representation.

In this chapter, we apply the theory of SK-languages to building a broadly applicable formal model of a linguistic database (LDB). This model describes the logical structure of LDB being the components of natural-language interfaces to intelligent databases as well as to other applied computer systems. The expressions of SK-languages enable us to associate with the lexical units the appropriate simple or compound semantic units.

7.2 Morphological Bases

Let's formally represent the information about the elements being the primary components of natural-language texts.

Morphology is a branch of linguistics studying the regularities of the alteration of words and word combinations (depending on grammatical number, case, tense, etc). A linguistic database (LDB) must include a morphological database (MDB) with the content depending on the considered language. In contrast to the English language, the Russian language (RL) and the German language (GL) are very flexible, that is, the words in these languages can be changed in many ways. That is why, though an MDB is rather simple for English, the situation is different for RL and GL.

There are many publications devoted to the formalization of morphology of Russian, German, and many other languages. However, in order to develop a structured algorithm of semantic-syntactic analysis of NL-texts (the input texts may be from Russian, English, and German languages), it was necessary to propose a new, more general look at morphology of Russian, English, German, and many other languages in comparison with the available approaches.

The goal was to indicate the role of morphological analysis as a part of semantic-syntactic analysis, avoiding too detailed treatment of the morphological problems. For achieving this goal, the notions of morphological determinant, morphological space, and morphological basis are introduced in this section.

Definition 7.1. *Morphological determinant (M-determinant)* is an arbitrary ordered triple of the form

$$(m, n, maxv), \tag{7.1}$$

where m, n are the positive integers; $maxv$ is a mapping from the set $\{1, 2, \ldots, m\}$ into the set of non-negative integers N^+.

Let *Det* be an M-determinant of the form (7.1), then m will be interpreted as the quantity of different properties (which are called morphological) of the words from the considered language; n be the maximal amount of different sets of the values of morphological properties associated with one word. If $1 \leq i \leq m$, then $maxv(i)$ is interpreted as the maximal numerical code of the value of the property with the ordered number i (see Fig. 7.1).

For example, three sequences of the values of morphological properties can be connected with the Russian word "knigi" ("book" or "books"): if "knigi" is a word in the singular form, then this word is in the genitive case; if "knigi" is a word in the plural form, then it can be both in nominative case and in accusative case. That's why $n \geq 3$.

Let us suppose that the morphological properties with the numerical codes 1 and 2 are the properties "a part of speech" and "a subclass of a part of speech." That is why every integer k, where $1 \leq k \leq maxv(1)$, will be interpreted as a value of a part of speech, and every r, such that $1 \leq r \leq maxv(2)$, will be interpreted as a value of a subclass of a part of speech.

Figure 7.2 illustrates the structure of one sequence of the values of morphological properties associated with one word.

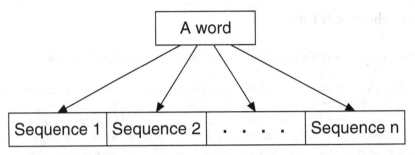

Fig. 7.1 The structure of the array of the values of morphological properties associated with one word

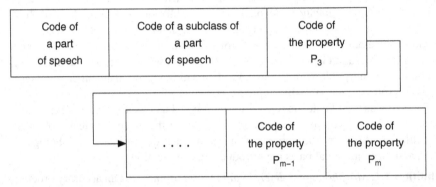

Fig. 7.2 The structure of one sequence of the values of morphological properties associated with one word

We will suppose that every word from the considered language can be associated with only one part of speech and with one subclass of a part of speech. On one hand, this assumption is true for a very large subset of Russian language and, for example, of German language. On the other hand, such assumption will allow for diminishing the complexity of the elaborated formal model of LDB (without any real harm for applications).

Definition 7.2. Let *Det* be a M-determinant of the form (7.1). Then the *morphological space defined by the M-determinant Det* is the set *Spmorph* consisting of all finite sequences of the form

$$(x_1, \ldots, x_m, x_{m+1}, \ldots, x_{2m}, x_{2m+1}, \ldots, x_{nm}), \qquad (7.2)$$

where (a) for every $k = 1, \ldots, n-1$, $x_{km+1} = x_1$, $x_{km+2} = x_2$; (b) for every $k = 1, \ldots, n$ and every such q that $(k-1)m+1 \leq q \leq km$, the following inequality is true: $0 \leq x_q \leq maxv(q-(k-1)m)$.

The conditions (a), (b) from this definition are interpreted in the following way: In the element of the form (7.2) from a morphological space, x_1 is the code of the part of speech. This code is located in every position separated by the distance $m, 2m, \ldots, (n-1)m$ from the position 1; x_2 is the code of the subclass of the part of speech, this code is located in all positions separated by the distance $m, 2m, \ldots,$ $(n-1)m$ from the position 2.

Let q be the numerical code of the letter designation of a morphological characteristic (or property). Then the mapping $maxv$ determines the diapason of the values of this characteristic $[1, maxv(q)]$. That is why for each position q, where $1 \leq q \leq m$, the inequality $0 \leq x_q \leq maxv(q)$ takes place for the sequence of the form (7.2).

If $1 \leq q \leq m\, x_q$ is a component of the sequence of the form (7.2), and $x_q = 0$, this means that a word associated with this element of the morphological space doesn't possess a morphological characteristic (or property) with the numerical code q. For instance, the nouns have no characteristic "time."

Every component x_s of an element of the form (7.2). of a morphological space, where $s = q+m, q+2m, \ldots, q+(n-1)m$, is interpreted as a possible value of the same morphological property as in case of the element x_q. That's why the inequality

$$0 \leq x_s \leq maxv(s-(k-1)m)$$

indicates the diapason of possible values of the element x_s, where the integer k in the borders from 1 to n is unambiguously defined by the condition

$$(k-1)m + 1 \leq s \leq km.$$

The definition of a morphological basis introduced below gives a new mathematical interpretation of the notion "a morphological database." Temporarily abstracting ourselves from mathematical details, we describe a morphological basis as an arbitrary system $Morphbs$ of the form

$$(Det, A, W, Lecs, lcs, fmorph, propname, valname), \qquad (7.3)$$

where Det is a morphological determinant, and the other components are interpreted as follows: A is arbitrary alphabet (a finite set of symbols); the symbols from A are used for forming the words of the considered sublanguage of natural language (English, Russian, etc.). Let A^+ be the set of all non empty (or non void) strings in the alphabet A (in other terms, over the alphabet A). Then W is a finite subset of A^+, the elements of this set are considered as words and fixed word combinations (for example, "has been received") used for constructing natural language texts. The elements of the set W will be called words.

The component $Lecs$ is a finite subset of the set W, the elements of $Lecs$ are called *the lexemes* and are interpreted as basic forms of the words and fixed word combinations (a noun in singular form and nominative case, a verb in the infinitive form, etc.).

The component lcs is a mapping from the set W into the set $Lecs$, associating a certain basic lexical unit with a word; the component $fmorph$ is a mapping associating an element of the morphological space $Spmorph(Det)$ with a word wd from W.

The component *propname* (it is an abbreviation from *property − name*) is a mapping linking a numerical code of a morphological characteristic (or property) with the letter designation of this morphological characteristic. For example, the following relationship can take place:

$$propname(1) = part\_of\_speech.$$

More exactly, it is a mapping

$$propname : \{1, 2, ..., m\} \longrightarrow A^+ \setminus W,$$

where $\setminus$ is the sign of set-theoretical difference.

The component *valname* is a partial mapping with two arguments. The first argument is the numerical code k of a morphological characteristic (or property). The second argument is the numerical code p of a certain possible value of this characteristic (property). The value of the mapping $valname(k, p)$ is the letter designation of p. For instance, the relationship $valname(1, 1) = verb$ can take place.

Definition 7.3. Let A, B be arbitrary non empty sets, and $f : A \longrightarrow B$ is a mapping from A into B. Then $Range(f)$ is the set of all such y that there is such element x from A that $f(x) = y$.

Definition 7.4. *Morphological basis* is an arbitrary 8-tuple $Morphbs$ of the form (7.3), where Det is an M-determinant of the form (7.1), A is an arbitrary alphabet, W is a finite subset of the set A^+ (the set of all non empty strings in the alphabet A), $Lecs$ is a finite subset of the set W, $lcs : W \longrightarrow Lecs$ is a mapping from W to $Lecs$, $fmorph : W \longrightarrow Spmorph(Det)$ is a mapping from W to the morphological space defined by the M-determinant Det, $propname$ is a mapping from the set $\{1, 2, ..., m\}$ to the set $A^+ \setminus W$, $valname$ is a partial mapping from the Cartesian product $N^+ \times N^+$ to the set $A^+ \setminus (W \cup Range(propname))$ defined for the pair (i, j) from $N^+ \times N^+ \Longleftrightarrow 1 \leq i \leq m, 1 \leq j \leq maxv(i)$.

Definition 7.5. Let $Morphbs$ be a morphological basis of the form (7.3). Then

$$Parts(Morphbs) = \{valname(1, 1), ..., valname(1, maxv(1))\},$$

$$Subparts(Morphbs) = \{valname(2, 1), ..., valname(2, maxv(2))\}.$$

Thus, $Parts(Morphbs)$ is the set of the letter designations of the parts of speech, $Subparts(Morphbs)$ is the set of the letter designations of the subclasses of the parts of speech for the morphological basis $Morphbs$.

For instance, it is possible to define an English-oriented morphological basis $Morphbs$ in such a way that the following relationships will take place:

$$Parts(Morphbs) \supseteq \{verb, noun, adjective, preposition, pronoun, participle,$$

$$adverb, cardinal\_numeral, ordinal\_numeral, conjunctive\},$$

$$Subparts(Morphbs) \supseteq \{common\_noun, proper\_noun\} \,.$$

Definition 7.6. Let *Morphbs* be a morphological basis of the form (7.3), $z \in$ *Spmorph(Det)* be an arbitrary element of morphological space, and $1 \le i \le mn$, then $z[i]$ is the i-th component of the sequence z (obviously, z has $m \cdot n$ components).

Definition 7.7. Let *Morphbs* be a morphological basis of the form (7.3). Then the mapping *prt* from the set of words W to the set *Parts(Morphbs)* and the mapping *subprt* from the set W to the set *Subparts(Morphbs)* $\cup \{nil\}$ are determined as follows: for arbitrary word $d \in W$,

$$prt(d) = valname(1, fmorph(d)[1]) \,,$$

if $fmorph(d)[2] > 0$,

$$subprt(d) = valname(2, fmorph(d)[2])$$

else

$$subprt(d) = nil \,.$$

Thus, the strings $prt(d)$ and $subprt(d)$ are respectively the letter designations of the part of speech and the subclass of the part of speech associated with the word d.

  Example. A morphological basis can be defined in such a way that

$$W \ni cup, France;\ prt(cup) = noun, subprt(cup) = common\_noun \,,$$

$$prt(France) = noun, subprt(France) = proper\_noun \,.$$

7.3 Text-Forming Systems

Natural language texts include not only words but also the expressions being the numerical values of different parameters, for example, the strings 90 km/h, 120 km, 350 USD. Let us call such expressions the *constructors* and suppose that these expressions belong to the class of elementary meaningful lexical units. It means that if we are building a formal model of linguistic database, we consider, for example, the expression 120 km as a symbol.

  Of course, while developing computer programs, we are to take into account that there is a blank between the elements "120" and "km," so "120 km" is a word combination consisting of two elementary expressions. However, the construction of every formal model includes the idealization of some entities from the studied domains, that is why we consider the constructors as symbols, i.e. as indivisible expressions.

  Except the words and constructors, NL-texts can include the *markers*, for instance, the point, comma, semi-colon, dash, etc, and also the expressions in inverted commas or in apostrophes being the names of various objects.

Definition 7.8. Let *Cb* be a marked-up conceptual basis. Then *a text-forming system* (*t.f.s.*) *coordinated with the basis Cb* is an arbitrary system *T form* of the form

$$(Morphbs, Constr, infconstr, Markers), \tag{7.4}$$

where

- *Morphbs* is a morphological basis of the form (7.3),
- *Constr* is a countable set of symbols not intersecting with the set of words *W*,
- *infconstr* is a mapping from the set *Constr* to the primary informational universe $X(B(Cb))$,
- *Markers* is a finite set of symbols not intersecting with the sets *W* and *Constr*,

and the following requirements are satisfied:

- for every *d* from the set *Constr*, the element $tp(infconstr(d))$ is a sort from the set $St(B)$;
- the sets *W*, *Constr*, *Markers* don't include the inverted commas and apostrophes.

The elements of the sets *W*, *Constr*, and *Markers* are called *the word forms (or words), constructors, and markers* of the system *T form*, respectively.

Obviously, every morphological basis *Morphbs* determines, in particular, an alphabet *A* and the set of words *W*.

Definition 7.9. Let *T form* be a text-forming system of the form (7.4), *Morphbs* be a morphological basis of the form (7.3). Then

$$Names(Tform) = Names1 \cup Names2,$$

where *Names*1 is the set of all expressions of the form "*x*," where *x* is an arbitrary string in the alphabet *A*, and *Names*2 is the set of all expressions of the form "*y*," where *y* is an arbitrary string in the alphabet *A*;

$$Textunits(Tform) = W \cup Constr \cup Names(Tform) \cup Markers;$$

Texts(*T form*) is a set of all finite sequences of the form $d_1, \ldots, d_n$, where $n \geq 1$, for $k = 1, \ldots, n, d_k \in Textunits(Tform)$.

Definition 7.10. Let *Cb* be a marked conceptual basis, *T form* be a text-forming system of the form (7.4) coordinated with the basis *Cb*. Then the mapping *tclass* from *Textunits*(*T form*) to the set *Parts*(*Morphbs*) $\cup$ {*constr, name*} and the mapping *subclass* from *Textunits*(*T form*) to *Subparts*(*Morphbs*)$\cup$ {*nil*}, where *nil* is an empty element, are determined by the following conditions:

- if $u \in W(Tform)$, then $tclass(u) = prt(u)$;
- if $u \in Constr$, then $tclass(u) = constr$;
- if $u \in Names(Tform)$, then $tclass(u) = name$;
- if $u \in Markers$, then $tclass(u) = marker$;
- if $u \in W(Tform)$, then $subclass(u) = subprt(u)$;

- if $u \in Constr$, then $subclass(u) = tp(inf constr(u))$, where $inf constr$ and tp are the mappings being the components of the text-forming system $Tform$ and of the primary informational universe $X(B(Cb))$ respectively;
- if $u \in Names(Tform) \cup Markers$, then $subclass(u) = nil$.

7.4 Lexico-semantic Dictionaries

Let us consider a model of a dictionary that establishes a correspondence between the elementary meaningful text units ("containers," "have prepared," etc.) and the units of semantic (or, in other words, informational) level. Lexico-semantic dictionary is one of the main components of a linguistic database. One part of the informational units corresponding to the words will be regarded as symbols; they are the elements of the primary informational universe $X(B(Cb))$, where Cb is a marked-up conceptual basis (m.c.b.) built for the considered domain, B is a conceptual basis (c.b.) being the first component of Cb.

The examples of such units are, in particular,

$$publication, entering1, entering2, station1, station2,$$

etc. Other informational units are compound. For example, the adjective "green" from W can be connected with the expression $Color(z1, green)$.

Definition 7.11. Let S be a sort system (s.s.) of the form

$$(St, P, Gen, Tol).$$

Then *the semantic dimension* of the system S is such maximal number $k > 1$ that one can find such sorts $u_1, \ldots, u_n \in St$ that for arbitrary $i, j = 1, \ldots, k$, where $i \neq j$, u_i and u_j are comparable for the compatibility (or tolerance) relation Tol, i.e. $(u_i, u_j \in Tol)$. This number k is denoted $dim(S)$.

Thus, $dim(S)$ is the maximal number of the different "semantic axes" used to describe one entity in the considered application domain.

Example 1. Let us consider the concepts *"a firm"* and *"a university"*. We can distinguish three *semantic contexts* of word usage associated with these concepts. First, a firm or a university can develop a tool, a technology etc., so the sentences with these words can realize *the semantic coordinate "intelligent system."* Second, we can say, "This firm is situated near the metro station 'Taganskaya,'" and then this phrase realizes *the semantic coordinate "spatial object."* Finally, the firms and institutes have the directors. We can say, for example, "The director of this firm is Alexander Semenov." This phrase realizes *the semantic coordinate "organization."*

In the considered examples, we'll presume that semantic dimension of the considered sort systems is equal to four or three.

A *lexico-semantic dictionary* is a finite set *Lsdic* consisting of the $k+5$-tuples of the form

$$(i, lec, pt, sem, st_1, \ldots, st_k, comment) , \qquad (7.5)$$

where k is the semantic dimension of the considered sort system, $i \geq 1$ is the ordered number of the $k + 5$-tuple (we need it to organize the loops in the algorithms of processing NL-texts), and the rest of the components are interpreted in the following way:

- lec is an element of the set of basic lexical units $Lecs$ for the considered morphological basis;
- pt is a designation of the part of speech for the basic lexical unit lec;
- the component sem is a string that denotes one of the possible meanings of the basic lexical unit lec.

The component sem for verbs, participles, gerunds is an informational unit connected with the corresponding verbal noun. For example, the verb *"enter"* has, in particular, the following two meanings: (1) entering a learning institution (in the sense "becoming a student of this learning institution"); (2) entering a space object ("John has entered the room," etc.).

So, for example, one system from a possible lexico-semantic dictionary will have, as the beginning, the sequence

$$i_1, enter, verb, entering1 ,$$

and the other will have, as the beginning, the sequence

$$i_2, enter, verb, entering2 .$$

Number k is the semantic dimension of the considered sort system, i.e. $k = dim(S(B(Cb)))$, where Cb is the considered marked-up conceptual basis; $st_1, \ldots, st_k$ are the different *semantic coordinates* of the entities characterized by the concept sem. For example, if $sem = firm$, then $st_1 = ints$, $st_2 = space.ob.$, $st_3 = org$, $k = 3$.

If an entity characterized by the concept sem has the various semantic coordinates $st_1, \ldots, st_p$, where $p < k$, then $st_{p+1}, \ldots, st_k$ is a special empty element nil. The component $comment$ is either a natural language description of a meaning associated with the concept sem or an empty element nil.

Definition 7.12. Let Cb be a marked-up conceptual basis of the form

$$(B, Qmk, Setmk, Cmk) ,$$

$Morphbs$ be a morphological basis of the form (7.3), Qmk be a questions marking-up of the form (5.1), and let the primary informational universe $X(B(Cb))$ and the set of variables $V(B(Cb))$ not to include the symbol nil (empty element).

Then *a lexico-semantic dictionary coordinated with the marked-up conceptual basis Cb and with the morphological basis Morphbs* is an arbitrary finite set $Lsdic$ consisting of the systems of the form (7.5), where

- $i \geq 1$, for each $lec \in Lecs$;

- $pt = prt(lec)$, $sem \in Ls(B(Cb)) \cup \{nil\}$;
- $k = dim(S(B(Cb)))$;
- for each $p = 1, \ldots, k$, $st_p \in St(B(Cb)) \cup \{nil\}$;
- $comment \in A^+ \cup \{nil\}$;

and the following conditions are satisfied:

- no two systems from *Lsdic* may have the same first component *i*;
- if two systems from *Lsdic* have different values of the *sem* component, then these two systems have different values of the *comment* component.

Example 2. A set *Lsdic* can be defined in such a way that *Lsdic* includes the following 8-tuples:

$$(112, container, noun, container1, dyn.phys.object, nil, nil, \text{``reservoir''}),$$

$$(208, enter, verb, entering1, sit, nil, nil, \text{``enter a college''}),$$

$$(209, enter, verb, entering2, sit, nil, nil, \text{``enter a room''}),$$

$$(311, aluminum, adj, Material(z1, aluminum), phys.ob., nil, nil, nil),$$

$$(358, green, adj, Color(z1, green), phys.ob, nil, nil, nil),$$

$$(411, Italy, noun, certn\, country * (Name1, {}'Italy'), space.ob, nil, nil, \text{``country''}),$$

$$(450, passenger, adj, sem1, dyn.phys.ob., nil, nil, nil),$$

where

$$sem1 = Purpose(z1, movement1 * (Object1,$$

$$certn\, set * (Qual - compos, person))).$$

7.5 Dictionaries of Verbal – Prepositional Semantic-Syntactic Frames

Verbs, participles, gerunds, and verbal nouns play the key role in forming sentences due to expressing the various relations between the entities from the considered application domain.

Thematic role is a conceptual relation between a meaning of a verbal form (a form with time, an infinitive, a participle, a gerund) or a verbal noun and a meaning of a word group depending on it in the sentence.

Thematic roles are also known as *conceptual cases, semantic cases, deep cases*, and *semantic roles*.

The concept of deep case was proposed by the world-known American linguist C. Fillmore in 1968. This concept very soon became broadly popular in computer linguistics and theoretical linguistics, because it underlies the basic procedures that find conceptual relationships between a meaning of a verbal form and a meaning of a word group dependending on it in a phrase.

Example 1. Let T1 = "The bulk carrier 'Mikhail Glinka' has arrived from Marseilles to Novorossiysk on the 27th of March." The compound verbal form "has arrived" denotes a certain event of the type "arrival" that can be connected with the label $e1$(event1). In the text T1, the following objects are mentioned: a certain ship $x1$; a certain city $x2$ named "Marseilles"; a certain city $x3$ named "Novorossiysk."

In the event $e1$, the object $x1$ plays the role "Agent of action" (Agent1), $x2$ plays the role "Initial place of movement" (*Place*1), $x3$ plays the role "Place of destination" (*Place*2). Then we can say that the text T1 realizes the thematic roles *Agent*1, *Place*1, *Place*2 as well as the thematic role *Time*.

Example 2. Let T2 = "The bulk carrier 'Mikhail Glinka' has arrived from Marseilles." The text T2 explicitly realizes only the thematic roles *Agent*1 and *Place*1, whereas the thematic roles *Time* and *Place*2 only are implied because of the semantics of the verb "arrive." Thus, the phrases with the same verb in the same meaning can explicitly realize the different subsets of thematic roles.

Formally, we will interpret thematic roles as the names of binary relations with the first attribute being a situation and second one being a real or abstract object playing a specific role in this situation. In this case, if an element $rel \in R_2(B)$, where B is a conceptual basis, and rel is interpreted as a thematic role, then its type $tp(rel)$ is a string of the form $\{(s, u)\}$, where s is a specification of the distinguished sort *sit* (situation), and u is a sort from the set $St(B)$.

The dictionaries of verbal – prepositional frames contain such templates (in other terms, frames) that enable us to represent the necessary conditions of realizing a specific thematic role in the combination

$$Verbal form + Preposition + Dependent word group,$$

where *Preposition* can be void (let *nil* be the sign of void preposition), and

Dependent word group is either a noun with dependent words or without them, or a construct, that is, a numeric value of a parameter.

For example, such expressions include the combinations "has arrived to the port," "left the city," "prepare 4 articles," "has bought the Italian shoes," "arrived before 16:30."

Definition 7.13. Let Cb be a marked-up conceptual basis of the form (5.4), $Tform$ be a text-forming system of the form (7.4) coordinated with Cb, $Morphbs$ be a morphological basis of the form (7.3), $Lsdic$ be a lexico-semantic dictionary coordinated with Cb and $Tform$.

Then *a dictionary of verbal – prepositional semantic-syntactic frames (d.v.p.f.) coordinated with Cb, Tform, and Lsdic* is an arbitrary finite set Vfr consisting of the ten-tuples of the form

$$(k, semsit, form, refl, vc, sprep, grcase, str, trole, expl), \qquad (7.6)$$

where

- $k \geq 1$, $semsit \in X(B)$, $form \in \{infin, ftm, nil\}$, $refl \in \{rf, nrf, nil\}$, $vc \in \{actv, passv, nil\}$,
- $sprep \in W \cup \{nil\}$, where nil is an empty element, W is the set of words from the text-forming system $Tform$; if $sprep \in W$, then $prt(sprep) = preposition$;
- $0 \leq grcase \leq 10$, $str \in St(B)$,
- $trole$ is a binary relational symbol from the primary informational universe $X(B)$, $tp(trole) = \{(s, u)\}$, where $s, u \in St(B)$, s being a concretization of the distinguished sort sit ("situation") (i.e. $sit \to s$);
- $expl \in A^{+} \cup \{nil\}$.

The components of an arbitrary 10-tuple of the form (7.6) from Vfr are interpreted in the following way:

- k is the ordered number of the collection;
- $semsit$ is a semantic unit identifying the type of situation (arrival, departure, receipt, etc.);
- $form$ is a verb form property;
- $infin$ is the indicator of the infinitive verb form;
- ftm is the indicator of a verb form with time, i.e., of the verb in indicative or subjunctive mood;
- $refl$ is the property of reflexivity of the verbs and participles, rf is the indicator of reflexive form, nrf is the indicator of non reflexive form;
- $actv$, $passv$ are the indicators of active and passive voices.

The components $semsit, form, refl, vc$ define the requirements to a verbal form, and the components $sprep, grcase, str$ formulate the requirements to a word or word group being dependent on the verbal form and used in a sentence for expressing a thematic role $trole$.

The string $sprep$ is a simple or compound preposition (for example, the preposition "during" is translated into Russian as "v techenie") or the sign of the void preposition nil; $grcase$ is the code of a grammatical case (that is why $1 \leq grc \leq 10$) or 0 (it is the sign of the lack of such information); str is a semantic restriction for the meaning of a dependent word group or word; $trole$ is such thematic role that the necessary conditions of its realization are represented by this collection (frame); $expl$ is an example in NL that explains the meaning of the thematic role or it is the empty example nil.

The maximal value 10 for the numerical code of a grammatical case is chosen with respect to the fact that the quantity of grammatical cases is 4 for the German language and 6 for the Russian language.

Example 3. Let us construct a certain dictionary $Vfr1$ helping us to find the conceptual relations in the sentences with the verb "to prepare." This verb has, in particular, the meanings $preparation1$ (the preparation of a report, article, etc.) and $preparation2$ (the preparation of the sportsmen of highest qualification, etc.).

In particular, the dictionary $Vfr1$ can be useful for the semantic analysis of the texts like T1 = "Professor Semenov prepared in June a report for the firm 'Sunrise' ";

T2 = "A report for the firm 'Sunrise' was prepared in June by Professor Semenov;"
T3 = "Professor Semenov prepared three Ph.D. scholars in chemistry during 2003–2008."

One can create such marked-up conceptual basis Cb that its first component is a conceptual basis B and the following relationships take place:

$$St(B) \supset (Sorts1 \cup Sorts2),$$

where

$$Sorts1 = \{org, ints, mom, inf.ob, dyn.phys.ob\},$$

$$Sorts2 = \{sit, event, qualif, space.ob, string\};$$

$$X(B) \supset (Units1 \cup Units2 \cup Units3 \cup Units4 \cup Units5),$$

where

$$Units1 = \{\#now\#, firm, university, professor, phd - scholar, learn.inst\},$$

$$Units2 = \{person, 'Semenov', preparation1, preparation2, report1\},$$

$$Units3 = \{Name, Surname, June, 3, 2003, 2008, Qual\},$$

$$Units4 = \{Agent1, Object1, Object2, Product1, Time\},$$

$$Units5 = \{Place1, Place2, Recipient1, Educ\_inst\};$$

$$sit \rightarrow event,$$

because the events are special cases of situations;

$$tp(preparation1) = tp(preparation2) = \uparrow event;$$

$$tp(firm) = tp(university) = \uparrow org * space.ob * ints;$$

$$tp(person) = \uparrow ints * dyn.phys.ob;$$

$$tp(professor) = tp(phd - scholar) = qualif;$$

$$tp(Agent1) = \{(event, ints)\}; tp(Recipient1) = \{(event, org)\},$$

$$tp(Time) = \{(event, mom)\};$$

$$tp(Place1) = tp(Place2) = \{(event, space.ob)\};$$

$$tp(Surname) = \{(ints, string)\}, tp('Semenov') = string;$$

$$tp(Object1) = \{(event, dyn.phys.ob)\}.$$

In that case, let $Vfr1$ be the set consisting of the following sequences:

$$(1, preparation1, ftm, nrf, actv, nil, 1, ints, Agent1, Expl1),$$

where $Expl1 = 'P.Somov\ prepared\ (a\ textbook)',$

$$(2, preparation1, ftm, nrf, passv, by, 5, ints, Agent1, Expl2),$$

where $Expl2 = ''(This book) was prepared by Professor Semenov'$,

$$(3, preparation1, ftm, nrf, actv, nil, 4, inf.ob, Product1, Expl3),$$

where $Expl3 = '(P.Somov) prepared a book'$,

$$(4, preparation1, ftm, nrf, passv, nil, 1, inf.ob, Product1, Expl4),$$

$Expl4 = 'This article was prepared (during three weeks)'$,

$$(5, preparation2, ftm, nrf, passv, nil, 4, qualif, Object2, Expl5),$$

where $Expl5 = 'Many masters of sport were prepared (by this school)'$,

$$(6, preparation1, nil, nil, in, 0, mom, Time, Expl6),$$

where $Expl6 = 'prepared in 2007'$,

$$(7, preparation2, nil, nil, in, 0, mom, Time, Expl7),$$

where $Expl7 = 'prepared in 2007'$.

In this example, the numerical codes of the grammatical cases are indicated for the Russian language, where six grammatical cases are distinguished. The reference to the Russian language helps to become aware of the significance of the component *grcase* of the elements of the dictionaries of verbal – prepositional frames for highly flexible languages.

7.6 The Dictionaries of Prepositional Frames

Let's consider in this section the following problem: how it would be possible to find one or several conceptual relationships realized in the word combinations of the form

$$Noun1 + Preposition + Noun2$$

or of the form

$$Noun1 + Noun2.$$

Example 1. Let us assume that $Expr1$ is the expression "an article by Professor Novikov," and a linguistic database includes a template of the form

$$(k1, 'by', inf.ob, ints, 1, Authors, 'a poem by Pushkin'),$$

where *ints* is the sort "intelligent system," 1 is the code of common case in English. We may connect the sorts *ints* and *dyn.phys.ob* (dynamic physical object) with the basic lexical unit "professor." We see that the expression $Expr1$ is compatible with this template having the number $k1$.

Definition 7.14. Let Cb be a marked-up conceptual basis of the form (5.4), $B = B(Cb)$, $Morphbs$ be a morphological basis of the form (7.3), $Tform$ be a text-forming system of the form (7.4) coordinated with m.c.b Cb; $Lsdic$ be a lexico-semantic dictionary consisting of the finite sequences of the form (7.5) coordinated with Cb and $Tform$.

Then *a dictionary of prepositional semantic-syntactic frames coordinated with Cb, $Tform$, and $Lsdic$* is an arbitrary finite set Frp consisting of the ordered 7-tuples of the form

$$(i, prep, sr1, sr2, grc, rel, ex), \tag{7.7}$$

where

- $i \geq 1$; $prep \in Lecs \cup \{nil\}$, where nil is the string denoting the void (empty) preposition; if $prep \in Lecs$, then $prt(prep) = preposition$;
- $sr1, sr2 \in St(B)$; $1 \leq grc \leq 10$;
- $rel \in R_2(B)$, where $R_2(B)$ is the set of binary relational symbols (thefore, it is a subset of the primary informational universe $X(B(Cb))$); $ex \in A^+$.

The components of the 7-tuples of the form (7.7) from the set Frp are interpreted as follows: The natural number $i \geq 1$ is the ordered number of the 7-tuple (it is used for organizing the loops while analyzing the data from the dictionary Frp), $prep$ is a preposition from the set of basic lexical units $Lecs$ or the void (or empty) preposition nil.

The elements $sr1$ and $sr2$ are interpreted as the sorts that may be associated respectively with the first and second nouns in the linguistically correct combination of the form "Noun1 + Preposition + Noun2"; grc (grammatic case) is the code of such grammatical case that the second noun must be in this grammatical case when it is a part of the correct combinations of the kind.

The component rel is a designation of such conceptual relation that this relation can be realized in such combinations when the specified conditions are satisfied; ex is an example being an expression where the same relation rel is realized.

Example 2. It is possible to build such marked-up conceptual basis Cb, a morphological basis $Morphbs$, a text-forming system $Tform$, a lexico-semantic dictionary $Lsdic$, and a dictionary of prepositional semantic-syntactic frames Frp that Frp includes the semantic-syntactic template (frame) with the number $k1$ considered in Example 1 and also the templates

$$(k2, \, 'for', \, substance, \, illness, \, 1, \, Against1, \, Expr1),$$

$$(k3, \, 'for', \, phys.ob, \, ints, \, 1, \, Addressee, \, Expr2),$$

where 1 is the code of common case in English, $ints$ is the sort "intelligent system," $Expr1 = $ "pills for flu," $Expr2 = $ "a letter for Mary."

Suppose that a dictionary of prepositional semantic-syntactic frames Frp contains no such 7-tuples where the components $prep \neq nil$, $sr1$, $sr2$, grc coincide but the components *rel* or *ex* don't coincide. In such cases the 4-tuple of the form (*prep*, *sr*1, *sr*2, *grc*) unambiguously determines the relation *rel*.

7.7 Linguistic Bases

Let's take two final steps for constructing a formal model of a linguistic database.

7.7.1 Semantic Information Associated with the Role Interrogative Words

Let's define the notion of *a dictionary of the role interrogative word combinations*. Consider the following pairs of the form (*prepqw*, *qwd*), where *prepqw* is a preposition or the void (or empty) preposition *nil*; *qwd* is an interrogative word being either a pronoun or adverb:

$$(nil, who), (by, whom), (for, whom),$$

$$(from, whom), (from, where), (nil, when).$$

Our language competence enables us to associate a certain rather general conceptual relation with each of such pairs:

$$(nil, who) \Rightarrow Agent;$$

$$(by, whom) \Rightarrow Agent$$

$$(for, whom) \Rightarrow Addressee;$$

$$(from, whom) \Rightarrow Source1$$

$$(from, where) \Rightarrow Place1$$

$$(nil, when) \Rightarrow Time.$$

We'll say that such pairs of the form (*prepqw*, *qwd*) are the role interrogative word combinations.

Taking this into account, we'll include one more dictionary into a linguistic database.

Definition 7.15. Let *Cb* be a marked-up conceptual basis of the form (5.4), *B* = *B*(*Cb*), *Morphbs* be a morphological basis of the form (7.3), *Tform* be a text-forming system of the form (7.4) coordinated with m.c.b *Cb*; *Lsdic* be a lexico-semantic dictionary consisting of the finite sequences of the form (7.5) coordinated with *Cb* and *Tform*.

Then *a dictionary of the role interrogative word combinations* coordinated with the marked-up conceptual basis *Cb*, the morphological basis *Morphbs*, and the lexico-semantic dictionary *Lsdic* is an arbitrary finite set *Rqs* consisting of the ordered 4-tuples of the form

$$(i, prepqw, qwd, relq),\tag{7.8}$$

where

- $i \geq 1$, $prepqw \in Lecs \cup \{nil\}$,
- $qwd \in W$, $prt(qwd) \in \{pronoun, adverb\}$,
- $relq \in R_2(B(Cb))$;
- if $prepqw \neq nil$, $prt(prepqw) = preposition$.

Example. It is possible to define B, Cb, $Morphbs$, $Lsdic$, Rqs in such a way that *Rqs* includes the 4-tuples

$$(1, nil, who, Agent), (2, nil, whom, Addressee),$$

$$(3, for, whom, Addressee), (4, from, whom, Source1),$$

$$(5, with, what, Tool) (6, nil, when, Time),$$

$$(7, from, where, Place1), (8, nil, where, Place2).$$

7.7.2 The Notion of a Linguistic Basis

The linguistic bases are formal models of linguistic databases (LDB).

Definition 7.16. The ordered 6-tuple *Lingb* of the form

$$(Cb, Tform, Lsdic, Vfr, Frp, Rqs)\tag{7.9}$$

is called *a linguistic basis (l.b.)* $\Leftrightarrow$ when Cb is a marked-up conceptual basis (m.c.b) of the form (5.4), $Tform$ is a text-forming system (t.f.s) of the form (7.4) coordinated with m.c.b Cb; $Lsdic$ is a lexico-semantic dictionary coordinated with m.c.b Cb and with t.f.s $Tform$, Vfr is a dictionary of verbal – prepositional semantic-syntactic frames coordinated with m.c.b Cb, t.f.s $Tform$, and lexico-semantic dictionary $Lsdic$; Rqs is a dictionary of the role interrogative word combinations coordinated with Cb, $Tform$, and $Lsdic$.

The structure of a linguistic basis is illustrated by Fig. 7.3. The introduced formal notion of a linguistic basis reflects the most significant features of broadly applicable logical structure of a linguistic database. This notion is constructive in the sense that it really can help to design LDB of practically useful linguistic processors, it is shown in the next chapters of this book.

The formal model of a linguistic database constructed above generalizes the author's ideas published, in particular, in [51, 54, 81, 85].

Problems

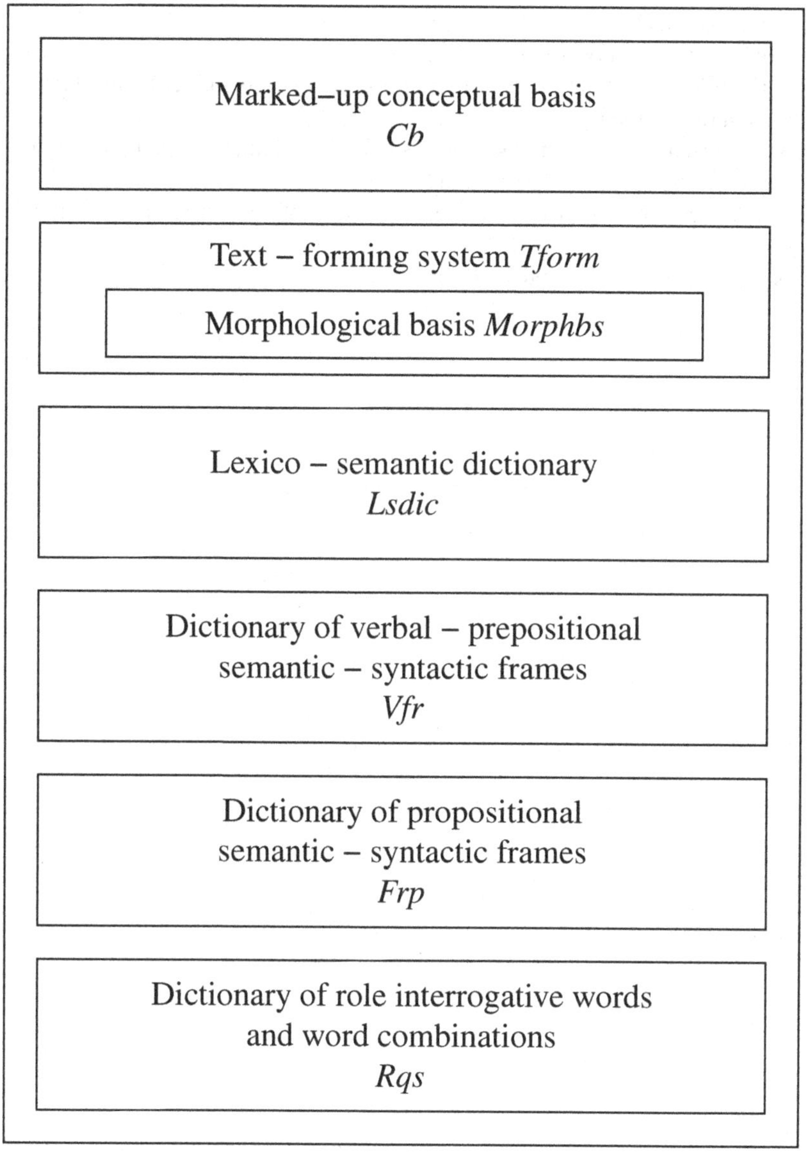

Marked–up conceptual basis
Cb

Text – forming system *Tform*

Morphological basis *Morphbs*

Lexico – semantic dictionary
Lsdic

Dictionary of verbal – prepositional
semantic – syntactic frames
Vfr

Dictionary of propositional
semantic – syntactic frames
Frp

Dictionary of role interrogative words
and word combinations
Rqs

Fig. 7.3 The structure of a linguistic basis

1. How are formally interpreted thematic roles (conceptual cases, deep cases)?
2. What is the semantic dimension of a sort system?
3. What is a morphological determinant (M- determinant)?
4. What is a morphological space?
5. What are the components of a morphological basis?
6. What are the components of a text-forming system?

7. What are constructs?

8. How is the subclass of constructs defined?

9. What is the role of the tolerance relation on the set of sorts in the definition of a lexico-semantic dictionary?

10. What is the structure of the finite sequences being the elements of a dictionary of semantic-syntactic verbal–prepositional frames?

11. What are the components of a dictionary of semantic-syntactic prepositional frames?

12. What is the structure of a linguistic basis?

Chapter 8
A New Method of Transforming Texts into Semantic Representations

Abstract This chapter sets forth a new method of describing the transformation of an NL-text (a statement, a command, or a question) into its semantic representation. According to this method, the transformation includes three phases: (a) Phase 1: The component-morphological analysis of the text; (b) Phase 2: The construction of a matrix semantic-syntactic representation (MSSR); (3) Phase 3: The assembly of a semantic representation of the text, proceeding from its MSSR.

8.1 A Component-Morphological Representation of an NL-text

Let's agree that in this and the next chapters we will consider as lexical units (or word forms) not only separate words but also compound verbal forms ("has been received" and so on), compound prepositions, compound terms ("Olympic games," "artificial intelligence" and so on). This approach allows for attracting the attention to central problems of developing the algorithms of semantic-syntactic analysis by means of abstracting from the details of text preprocessing (it is reasonable to consider such details at the level of program implementation).

Let's say that *elementary meaningful units of texts* are all lexical units, the constructs (the designations of the values of different numeric parameters: 780 km, 12 kg, 7 percent, and so on), the markers (punctuation marks), and the expressions in quotes or apostrophes – the names of various objects.

In order to determine the notion of a matrix semantic-syntactic representation (MSSR) of an NL-text, we will introduce a number of additional data structures associated with the input texts of applied intelligent systems with respect to a considered linguistic basis.

8.1.1 Morphological Representation

Temporarily skipping a number of mathematical details, we'll suppose that *a morphological representation* of a text T with the length nt is a two-dimensional array

V.A. Fomichov, *Semantics-Oriented Natural Language Processing*, IFSR International Series on Systems Science and Engineering 27, DOI 10.1007/978-0-387-72926-8_8, © Springer Science+Business Media, LLC 2010

Rm with the names of columns *base* and *morph*, where the elements of the array rows are interpreted in the following way.

Let *nmr* be the number of the rows in the array *Rm* that was constructed for the text *T*, and *k* be the number of a row in the array *Rm*, i.e. $1 \leq k \leq nmr$. Then *Rm*[*k, base*] is the basic lexical unit (the lexeme) corresponding to the word in the position *p* from the text *T*. Under the same assumptions, *Rm*[*k, morph*] is a sequence of the collections of the values of morphological characteristics (or features) corresponding to the word in the position *p*.

Definition 8.1. Let *T form* be a text-forming system of the form (7.4), *Morphbs* be a morphological basis of the form (7.3), $T \in Texts(T form)$, *nt* be the length of the text *T*. Then *a morphological representation of the text T* is such two-dimensional array *Rm* with the indices of the columns *base* and *morph* that the following conditions are satisfied:

1. Every row of the array *Rm* contains information about a certain word from the input text *T*, i.e. if *nmr* is the number of the rows in the array *Rm*, then for each *i* from 1 to *nmr*, such position *p* can be found in the text *T*, where $1 \leq p \leq nt$, that

$$t_p \in W, \ Rm[i, base] = lcs(t_p),$$

$$Rm[i, morph] = fmorph(t_p).$$

2. For each word from the text T, there is a row in *Rm* representing morphological information about this word, i.e., for each position *p* in the text *T*, where $1 \leq p \leq nt$, $t_p \in W$, there is such *k*, where $1 \leq k \leq nmr$, that

$$lcs(t_p) = Rm[k, base],$$

$$fmorph(t_p) = Rm[k, morph].$$

3. Every two rows in the array *Rm* differ either due to different basic lexical units in the column *base* or due to different collections of the values of morphological properties in the column with the index *morph*. This means that if

$$1 \leq k \leq nmr,$$

$$1 \leq q \leq nmr, \ k \neq q,$$

then either $Rm[k, base] \neq Rm[q, base]$
or $Rm[k, morph] \neq Rm[q, morph]$.

Thus, any row from *Rm* points a basic lexical form and a collection of the values of morphological properties connected with a certain lexical unit from the text *T*. At the same time, for each lexical unit from *T*, a corresponding row can be found in *Rm*.

Example 1. Let T1 be the question "What (1) Russian (2) publishing (3) house (3) released (4) in (5) the (6) year (6) 2007 (6) the (7) work (7) on (8) multi-agent systems (9) 'Mathematical Foundations of Representing the Content of Messages Sent by Computer Intelligent Agents' (10) by (11) professor (12) Fomichov (13) ?

(14)." The text T1 is marked-up in the following way: every elementary expression from the text is followed by the number of elementary meaningful unit of text including this expression. Then a morphological representation Rm of the text T1 may have the following form:

base	morph
what	md_1
Russian	md_2
publishing house	md_3
released	md_4
in	md_5
the work	md_6
on	md_7
multi-agent systems	md_8
by	md_9
Professor	md_{10}
Fomichov	md_{11}

Here $md_1, \ldots, md_{11}$ are the numerical codes of morphological features collections, that is connected in corresponding the words from input the text T1. In particular, md_3 encodes the next information: the part of speech – noun, the subclass of part of speech – common noun, the number – singular, the case – common.

The collection of non-negative integers md_{11} encodes the following information: the part of speech – noun, the subclass of part of speech – proper noun, the number – singular, the case – common.

8.1.2 Classifying Representation

Let $Tform$ be a text-forming system of the form (7.4),

$$T \in Texts(Tform), \quad nt = length(T).$$

Then, from an informal point of view, we will say that *a classifying representation of the text T coordinated with the morphological representation Rm of the text T*, is a two-dimensional array Rc with the number of the rows nt and the column with the indices *unit, tclass, subclass, mcoord*, in which its elements are interpreted in the following way.

Let k be the number of any row in the array Rc i.e. $1 \le k \le nt$. Then $Rc[k, unit]$ is one of elementary meaningful units of the text T, i.e. if $T = t_1 \ldots, t_{nt}$, then $Rc[k, unit] = t_k$.

If $Rc[k, unit]$ is a word, then $Rc[k, tclass], Rc[k, subclass], Rc[k, mcoord]$ are correspondingly part of speech, subclass of part of speech, a sequence of the collections of morphological features' values.

If $Rc[k, unit]$ is a construct (i.e. a value of a parameter), then $Rc[k, tclass]$ is the string *constr*, $Rc[k, subclass]$ is a designation of a subclass of informational unit that corresponds to this construct, $Rc[k, mcoord] = 0$.

Example 2. Let T1 be the question "What Russian publishing house released in the year 2007 the work on multi-agent systems 'Mathematical Foundations of Representing the Content of Messages Sent by Computer Intelligent Agents' by professor Fomichov?" Then a classifying representation Rc of the text $T1$ coordinated with the morphological representation Rm of $T1$ may have the following form:

unit	tclass	subclass	mcoord
What	pronoun	nil	1
Russian	adject	nil	2
publishing house	noun	common-noun	3
released	verb	verb-in-indic-mood	4
in	prep	nil	5
the year 2007	constr	nil	0
the work	noun	common-noun	6
on	prep	nil	7
multi agent systems	noun	common-noun	8
book-title	name	nil	0
by	prep	nil	9
Professor	noun	common-noun	10
Fomichov	noun	proper-noun	11
?	marker	nil	0

Here the element *book-title* is the name of the monograph "Mathematical Foundations of Representing the Content of Messages Sent by Computer Intelligent Agents."

Definition 8.2. Let $Tform$ be a text-forming system of the form (7.4), $Morphbs$ be a morphological basis of the form (7.3), $nt = length(T)$, $T \in Texts(Tform)$, Rm be a morphological representation of T. Then *a classifying representation of the text T coordinated with Rm* is a two-dimensional array Rc with the indices of the columns *unit, tclass, subclass, mcoord*, and the number of the rows nt, satisfying the following conditions:

1. For $k = 1,\ldots, nt$, $Rc[k, unit] = tk$.
2. If $1 \leq k \leq nt$ and $t_k \in W$, then

$$Rc[k, tclass] = prt(t_k), \ Rc[k, subclass] = subprt(t_k),$$

and it is possible to find such q, where $1 \leq q \leq nrm$, nmr is the number of the rows in Rm, that

$$Rc[k, mcoord] = q, \ Rm[q, base] = lcs(t_k),$$

$$Rm[q, morph] = fmorph(t_k).$$

3. If $1 \leq k \leq nt$ and $t_k \in Constr$, then

$$Rc[k, tclass] = constr, \ Rc[k, subclass] = tp(infconstr(t_k)),$$

$$Rc[k, mcoord] = 0.$$

4. If $1 \leq k \leq t$ and $t_k \in Names(Tform)$, then

$$Rc[k, tclass] = name, Rc[k, subclass] = nil,$$

$$Rc[k, mcoord] = 0.$$

5. If $1 \leq k \leq nt$ and $t_k \in Markers$, then

$$Rc[k, tclass] = marker, Rc[k, subclass] = nil,$$

$$Rc[k, mcoord] = 0.$$

Thus, a classifying representation of the text T sets the following information:

1. For each lexical unit, it indicates a part of speech, a subclass of part of speech (if it is defined), and the number of a row from the morphological representation Rm containing the numerical codes of morphological characteristics corresponding to this lexical unit.
2. For each construct it indicates the class $constr$ and a subclass, that is the sort of information unit corresponding to this construct.
3. For each element from the set $Names(Tform)$, it indicates the class $name$, the subclass nil, and the number 0 in the column $mcoord$.
4. For each separator (the punctuation marks), it indicates the class $marker$, the subclass nil, and 0 in the column $mcoord$.

Definition 8.3. Let $Tform$ be a text-forming system of the form (7.4), $T \in Texts(T form)$. Then *a component-morphological representation (CMR) of the text T* is an ordered pair of the form

$$(Rm, Rc),$$

where Rm is a morphological representation of the text T, Rc is a classifying representation of the text T coordinated with Rm.

8.2 The Projections of the Components of a Linguistic Basis on the Input Text

Let *Lingb* be a linguistic basis of the form (7.9), and *Dic* be one of the following components of *Lingb* : the lexico-semantic dictionary *Lsdic*, the dictionary of verbal – prepositional semantic-syntactic frames *Vfr*, the dictionary of prepositional semantic-syntactic frames *Frp*. Then *the projection of the dictionary Dic on the input text $T \in Texts(Tform)$* is a two-dimensional array whose rows represent all data from *Dic* linked with the lexical units from T.

Let's introduce the following denotations to be used in this and next chapters:

- *Arls* is *the projection of the lexico-semantic dictionary Lsdic on the input text* $T \in Texts(T form)$;
- *Arvfr* is *the projection of the dictionary of verbal – prepositional frames Vfr on the input text* $T \in Texts(T form)$;
- *Arfrp* is *the projection of the dictionary of prepositional frames Frp on the input text* $T \in Texts(T form)$.

Example 1. Let T1 be the question "What (1) Russian (2) publishing (3) house (3) released (4) in (5) the (6) year (6) 2007 (6) the (7) work (7) on (8) multi-agent systems (9) 'Mathematical Foundations of Representing the Content of Messages Sent by Computer Intelligent Agents' (10) by (11) professor (12) Fomichov (13) ? (14)" Then the array *Arls* may have the following form:

ord	sem	st1	st2	st3	comment
2	Country(z1, Russia)	space.ob	nil	nil	nil
3	publish-house	org	ints	space.ob	nil
4	releasing1	sit	nil	nil	comment1
4	releasing2	sit	nil	nil	comment2
7	work1	sit	nil	nil	comment3
7	work2	inf.ob	dyn.phys.ob	nil	comment4
9	sem1	field-of-activ	nil	nil	comment5
12	sem2	ints	dyn.phys.ob	nil	nil
13	sem3	ints	dyn.phys.ob	nil	nil

where

$$sem1 = multi\_agent\_systems,$$

$$sem2 = certn\,person * (Qualif, professor),$$

$$sem3 = certn\,person * (Surname, "Fomichov"),$$

*comment*1 = "This film was released in 2005,"
*comment*2 = "Yves released her hand,"
*comment*3 = "This work took 3 h,"
*comment*4 = "This work was sent via DHL,"
*comment*5 = "a scientific – technical field of studies."

The elements of the column *ord* (ordered number) are the ordered numbers of the rows from the classifying representation *Rc*, that is, the ordered numbers of elementary meaningful units (or tokens) of the text *T*.

The number of the rows of the array *Arls* corresponding to one elementary meaningful lexical unit (i.e., corresponding to one row of the classifying representation *Rc*) is equal to the number of different meanings of this lexical unit.

The purpose of considering a two-dimensional array *Arvfr* is as follows: for each verbal form from the text *T*, this array contains all templates (in other terms, frames) from the dictionary *Vfr* enabling a linguistic processor to find the possible conceptual (or semantic) relations between a meaning of this verbal form and a meaning of a word or word group depending on this verbal form in a sentence from the text T.

Example 2. Let T1 be the question "What (1) Russian (2) publishing (3) house (3) released (4) in (5) the (6) year (6) 2007 (6) the (7) work (7) on (8) multi agent systems (9) 'Mathematical Foundations of Representing the Content of Messages Sent by Computer Intelligent Agents' (10) by (11) professor (12) Fomichov (13) ? (14)" Then a fragment of the array $Arvfr$ may have the following form:

nb	semsit	fm	refl	vc	trole	sprep	grc	str	expl
4	releasing1	ftm	nrf	actv	Agent2	nil	1	org	expl1
4	releasing1	ftm	nrf	passv	Agent2	by	1	org	expl2
4	releasing1	ftm	nrf	actv	Product1	nil	1	inf.ob	expl3
4	releasing2	ftm	nrf	actv	Agent1	nil	1	ints	expl4
4	releasing2	ftm	nrf	actv	Object1	nil	1	dyn.phys.ob	expl5

Here 4 is the position of the verb "released" in the considered text T1; ftm (form with time) is the indicator of the verbs in indicative and subjunctive mood; nrf is the indicator of non reflexive verbs; $actv$ and $passv$ are the values "active" and "passive" of the voice; $Agent1$, $Agent2$, $Product1$, $Object1$ are the designations of thematic roles, 1 is the numeric code of the common grammatical case in English; org, $inf.ob$, $ints$, $dyn.phys.ob$ are the sorts "organization," "informational object," "intelligent system," "dynamic physical object,"

$expl1$ = "The studio released (this film in 2005),"

$expl2$ = "(This film) was released by the studio (in 2005),"

$expl3$ = "(The studio) released this film (in 2005),"

$expl4$ = "Yves released (her hand for several seconds),"

$expl5$ = "Yves (released her hand) for several seconds,"

where the auxiliary parts of the examples are surrounded by brackets.

The connection of the array $Arvfr$ with the array $Arls$ is realized by means of the column $semsit$. A template (frame) from the array $Arvfr$ being the m-th row of $Arvfr$ is associated with the row k of the array $Arls$ ⇔ when this template and the row k correspond to the same lexical unit from the text, and

$$Arls[k, sem] = Arvfr[m, semsit].$$

In the same way the array $Arfrp$ can be built, it is called the projection of the dictionary of prepositional frames Frp on the input text. This array is intended for representing all data from the dictionary Frp relating to the prepositions from the text T and to the empty preposition nil (in case the text T contains the word combinations of the form "Noun1 + Noun2").

Example 3. Let T1 be the question "What (1) Russian (2) publishing (3) house (3) released (4) in (5) the (6) year (6) 2007 (6) the (7) work (7) on (8) multi-agent systems (9) 'Mathematical Foundations of Representing the Content of Messages Sent by Computer Intelligent Agents' (10) by (11) professor (12) Fomichov (13) ? (14)" Then a fragment of the array $Arfrp$ may have the following form:

prep	sr1	sr2	grc	rel	ex
on	inf.ob	field-of-activ	1	Field1	"a book on art"
on	phys.ob	phys.ob	1	Location1	"a house on the hill"

8.3 Matrix Semantic-Syntactic Representations of NL-Texts

Let's consider a new data structure called *a matrix semantic-syntactic representation (MSSR) of a natural language input text T*. This data structure will be used for representing the intermediate results of semantic-syntactic analysis on an NL-text.

An MSSR of an NL-text T is a string-numerical matrix $Matr$ with the indices of columns or the groups of columns

$$locunit, nval, prep, posdir, reldir, mark, qt, nattr,$$

it is used for discovering the conceptual (or semantic) relations between the meanings of the fragments of the text T, proceeding from the information about linguistically correct, short, word combinations. Besides, an MSSR of an NL-text allows for selecting one among several possible meanings of an elementary lexical unit.

The number of the rows of the matrix $Matr$ equals nt – the number of the rows in the classifying representation Rc, i.e., it equals the number of elementary meaningful text units in T.

Let's suppose that k is the number of arbitrary row from MSSR $Matr$. Then the element $Matr[k, locunit]$, i.e., the element on the intersection of the row k and the column with the index $locunit$, is *the least number* of a row from the array $Arls$ (it is the projection of the lexico-semantic dictionary $Lsdic$ on the input text T) corresponding to the elementary meaningful lexical unit $Rc[k, unit]$.

It is possible to say that the value $Matr[k, locunit]$ for the k-th elementary meaningful lexical unit from T is *the coordinate of the entry to* the array $Arls$ corresponding to this lexical unit.

The column $nval$ of $Matr$ is used as follows. If k is the ordered number of arbitrary row in Rc and $Matr$ corresponding to an elementary meaningful lexical unit, then the initial value of $Matr[k, nval]$ is equal to the quantity of all rows from $Arls$ corresponding to this lexical unit, that is, corresponding to different meanings of this lexical unit.

When the construction of $Matr$ is finished, the situation is to be different for all lexical units with several possible meanings: for each row of $Matr$ with the ordered number k corresponding to a lexical unit, $Matr[k, nval] = 1$, because a certain meaning was selected for each elementary meaningful lexical unit.

For each row of $Matr$ with the ordered number k associated with a noun or an adjective, the element in the column $prep$ (preposition) specifies the preposition (possibly, the void, or empty, preposition nil) relating to the lexical unit corresponding to the k-th row.

Let's consider the purpose of introducing the column group

$$posdir(posdir_1, posdir_2, \ldots, posdir_n),$$

where n is a constant between 1 and 10 depending on program implementation. Let $1 \leq d \leq n$. Then we will use the designation $Matr[k, posdir, d]$ for an element located at the intersection of the k-th row and the d-th column in the group $posdir$.

If $1 \le k \le nt$, $1 \le d \le n$, then $Matr[k, posdir, d] = m$, where m is either 0 or the ordered number of the d-th lexical unit wd from the input text T, where wd governs the text unit with the ordered number k.

There are no governing lexical units for the verbs in the principal clauses of the sentences, that is why for the row with the ordered number m associated with a verb, $Matr[m, posdir, d] = 0$ for any d from 1 to n.

Let's agree that the nouns govern the adjectives as well as govern the designations of the numbers (e.g. "5 scientific articles"), cardinal numerals, and ordinal numerals.

The group of the columns *reldir* consists of semantic relations whose existence is reflected in the columns of the group *posdir*. For filling in these columns, the templates (or frames) from the arrays $Arls$, $Arvfr$, $Arfrp$ are to be used (the method can be grasped from the analysis of the algorithm of constructing a matrix semantic – syntactic representation of an input NL-text stated in the next chapter).

The column with the index *mark* is to be used for storing the variables denoting the different entities mentioned in the input text (including the events indicated by verbs, participles, gerunds, and verbal nouns).

The column *qt* (quantity) equals either zero or the designation of the number situated in the text before a noun and connected to a noun.

The column *nattr* (number of attributes) equals either zero or the quantity of adjectives related to a noun presented by the k-th row, if we suppose that $Rc[k, unit]$ is a noun.

Example. Let T1 be the question "What (1) Russian (2) publishing (3) house (3) released (4) in (5) the (6) year (6) 2007 (6) the (7) work (7) on (8) multi-agent systems (9) 'Mathematical Foundations of Representing the Content of Messages Sent by Computer Intelligent Agents' (10) by (11) professor (12) Fomichov (13) ? (14)." The morphological and classifying representations Rm and Rc of T1, the possible projections of the dictionaries $Lsdic$, Vfr, Frp on the input text T1 are considered in two preceding sections.

With respect to the arrays Rm, Rc, $Lsdic$, Vfr, Frp constructed for the text T1, its MSSR can have the following form:

locunit	nval	prep	posdir	reldir	mark	qt	nattr
0	1	nil	0, 0	nil, nil	nil	0	0
1	1	nil	3, 0	conc1, nil	nil	0	1
2	1	nil	4, 0	Agent2, nil	x1	0	1
3	1	nil	0, 0	nil, nil	e1	0	0
0	1	in	0, 0	nil, nil	nil	0	0
0	1	in	4, 0	Time, nil	nil	0	0
6	1	nil	4, 0	Product1, nil	x2	0	0
0	1	on	0, 0	nil, nil	nil	0	0
7	1	on	7, 0	Field1, nil	nil	0	0
0	1	0	7, 0	Name1, nil	x2	0	0
0	1	by	0, 0	nil, nil	nil	0	0
8	1	by	7, 0	conc2, nil	x3	0	0
9	1	by	7, 0	conc3, nil	x3	0	0
0	1	nil	0, 0	nil, nil	nil	0	0

where

$$conc1 = Country(z1, Russia),$$

$$conc2 = Qualif(z1, Professor),$$

$$conc3 = Surname(z1, \text{``}Fomichov\text{''}).$$

The constructed matrix reflects the final configuration of the MSSR *Matr*. It means that all semantic relations between the text units were found.

8.4 A New Method of Transforming NL-Texts into Semantic Representations

The concepts introduced above and the stated principles provide the possibility to formulate a new method of transforming the NL-texts (in particular, the requests, statements, or commands) into semantic representations (SRs) of texts.

8.4.1 Formulation of the Method

The proposed method is intended for designing the dialogue systems and includes the following three stages of transformation:

Transformation 1: A component-morphological analysis of the input text.

The essence of the first transformation is as follows. Proceeding from an NL-text *T*, one constructs one or several component-morphological representations (CMR) of the text *T*. This means that one constructs one or several pairs of the form (Rm, Rc), where *Rm* is a morphological representation of the text, i.e., a representation of possible values of the morphological properties for the components of the text T being lexical units (contrary to the numerical values of the properties, to the markers, and to the expressions in apostrophes or inverted commas); *Rc* is a classifying representation of the text.

In other words, the first transformation consists in (a) distinguishing such fragments of the text (called further the elementary meaningful textual units) that each of these fragments either is a marker (comma, semi-colon, etc.) or an expression in inverted commas or in apostrophes or is associated with certain meaning (or meanings); (b) associating one or several collections of the values of morphological properties (a part of speech, a number, a grammatical case, etc.) with each elementary textual unit being a word or a word group, (c) associating a semantic item (a sort) with each elementary meaningful textual unit belonging to the class of constructs: the numbers and the expressions like "1200 km," "70 km/h," etc. For instance, the combination "were delivered" is associated with the part of speech "a verb," the plural, and the past simple tense.

In the major part of cases, only the CMR will correspond to the separate phrases from the input text. If there are several variants of dividing the input text T into the elementary meaningful units or several parts of speech can be associated with any text unit, then the computer system puts the questions to the end user of the dialogue system, and ambiguities are eliminated after the processing of the answers of the end user to these questions.

Transformation 2: The construction of a matrix semantic-syntactic representation (MSSR) of the text.

The first goal of this transformation is to associate with each elementary textual unit that is not a marker or not an expression in inverted commas or in apostrophes one definite meaning from the collection of several meanings linked with this unit. For instance, the verb "to deliver" has, in particular, the meanings "to deliver a lecture" and "to deliver a thing," and the noun "a box" is linked with two different meanings "a box as a container" and "a box as a theater concept."

The second goal of the Transformation 2 is to establish the conceptual relationships between various elementary textual items and, in some cases, between larger items (e.g., between the meaning of a noun and the meaning of an attributive clause).

Since it is done step by step, the MSSR initially is underspecified. In order to eliminate the ambiguities, the system can apply to the end users with diverse questions. Each new step is able to modify the current configuration of the built MSSR as a consequence of obtaining new information from the analysis of a text's fragment and of inscribing this new piece of information into the MSSR.

During this process, mainly the data from the considered linguistic database (LDB) are used and, besides, the knowledge about the admissible manners to combine the various text units into linguistically correct combinations.

Transformation 3: The assembly of a semantic representation of the input NL-text.

The purpose of this transformation is to "assemble" a semantic representation (SR) of the considered text T, proceeding from the information stored in its MSSR. It is important to note that such SR of T is an expression of an SK-language. That is, it is a K-representation of the text T.

An algorithm transforming an MSSR *Matr* of an input NL-text T into a formal expression *Semrepr* $\in Ls(B)$, where B is the conceptual basis being the first component of the considered marked-up conceptual basis Cb, and $Ls(B)$ is the SK-language in the basis B, will be called *an algorithm of semantic assembly*.

Example 1. Let T1 be the question "What (1) Russian (2) publishing (3) house (3) released (4) in (5) the (6) year (6) 2007 (6) the (7) work (7) on (8) multi-agent systems (9) 'Mathematical Foundations of Representing the Content of Messages Sent by Computer Intelligent Agents' (10) by (11) professor (12) Fomichov (13) ? (14)" The morphological and classifying representations *Rm* and *Rc* of T1, the possible projections of the dictionaries *Lsdic*, *Vfr*, *Frp* on the input text T1, and a matrix semantic–syntactic representation *Matr* of T1 are considered in three preceding sections.

With respect to the previous examples concerning the analysis of the question T1, its possible K-representation *Semrepr*1 can be as follows:

$$Question(x1, Situation(e1, releasing1 * (Time, certn\,mom*$$

$$(Earlier, \#now\#) : t1)(Agent2, certn\,publish-house*(Country,$$

$$Russia) : x1)(Product1, certn\,work2*(Field1, multi-agent\,systems)$$

$$(Name1, Title1)(Authors, certn\,person*(Qualif, professor)$$

$$(Surname, ``Fomichov") : x3) : x2))),$$

where *Title*1 is the string "Mathematical Foundations of Representing the Content of Messages Sent by Computer Intelligent Agents."

Figure 8.1 illustrates the proposed method of transforming NL-texts into their semantic representations.

8.4.2 The Principles of Selecting the Form of a Text's Semantic Representation

The form of a semantic representation of an NL-text T to be assembled from the data stored in the MSSR *Matr*, in the classifying representation *Rc*, and in the two-dimensional array *Arls* – the projection of the lexico-semantic dictionary *Lsdic* on the input text *T* is to depend on the kind of *T*.

Let's consider the examples illustrating the recommendations concerning the choice of the form of an SR being an expression of a certain SK-language, that is, being a K-representation of *T*. In these examples, the SR of the input text *T* will be the value of the string variable *Semrepr* (semantic representation).

Example 2. Let T1 = "Professor Igor Novikov teaches in Tomsk." Then

$$Semrepr = Situation\,(e1, teaching*(Time, \#now\#)$$

$$(Agent1, certn\,person*(Qualification, professor)$$

$$(Name, 'Igor')(Surname, 'Novikov') : x2)$$

$$(Place1, certn\,city*(Name, 'Tomsk') : x3)).$$

Example 3. Let T2 = "Deliver a box with details to the warehouse 3." Then

$$Semrepr = (Command\,(\#Operator\#, \#Executor\#, \#now\#, e1)$$

$$\wedge Target\,(e1, delivery1*(Object1, certn\,box1*$$

$$(Content1, certn\,set*(Qual-compos, detail)) : x1)$$

$$(Place2, certn\,warehouse*(Number, 3) : x2)).$$

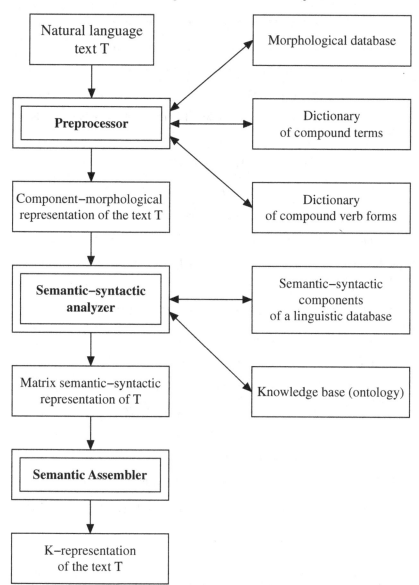

Fig. 8.1 The scheme of transforming NL-texts into their K-representations

Example 4. Let T3 = "Did the international scientific conference 'COLING' take place in Asia?" Then

$$Semrepr = Question\,(x1,\,(x1 \equiv$$

$$Truth - value\,(Situation\,(e1,\,taking\_place*$$

$$(Time, certn\,moment * (Earlier, \#now\#) : t1)$$

$$(Event1, certn\,conference * (Type1, international)$$

$$(Type2, scientific)\,(Name, 'COLING') : x2)$$

$$(Place, certn\,continent * (Name, 'Asia') : x3))))).$$

Example 5. Let T4 = "What publishing house has released the novel 'Winds of Africa'?" Then

$$Semrepr = Question\,(x1, Situation\,(e1, releasing1*$$

$$(Time, certn\,moment * (Earlier, \#now\#) : t1)$$

$$(Agent2, certn\,publ - house : x1)$$

$$(Product1, certn\,novel1 * (Name1, 'Winds\,of\,Africa') : x2))).$$

Example 6. Let T5 = "What foreign publishing houses the writer Igor Somov is collaborating with?" Then

$$Semrepr = Question\,(S1, (Qual - compos\,(S1, publish - house *$$

$$(Type - geographic, foreign)) \wedge$$

$$Description(arbitrary\,publish - house * (Element, S1) : y1,$$

$$Situation\,(e1, collaboration * (Time, \#now\#)$$

$$(Agent1, certn\,person * (Occupation, writer)$$

$$(Name, 'Igor')\,(Surname, 'Somov') : x1)$$

$$(Organization1, y1))))).$$

Example 7. Let T6 = "Who produces the medicine 'Zinnat'?" Then

$$Semrepr = Question\,(x1, Situation\,(e1, production1 *$$

$$(Time, \#now\#)\,(Agent2, x1)$$

$$(Product2, certn\,medicine1 * (Name1, 'Zinnat') : x2))).$$

Example 8. Let T7 = "For whom and where the three-ton aluminum container has been delivered from?"

$$Semrepr = Question\,((x1 \wedge x2),$$

$$Situation\,(e1, delivery2 *$$

$$(Time, certn\,moment * (Earlier, \#now\#) : t1)$$

$$(Recipient, x1)\,(Place1, x2)$$

$$(Object1, certn\,container1 * (Weight, 3/ton)$$

$$(Material, aluminum) : x3))).$$

Example 9. Let T8 = "How many people did participate in the creation of the textbook on statistics?" Then

$$Semrepr = Question\,(x1, ((x1 \equiv Numb(S1))$$

$$\wedge\,Qual - compos\,(S1, person) \wedge$$

$$Description\,(arbitrary\,person * (Element, S1) : y1,$$

$$Situation\,(e1, participation1 *$$

$$(Time, certn\,moment * (Earlier, \#now\#) : t1)$$

$$(Agent1, y1)\,(Type - of - activity, creation1 *$$

$$(Product1, certn\,textbook * (Area1, statistics) : x2))))).$$

Example 10. Let T9 = "How many times Mr. Stepan Semenov flew to Mexico?" Then

$$Semrepr = Question\,(x1, ((x1 \equiv Numb\,(S1))$$

$$\wedge\,Qual - compos\,(S1, sit) \wedge$$

$$Description\,(arbitrary\,sit * (Element, S1) : e1,$$

$$Situation\,(e1, flight * (Time, certn\,moment *$$

$$(Earlier, \#now\#) : t1)\,(Agent1, certn\,person *$$

$$(Name, 'Stepan')(Surname, 'Semenov') : x2)$$

$$(Place2, certn\,country * (Name, 'Mexico') : x3))))).$$

Problems

1. What is the difference between the morphological and classifying representations of the input NL-text?
2. What is the connection between the quantity of the rows of the two-dimensional array *Arls* – the projection of the lexico-semantic dictionary *Lsdic* on the input text corresponding to a lexical unit and the quantity of the meanings associated with this lexical unit?
3. How are the following interpreted: (a) the columns *locunit* and *nval*, (b) the groups of columns *posdir* and *reldir* of a matrix semantic-syntactic representation?
4. What new expressive mechanisms of SK-languages are to be used for building K-representations of the questions about the number of objects characterized with the help of the verbs with dependent words?

Chapter 9
Algorithm of Building a Matrix Semantic-Syntactic Representation of a Natural Language Text

Abstract This chapter sets forth an original algorithm of constructing a matrix semantic-syntactic representation of a natural language text. This algorithm, called *BuildMatr*1, is multilingual: the input texts (the statements, commands, and questions) may belong to the sublanguages of English, German, and Russian languages (a Latin transcription of Russian texts is considered). A pure syntactic representation of an analyzed text isn't used: the proposed algorithm is oriented at directly finding the conceptual relations between the meanings of the fragments of an NL-text.

9.1 Task Statement

9.1.1 Purpose of Development and Non-detailed Structure of the Algorithm

During the last decade, the Internet has become a multilingual system. That is why one has been able to observe a permanently growing interest in multilingual computer programs for processing NL-texts. It seems that the complete potential of semantics-oriented approaches to developing multilingual algorithms of processing NL-texts is far from being realized.

The purpose of this chapter and Chap. 10 is to make a new step (after the author's monograph [85]) in the realization of this potential. With this aim, a new algorithm of semantic-syntactic analysis of the NL-texts is developed. The input NL-texts may belong to the sublanguages of English, German, and Russian. The proposed algorithm implements a new method of explicating the semantic-syntactic structure of NL-texts being a part of a new method of fulfilling the transformation "Natural Language (NL) text – Semantic Representation (SR) of the text" described in Chap. 8.

The formal notion of a linguistic basis introduced in Chap. 7 is interpreted here as a description of the structure of linguistic database (LDB) used by the algorithm.

V.A. Fomichov, *Semantics-Oriented Natural Language Processing*, IFSR International Series on Systems Science and Engineering 27, DOI 10.1007/978-0-387-72926-8_9,

The considered texts may express the statements, commands, specific questions (i.e. the questions with interrogative words) of several kinds, general questions (i.e. the questions with the answers "Yes" / "No").

To realize the new method of fulfilling the transformation "NL-text $\Rightarrow$ SR of the text" suggested in Chap. 8, the following task is stated: to develop an algorithm *SemSynt*1 being the composition of the algorithms *BuildMatr*1 and *BuildSem*1 meeting the following requirements:

1. *BuildMatr*1 is an algorithm of transforming the texts from some sublanguages of English, German, and Russian languages being of practical significance to their matrix semantic-syntactic representations (MSSR);
2. *BuildSem*1 is an algorithm constructing a semantic representation of the NL-text from its MSSR. Moreover, this SR of the text is an expression of a certain SK-language (i.e. it is a K-representation of the input text).

One of the input data of the algorithm *BuildMatr*1 is the string variable *lang* with the values *English, German, Russian*.

The main output data of the algorithm are the string *kindtext* defining the form of the input text (i.e. classifying this text) and the string-numerical matrix *Matr* – an MSSR of the input text.

An important peculiarity of the algorithm *BuildMatr*1 is that it directly discovers the semantic relationships between the meanings of the text's fragments without fulfilling the traditional syntactic analysis of the input text.

The starting point for developing the algorithm *BuildMatr*1 was the analysis of the surface and semantic structure of the texts from the following sublanguages of English, German, and Russian languages being of practical interest:

- the questions and statements in natural language regarding the scientific publications and specialists' participation in scientific conferences (it is supposed that these questions and statements are addressed to an online searching system of a new generation);
- the commands and questions to a transport-loading intelligent robot, in particular, to a robot operating in an automated warehouse and to a robot acting in an airport;
- the questions and statements concerning the manufacture, export, and import of certain products (it is supposed that these questions and statements are addressed to an intelligent database);
- the questions of an automated warehouse operator addressed to an intelligent database;
- the questions of potential clients to an intelligent database of an online shop.

9.1.2 Some Peculiarities of the Approach

It is known that the problem of computer semantic-syntactic analysis (SSA) of NL-texts includes many aspects being not equally worked out. The initial stage of semantic-syntactic analysis of the NL-texts includes finding the basic forms of the words of the input text, finding the possible values of morphological characteristics (number, case, person, tense, etc.) for these words and splitting the text into the segments corresponding to certain elementary semantic units. These segments include, for example, the following expressions: "were shipped," "will be prepared," "Olympic games," "840 km," "1999th year," "2 h," "five percents."

The questions of designing the morphological analyzers of NL-texts are investigated in many publications.

The problems concerning the automatic selection of the short text segments designating the elementary units of semantic level are not very complicated from the logical standpoint. These problems can be solved directly at the level of programming an algorithm.

A literature analysis and own experience of the author show that the main logical difficulties concerning the automation of SSA of NL-texts are related to the search of semantic relationships between the components of the input text.

That is why the algorithm *BuildMatr*1 is focused on formalization and algorithmization of the process of searching the semantic relationships between the components of the input text. This focus is achieved as follows:

1. In many cases the input text T of the algorithm is a certain abstraction as compared with the real input text of a linguistic processor. That is because the units of the input text of the algorithm could be represented not only by the words but also by the short phrases designating the elementary semantic, or conceptual, units ("were shipped," "will be prepared," "Olympic games," "840 km," "1999th year," "2 h").
2. It is assumed that there is a mapping associating the words and their basic forms with the sets of values of morphological characteristics. A mapping of the kind also associates each construct (a numerical value of a certain parameter) with a certain semantic, or conceptual, unit. The concept of a text-forming system was introduced in Chap. 7 just with this aim.

There are two main approaches to the algorithm development: top-to-bottom (descending) and bottom-to-top (ascending) design. It seemed to be expedient to combine both these methods during the development of a strongly structured algorithm *Semsynt*1. This combination of descending and ascending methods of algorithm design is reflected in the description of the algorithm *BuildMatr*1 in the next sections of this book and in the description of the algorithm *BuildSem*1 in Chap. 10.

In cases when an auxiliary algorithm is very simple (can be directly programmed) or its development does not represent any theoretical difficulties considering existing scientific publications, only an external specification of such algorithm is included in the structured descriptions of the algorithms *BuildMatr*1 and *BuildSem*1, i.e., a description of the purpose of development, input and output data of an algorithm.

One of possible variants of a language for algorithm development (or pseudocode, see [140]) is used for describing the algorithms in this and next chapters. The following service words are used in the algorithms: *begin, end, if, then, else, end − if, loop, loop − until, end − of − loop, case − of, end − case − of.*

9.2 Initial Stages of Developing the Algorithm BuildMatr1

This section contains an external specification and a plan of the algorithm *Build Matr1* to be developed in the sections of this chapter.

9.2.1 External Specification of the Algorithm BuildMatr1

Input:
Lingb – linguistic basis (see Sect. 7.7);
T – input text;
lang – string variable with the values *English, German, Russian.*
Output:
nt – integer – the quantity of elementary meaningful text units;
Rc – classifying representation of the input text *T* (see Sect. 8.1);
Rm – morphological representation of the input text (see Sect. 8.1);
kindtext – string variable; its value allows to attribute the input text to one of the text subclasses;
Arls – two-dimensional array – projection of the lexico-semantic dictionary *Lsdic* on the input text *T*;
Arvfr – two-dimensional array – projection of the dictionary of verbal-prepositional frames *Vfr* on the input text *T*;
Arfrp – two-dimensional array – projection of the dictionary of prepositional semantic-syntactic frames *Frp* on the input text *T* (see Sect. 8.2);
Matr – matrix semantic-syntactic representation (MSSR) of the input text (see Sect. 8.3).

9.2.2 Development of a Plan of the Algorithm BuildMatr1

Plan of the Algorithm BuildMatr1
> *Begin*
>> $Building - compon - morphol - representation\,(T, Rc, nt, Rm)$
>> $Building - projection - of - lexico - semantic - dictionary$
>> $(Lsdic, nt, Rc, Rm, Arls)$
>> $Building - projection - of - verbal - frames - dictionary$
>> $(Arls, Vfr, nt, Rc, Rm, Arvfr)$
>> $Building - projection - of - prepos - frames - dictionary$
>> $(Arls, Frp, nt, Rc, Rm, Arfrp)$
>> $Forming - initial - values - of - data$
>> $Defining - form - of - text$
>> $(nt, Rc, Rm, leftprep, mainpos, kindtext, pos)$
>> $loop - until$

$$pos := pos + 1$$
$$Class := Rc\,[pos, tclass]$$
$$case\ class\ of$$
$$\quad preposition:\ Processing - preposition$$
$$\quad adjective:\ Processing - adjective$$
$$\quad cardinal - number:\ Processing - cardinal - number$$
$$\quad noun:\ Processing - noun$$
$$\quad verb:\ Processing - verbal - form$$
$$\quad conjunction:\ Empty - operator$$
$$\quad construct:\ Processing - construct$$
$$\quad name:\ Processing - name$$
$$\quad marker:\ Empty\ operator$$
$$end - case - of$$
$$exit - when\ (pos = nt)$$
$$end$$

$$Algorithm\ ``Forming - initial - values - of - data"$$

Begin

Set to null all integer variables.

Assign the value *nil* (empty element) to all string variables.

Set to null all numeric columns and fill in with the string *nil* all string positions of the used one-dimensional and two-dimensional arrays.

For each row k of the classifying representation *Rc* corresponding to the word of the input text carry out the following actions:

$Matr\,[k, locunit] :=$ the minimal number of the row of array *Arls* (projection of the lexico-semantic dictionary *Lsdic* on the considered text) containing the information corresponding to the considered word;

$Matr\,[k, nval] :=$ quantity of rows of *Arls* corresponding to this word

end

The next sections of this chapter are dedicated to the detalization of this plan, i.e. to the development of the first part of the algorithm of semantic-syntactic analysis of the texts from sublanguages of English, German, and Russian being of practical interest.

9.3 Description of the Algorithm Classifying the Input Texts

9.3.1 The Purpose of the Algorithm

The algorithm "Defining-form-of-text" is designed for attributing the text to a certain class. The type of the class is a value of the output variable *kindtext*. The range of possible values of the variable *kindtext* can be described as follows:

- Statements – the value *stat* – Example (English): "The joint-stock company 'Sail' has been exporting the goods to Bulgaria since 1999."
- Commands – the value *imp* – Example (English): "Ship the containers to the factory 'Dawn' by 16:30."
- Closed (general) questions – the value *genqs* – Example (English): "Does the joint-stock company 'Sail' export the goods to Bulgaria?"
- Questions about the quantity of objects – the value *specqs – quant*1 – Example (English): "How many articles on organic chemistry by Professor Igor Somov were published last year?"
- Questions about the quantity of events – the value *specqs – quant*2 – Example (English): "How many times this year was a textbook by Korobov requested?"
- Role questions – the value *specqs – role* - Example (English): "Where three-ton containers came from?"
- Questions with an interrogative word attached to a noun in singular, where the sought-for value is an individual – the value *specqs – relat*1 – Example (English): "What institute does Professor Igor Somov work at?"
- Questions with an interrogative word attached to a noun in plural, where the sought-for value is a set of individuals – the value *specqs – relat*2 – Example (English): "What countries does the joint-stock company 'Rainbow' import the component parts from?"

The value of the variable *kindtext* is used in the algorithm building the semantic representation (SR) of the text from its matrix semantic-syntactic representation. For example, let Qs1 = "Where three-ton containers came from?" Qs2 = "How many times this year was the textbook by Korobov requested?"

For the question Qs1, *kindtext = specqs – role* and SR of the question Qs1 may be the following K-formula:

$$Question\,(x1,\,(Situation\,(e1,\,delivery2 * (Place1,\,x1)$$

$$(Time,\,x2)\,(Object1,\,certn\,set * (Compos,\,container *$$

$$(Weight,\,3/ton)) : S1)) \wedge Earlier\,(x2,\,\#now\#))).$$

In case of question Q2 *kindtext = specqs – quant*2 and SR of the question Q2 may be the following K-formula:

$$Question\,(x1,\,(x1 \equiv Quant - elem\,(all\,inquiry1 *$$

$$(Time,\,current - year)\,(Subject - of - inquiry,\,certn\,textbook *$$

$$(Author,\,certn\,person * (Surname,\,"Korobov")) : x2))))).$$

Besides the variable *kindtext*, the output data of the algorithm "Defining-form-of-text" also include the variables *mainpos* and *pos*. The value of the variable *mainpos* is the position number of an interrogative word in the text.

For example, for the question Qs3 = "From which countries the joint-stock company 'Rainbow' imports the component parts?" and Qs1 = "Where three-ton containers came from?" the variable *mainpos* takes the values 2 and 1 correspondingly. For the commands, statements, and general questions the variable *mainpos* takes the value 1.

9.3.2 Compound Designations of the Subclasses of Lexical Units

The proposed algorithm of building a matrix semantic-syntactic representation (MSSR) of an input NL-text is a modification of the algorithm *BuildMatr* described in Russian in the monograph [85], taking into account some peculiarities of the Russian language.

In order to develop a multilingual algorithm of constructing an MSSR of an input NL-text from the sublanguages of English, German, and Russian, it turned out to be reasonable to introduce the compound designations of the subclasses of lexical units.

The principal impulse was given by the following observation. The interrogative word "what" in English can be used at least in two different contexts illustrated by the questions Qs1 = "What has John received" and Qs2 = "What books are being edited by John?".

Let's agree to say that in the question Qs1, "what" is a role interrogative word, and in the question Qs2, "what" is a noun-attaching interrogative word.

That is why it is proposed to consider the following interlingual subclasses of interrogative words:

- *roleqswd* ("who," "whom" in English; "wer", "wem", "woher" in German; "kto", "komu," etc., in Russian);
- *noun_attach_qswd* ("welche" in German; "kakoy", "kakie" in Russian);
- *roleqswd | noun_attach_qswd* ("what" in English);
- *qswd_quant*1 *| qswd_quant*2 ("how many" in English; "wieviel" in German; "skol'ko" in Russian), where the component *qswd_quant*1 is realized in the questions about the quantity of events ("How many times" in English; "Wieviel Mal" in German; "Skol'ko raz" in Russian), and the component *qswd_quant*2 is realized in the questions about the quantity of objects.

Let *Lexsubclasses* be the set of strings denoting all subclasses of considered lexical units. Then let the mapping *Setlextypes* be defined as follows:

- for every simple (i.e. containing no symbol |) d from *Lexsubclasses*,

$$Setlextypes(d) = \{d\};$$

- for every compound string h of the form

$$d_1 \mid d_2 \mid \ldots \mid d_n$$

from *Lexsubclasses*,

$$Setlextypes(h) = \{d_1, d_2, \ldots d_n\}.$$

Example. *Setlextypes(roleqswd | noun_attach_qswd)*
= {*roleqswd, noun_attach_qswd*}.

9.3.3 External Specification of the Algorithm "Defining-form-of-text"

Input:
Lingb – linguistic basis,
Rc and *Rm* – classifying and morphological representations of the input text *T* (see Sect. 8.1);
lang – string variable with the values *English, German, Russian*.
Output:
kindtext – string variable designating the form of the input text;
leftprep – string variable designating the preposition in the beginning of the text;
mainpos – integer variable designating the rightest position of an interrogative word;
pos – integer variable designating such position in the input text that it is required to continue processing of the text after this position.

External specification of auxiliary algorithms

Specification of the algorithm − function "Number"

Input:
d – element of the morphological space *Spmorph* (*T form* (*Lingb*)) corresponding to the word which may have a singular or a plural number.
Output:
*number*1 – value 1 for singular number, 2 for plural number, 3 in the cases when the word could be attributed to both singular and plural.

Specification of the algorithm "Form − of − verb"

Input:
d – element of the morphological space *Spmorph* (*T form* (*Lingb*)) corresponding to the verb.
Output:
*form*1 – string with the value "infinitive" if *d* corresponds to the verb in infinitive form; the value "indicative" for representing the verb in indicative mood; the value "imperative" if *d* corresponds to the verb in imperative mood.

Specification of the algorithm

$$\text{``}Right\_noun\,(pos1,\,posnoun1)\text{''}$$

Input:

Rc and *Rm* – classifying and morphological representations of the input text *T* (see Sect. 8.1);

*pos*1 – integer – position of a lexical unit in the input text, i.e. position in the classifying representation *Rc*

Output:

*posnoun*1 – integer – either 0 or the position of a noun from the input text to the right from *posnoun*1 having the minimal distance from *pos*1.

$$Auxiliary\,Algorithm$$

$$Right\_noun\,(pos1,\,posnoun1)$$

Comment (condition of calling the algorithm):

$$Setlextypes\,(Rc[pos1,\,subclass]) \ni noun\_attach\_qswd$$

```
begin posnoun1 := 0; k := pos1
      loop_until
      k := k + 1
      part1 := Rc[k, tclass])
      if part1 = noun then posnoun1 := k end − if
      until ((k = nt) or (part1 = verb)
      or (part1 = adverb) or (part1 = preposition)
      or (part1 = conjunction) or (part1 = marker)
      or (part1 = constr) or (part1 = name))
      end − of − loop
end
```

9.3.4 Algorithm "Defining-form-of-text"

```
Begin        unit1 := Rc[1, unit]
             part1 := Rc[1, tclass]
             unit2 := Rc[2, unit]
             part2 := Rc[2, tclass]
             last_unit := Rc[nt, unit]
             last_class := Rc[nt, tclass]
             before_last_unit := Rc[nt − 1, unit]
             before_last_class := Rc[nt − 1, tclass]
             prep_end := nil
             if (lang = English) and (last_unit := '?')
```

thenprep_end := *before_last_unit*
if(part1 = *verb) and*
(*Rc*[1, *subclass*] ∈ {*imp, infin*})
and (last_unit ≠ '?')
then kindtext := *imp*

Comment

The input text is a command.

Example: "Ship the containers to the factory 'Dawn' by 16:30."

End-of-comment

if(part1 = *verb) and*
(*Rc*[1, *subclass*] = *ftm*)

Comment

ftm means "form with time," it is the introduced (non traditional) property of the verbs in indicative and subjunctive moods.

End-of-comment

and (last_unit = '?')
then kindtext := *genqs*

Comment

The input text is a general question, i.e., it is a question with the answer "Yes/No."

Example: "Does the joint-stock company 'Sail' export the goods to Bulgaria?"

End-of-comment

loghelp1 := *false*
loghelp2 := *false*
loghelp3 := *false*
if qswd_quant1 | qswd_quant2
belongs toSetlextypes(Rc[1, *subclass*])
then if (lang = *English) and (unit2* = '*times*')
then loghelp1 := *true*
end − if
if(lang = *German)and(unit2* = '*Mal*')
then loghelp2 := *true*
end − if
if(lang = *Russian)and(unit2* = '*raz*')
then loghelp3 := *true*
end − if
if(loghelp1 = *true)or(loghelp2* = *true)*
or(loghelp3 = *true)*
then begin mainpos := 1
pos := 2
kindtext := *specqs-quant1*
end

Comment

Example: "How many times this year the textbook by Korobov was requested?"

End-of-comment

$elsebegin\ mainpos := 1$
$\quad pos := 1$
$\quad kindtext := specqs - quant2$
$\quad var1 := {}'S1'$
$\quad end$
$end - if$

Comment

Example: "How many articles on organic chemistry by professor Igor Somov were published last year?"

End-of-comment

$if\ kindtext := nil$
$then$
$begin$
$loghelp7 := ((part1 = pronoun)$
$or\ ((part1 = preposition)\ and\ (part2 = pronoun)));$
$if(part1 = pronoun)$
$thenmainpos := 1$
$end - if$
$if(part1 = preposition)\ and\ (part2 = pronoun))$
$thenmainpos := 2$
$end - if$
$Set1 := Setlextypes(Rc[mainpos, subclass]);$
$n1 := Number\_of\_elem(Set1);$
$if\ (roleqswd \in Set1)and(n1 = 1)$
$thenkindtext := specqs - role$
$\quad Processing - of - role - interrog - words$

Comment

The algorithm $Processing - of - role - interrog - words$ is described in next section.

End-of-comment

$\quad (pos, mainpos)$
$end - if$
$if\ (roleqswd \in Set1)\ and\ (noun\_attach\_qswd \in Set1)$
$then$

Comment

Recognize one of two cases:

Case 1: Example – "What has he received?"

Case 2: Example – "What scholars from France did participate in this conference?"

End-of-comment

$Calltheprocedure"Right - noun(mainpos, posnoun)$
$If\ (posnoun = 0)$
$thenkindtext := specqs - role$
$\quad Processing - of - role - interrog - words$

Comment

The algorithm *Processing − of − role − interrog − words* is described in next section.

End-of-comment

$$(pos, mainpos)$$

Comment

Example: "Where did three-ton containers come from?"

Example (German): "Wo arbeitet Herr Professor Dr. Schulz?"

End-of-comment

    else

Comment: $(posnoun > 0)$ end-of-comment

    $noun1 = Rc[posnoun, unit]$

    if $(noun1\ is\ associated\ with\ singular)$

    then $var1 := 'x1'$

    $kindtext := specqs − relat1$

Comment

Example: "What institute does Professor Igor Somov work at?"

End-of-comment

    else $var1 := 'S1'$

    $kindtext := specqs − relat2$

    end − if

Comment

"What countries does the joint-stock company 'Rainbow' import the component parts from?"

End-of-comment

End − if

End − if

    if$(kindtext = nil)\ and\ (nbqswd = 0)$

    then $kindtext := stat$

    End − if

Comment

"The joint-stock company 'Rainbow' has been exporting the goods to Bulgaria since 1999"

End-of-comment

end

9.4 Principles of Processing the Role Interrogative Words

Let's agree to say that a sentence starts with an interrogative word *wd* if the first word of this sentence is either *wd* or a certain preposition followed by *wd*. For example, let's agree that each of the questions Q1 = "Whom have you sent on a business trip to Prague?" and Q2 = "Whom did 3 two-ton containers come for?" starts with the interrogative word "whom."

Let's split all interrogative words the considered questions could begin with into two groups. The first group includes, in particular, the expressions "how many," "how many times," and "what." The second group includes, in particular, the pronouns "who," "whom," "what," etc., and the adverbs "where," "when." Each word of this group together with a certain preposition is used to express a certain thematic role, i.e., semantic relationship between the verb and its dependent expression.

If an interrogative word *wd* does not require a preposition, then let's say that this word is used with the void (or empty) preposition *nil*. For example, the pair (*nil, whom*) in a question Q1 is used to describe the thematic role "Object of action." The pair (*for, whom*) in the question Q2 expresses the thematic role "Recipient." With respect to these observations, the words of the second group will be called *the role interrogative word combinations*.

An initial processing of the interrogative pronouns from the first group is done by the algorithm "Defining-form-of-text." A position of an interrogative word is the value of the output variable *mainpos*. The value of the variable *kindtext* indicates a subclass the considered interrogative word belongs to.

The role interrogative words (i.e. the words of the second group) together with the prepositions may form the sequences being the left segments of the questions. The question Q3 = "When and where three aluminum containers came from?" could serve as an example. That is why a special algorithm "Processing-of-role-interrog-words" is required. This algorithm called by the algorithm "Defining-form-of-text" uses a dictionary of the role interrogative words and word combinations *Rqs* being a part of the considered linguistic basis.

Let's agree to say that a sentence begins with an interrogative word *wd* if the first word of this sentence is either *wd* or a certain preposition followed by *wd*. For example, let's agree that each of the questions Q1 = "Whom have you sent on a business trip to Prague?" and Q2 = "Whom did 3 two-ton containers come for?"

Algorithm "Processing-of-role-interrog-words" is used for processing the role interrogative words located in the beginning of many questions. Such words are the interrogative pronouns ("who," "whom," "what," etc.) or pronominal adverbs ("where," "when").

External specification of the algorithm

"Processing − of − role − interrog − words"

Input:
nt − integer − the quantity of the text units;
Rc − classifying representation of the input text *T* (see Sect. 7.1);
pos − integer − the position of the interrogative word in *Rc*;
Rqs − dictionary of the role interrogative words (one of the components of *Lingb*);
Matr − matrix semantic-syntactic representation (MSSR) of the input text (see Sect. 8.3);
leftprep − string − the value of the preposition on the left;
numbent − integer − the quantity of objects mentioned in the text;

numbqswd – integer – quantity of the interrogative words that have been already found in the text;

posqswd – one-dimensional array of the length *nt*, where for $k \geq 1$, *posqswd* [*k*] is either a position of the interrogative word having the number *k* in *Rc* or 0.

Output:
Matr, numbent, numbqswd, posqswd, leftprep.

$$Algorithm\ "Processing - of - role - interrog - words"$$

Begin
Comment (condition of application):

$$(roleqswd \in Setlextypes(Rc[mainpos, subclass])$$

$$and\ (kindtext = 'roleqs')$$

End-of-comment
$\qquad k1 := mainpos - 1$
$\qquad loop - until$
$\qquad k1 := k1 + 1$
$\qquad class1 := Rc[k1, tclass]$
The new book of the fragment to be obtained:
$\qquad k1 := k1 + 1$
$\qquad class1 := RC[k1, tclass]$
$\qquad if\ class1 = pronoun\ then\ numbqswd := numbqswd + 1$
$\qquad numbqswd := numbqswd + 1$
$\qquad if class1 = pronoun\ then$
Comment
Quantity of interrogative words already encountered in the text
End-of-comment
$\qquad posqswd\,[numbqswd] := k1$
$\qquad end - if$
$\qquad until\ class1 \neq \notin \{pronoun, preposition, conjunction, marker\}$
$\qquad end - of - loop$
$\qquad loop\ for\ m1\ from\ 1\ to\ numbqswd$
$\qquad p1 := posqswd[m1]$
$\qquad word1 := Rc[p1, unit]$
$\qquad if Rc[p1, tclass] = pronoun$
$\qquad then if (Rc[p1 - 1, tclass] = preposition$
$\qquad\quad then leftprep = Rc[p1 - 1, unit]$
$\qquad\quad else$
$\qquad\quad begin$
$\qquad\quad if (lang = 'German')or(lang = 'Russian')$
$\qquad\quad then leftprep = nil$
$\qquad\quad end - if$
$\qquad\quad if (lang = 'English')\ and\ (m1 < numbqswd)$
$\qquad\quad then\ leftprep := nil$

$end-if$
 $if\ (lang\ =\ 'English')\ and\ (m1\ =\ numbqswd)$
 $then\ if\ prep\_end\ \neq\ nil$
 $then\ leftprep\ =\ prep\_end$
 $end-if$
 $end-if$
 end
$end-if$
$else\ leftprep\ :=\ nil$ Comment: "when"in English, "wohin" in German
$end-if$
$if(Rc[p1, tclass]\ =\ pronoun)$
Find in the dictionary of the role interrogative word combinations Rqs
being a component of the considered linguistic basis $Lingb$
such ordered four-tuple with the least possible number of a row $n1$ that

$$Rqs[n1, prep]\ =\ leftprep$$

$$Rqs[n1, qswd]\ =\ word1$$

$else\ \{the case\ of\ Rc[p1, tclass]\ =\ adverb\}$
Find in the dictionary of the role interrogative word combinations Rqs
being a component of the considered linguistic basis $Lingb$
such ordered four-tuple with the least possible number of a row $n1$ that
 $Rqs[n1, qswd]\ =\ word1$
 $end-if$
 $role\ :=\ Rqs[n1, relq]$
 $Matr[p1, reldir, m1]\ :=\ role$
 $numbent\ :=\ m1$
Comment
Quantity of entities mentioned in already analyzed fragment of the text
End-of-comment
 $var1\ :=\ var('x', numbent)$
 $Matr[p1, mark]\ :=\ var1$
 end
 $end-of-loop\ \{for\ m1\}$
 $end\ (of\ algorithm)$

9.5 An Algorithm of Searching for Semantic Connections of the Verbs

Let us start to formalize the conditions required for the existence of a semantic relationship between a meaning of a verbal form and a meaning of a word or word combination depending in a sentence on this verbal form.

9.5.1 Key Ideas of the Algorithm

Let's agree to use the term "noun group" for designation of the nouns and the nouns together with the dependent words representing the concepts, objects, and sets of objects. For example, let T1 = "When and where two aluminum containers with ceramic tiles have been delivered from?", T2 = "When the article by Professor P. Somov was delivered?" and T3 = "Put the blue box on the green case." Then the phrases "two aluminum containers," "the article by professor P. Somov," "blue box" are the noun groups.

Let's call "a verbal form" either a verb in personal or infinitive form or a participle. A discovery of possible semantic relationships between a verbal form and a phrase including a noun or an interrogative pronoun plays an important role in the process of semantic-syntactic analysis of NL-text.

Let's suppose that $posvb$ is the position of a verbal form in the representation Rc, $posdepword$ is the position of a noun or an interrogative pronoun in the representation Rc.

The input data of the algorithm "Find-set-of-thematic-roles" are real numbers $posvb$, $posdepword$, and two-dimensional arrays $Arls$, $Arvfr$, where $Arls$ is the projection of the lexico-semantic dictionary $Lsdic$ on the input text, $Arvfr$ is the projection of the dictionary of verbal-prepositional frames Vfr on the input text.

The purpose of the algorithm "Find-set-of-thematic-roles" is firstly to find the integer number $nrelvbdep$ – the quantity of possible semantic relationships between the values of the text units with the numbers $p1$ and $p2$ in the representation Rc.

Secondly, this algorithm should build an auxiliary two-dimensional array $Arrelvbdep$ keeping the information about possible semantic connections between the units of Rc with the numbers $p1$ and $p2$. The rows of this array represent the information about the combinations of a meaning of the verbal form and a meaning of the dependent group of words (or one word).

The structure of each row of the two-dimensional array $Arrelvbdep$ with the indices of columns

$$linenoun, \quad linevb, \quad trole, \quad example$$

is as follows.

For the filled-in row with the number k of the array $Arrelvbdep$ ($k \geq 1$)

- $linenoun$ is the ordered number of the row of the array $Arls$ corresponding to the word in the position $p1$;
- $linevb$ is the ordered number of the row of the array $Arls$ corresponding to the verbal form in the position $p2$;
- $trole$ is the designation of the semantic relationship (thematic role) connecting the verbal form in the position $p2$ with the dependent word in the position $p1$;
- $example$ is an example of an expression in NL realizing the same thematic role.

The search of the possible semantic relationships between a meaning of the verbal form (VF) and a meaning of the dependent group of words (DGW) is done with the help of the projection of the dictionary of verbal-prepositional frames (d.v.p.f.) $Arvfr$ on the input text. In this dictionary such a template (or templates) is searched that it would be compatible with the certain semantic-syntactic characteristics of the VF in the position *posvb* and the DGW with the number *posdepword* in *Rc*.

Such characteristics include, first of all, the set of codes of grammatic cases *Grcases* associated with the text-forming unit having the ordered number – value *posdepwd* ("the position of dependent word") in *Rc*.

Let's suppose that *Rc* [*posvb, tclass*] = *verb*. Then *Grcases* is a set of grammatic cases corresponding to the noun in the position *posdepword*.

9.5.2 Description of an Algorithm of Searching for Semantic Connections Between a Verb and a Noun Group

9.5.2.1 Purpose of the Algorithm "Find-set-relations-verb-noun"

The algorithm is to establish a thematic role connecting a verbal form in the position *posvb* with a word (noun or connective word) in the position *posdepword* taking into account a possible preposition before this word. As a consequence, to select one of the several possible values of a verbal form and one of the several possible values of a word in the position *posdepword*, three enclosed loops are required: (1) with the parameter being a possible value of the word in the position *posdepword*; (2) with the parameter being a possible value of the verbal form; (3) with the parameter being a verbal-prepositional frame connected with this verbal form.

9.5.2.2 External Specification of the Algorithm "Find-set-relations-verb-noun"

Input:

Rc – classifying representation;

nt – integer – quantity of the text units in the classifying representation *Rc*, i.e. the quantity of rows in *Rc*;

Rm – morphological representation of the lexic units of *Rc*;

posvb – integer – position of a verbal form (a verb in a personal or infinitive form, or a participle);

posdepword – integer – position of a noun;

Matr – initial value of MSSR of the text;

Arls – array – projection of the lexico-semantic dictionary *Lsdic* on the input text *T*;

Arvfr – array – projection of the dictionary of verbal-prepositional frames *Vfr* on the input text *T*.

Output:

arrelvbdep – one-dimensional array designed to represent the information about (a) a meaning of a dependent word, (b) a meaning of a verbal form, and (c) about a semantic relationship between the verbal form in the position *posvb* and the dependent word in the position *posdepword*;

nrelvbdep – integer – quantity of meaningful rows in the array *arrelvbdep*.

9.5.2.3 External Specifications of Auxiliary Algorithms

Specification of the algorithm

"Characteristics − of − verbal − form"

Input:
$p1$ − number of a row of Rc corresponding to a verb or a participle.
Output:
$form1$, $refl1$, $voice1$ − strings; their values are defined in the following way:

If $p1$ is the position of a verb, then $form1$ may have one of the following values: *indicat* (the sign of the indicative mood), *infinit* (the sign of the infinitive form of a verb), *imperat* (the sign of the imperative mood).

If $p1$ is the position of a participle, then $form1 := indicat$.

The string $refl1$ represents a value of the property "reflexivity"

The string $voice1$ takes the value *active* (the sign of the active voice) or *passive* (the sign of the passive voice).

The values of the parameters $form1$, $refl1$, $voice1$ are calculated based on the set of the numeric codes of the values of the morphological characteristics of the text unit with the ordered number $p1$.

Specification of the algorithm "Range − of − sort"

Input:
z − sort, i.e., an element of the set $St\,(B\,(Cb\,(Lingb)))$, where $Lingb$ is a linguistic basis.
Output:
$spectrum$ − set of all sorts being the generalizations of the sort z including the sort z itself.

9.5.2.4 Algorithm "Find-set-relations-verb-noun"

Algorithm

Begin
        $Characteristics − of − verbal − form\,(posvb, form1, refl1, voice1)$
        $nrelvbdep := 0$
Comment
Now the preposition is being defined
End-of-comment
        $prep := leftprep$
Comment
Calculation of $posn1$ − position of the noun that defines the set of sorts of the text unit in the position $posdepword$

End-of-comment
$\quad\quad posn1 := posdepword$
Comment

Then the set of grammatic cases $Grcases$ is being formed. This set will be connected with the word in the position $posdepword$ in order to find a set of semantic relationships between the words in the positions $posvb$ and $posdepword$.

End-of-comment
$\quad\quad\quad t1 := Rc\,[posvb, tclass]$
$\quad\quad\quad t2 := Rc\,[posvb, subclass]$
$\quad\quad\quad p1 := Rc\,[posdepword, mcoord]$
$\quad\quad\quad Grcases := Cases\,(Rm\,[p1, morph])$
$\quad\quad\quad line1 := Matr\,[posn1, locunit]$
$\quad\quad\quad numb1 := Matr\,[posn1, nval]$
Comment

The quantity of the rows with the noun meanings in $Arls$

End-of-comment
$\quad\quad\quad loop\ for\ i1\ from\ line1\ to\ line1 + numb1 - 1$
Comment

A loop with the parameter being the ordered number of the row of the array $Arls$ corresponding to the noun in the position $posn1$

End-of-comment
$\quad\quad\quad Set1 := empty\ set$
$\quad\quad\quad loop\ for\ j\ from\ 1\ to\ m$
Comment

m – semantic dimension of the sort system $S\,(B\,(Cb\,(Lingb)))$, i.e., the maximal quantity of incomparable sorts that may characterize one essence

End-of-comment
$\quad\quad\quad current - sort := Arls\,[i1, st_j]$
$\quad\quad\quad if\ current - sort \neq nil$
$\quad\quad\quad then\ Range - of - sort\,(current - sort, spectrum)$
$\quad\quad\quad\quad Set1 := Set1 \cup spectrum$
$\quad\quad\quad end - if$
Comment

For an arbitrary sort z the value $spectrum$ is the set of all sorts being the generalizations of the sort z including the sort z itself

End-of-comment
$\quad\quad\quad end - of - loop$
Comment

End of the loop with the parameter j
End-of-comment
Comment

Then the loop with the parameter being a value of the verbal form follows
End-of-comment
$\quad\quad\quad line2 := Matr\,[posvb, locunit]$
$\quad\quad\quad numQ2 := Matr\,[posvb, nval]$

Comment

Quantity of rows with the values of the verbal form in *Arls*

End-of-comment

$$loop \; for \; i2 \; from \; line2 \; to \; line2 + numQ2 - 1$$

Comment

A loop with the parameter being the ordered number of the row of the array *Arls* corresponding to the verb in the position *posvb*

End-of-comment

$$current - pred := Arls\,[i2, sem]$$
$$loop \; for \; k1 \; from \; 1 \; to \; narvfr$$
$$if \; Arvfr\,[k1, semsit] = current - pred$$
$$then \; begin \; s1 := Arvfr\,[k1, str]$$
$$if \; ((prep = Arvfr\,[k1, sprep]) \; and \; (s1 \in Set1) \; and \; (form1 = Arvfr\,[k1, form]) \; and \; (refl1 = Arvfr\,[k1, refl]) \; and \; (voice1 = Arvfr\,[k1, voice]))$$
$$then \; grc := arvfr\,[k1, grcase]$$
$$if \; (grc \in Grcases)$$
$$then$$

Comment

The relationship exists

End-of-comment

$$nrelvbdep := nrelvbdep + 1$$
$$arrelvbdep\,[nrelvbdep, linevb] := i2$$
$$arrelvbdep\,[nrelvbdep, linenoun] := i1$$
$$arrelvbdep\,[nrelvbdep, gr] := grc$$
$$arrelvbdep\,[nrelvbdep, role] := arvfr\,[k1, trole]$$
$$end - if$$
$$end - if$$
$$end$$
$$end - if$$
$$end - of - loop$$
$$end - of - loop$$
$$end - of - loop$$

Comment

End of loops with the parameters *i*1, *i*2, *k*1

End-of-comment

$$end$$

9.5.2.5 Commentary on the Algorithm "Find-set-relations-verb-noun"

The quantity *nrelvbdep* of the semantic relationships between the verbal form and a noun depending on it in the considered sentence is found. Let's consider such sublanguages of English, German, and Russian languages that in all input texts a verb is always followed (at certain distance) by at least one noun.

The information about such combinations of the meanings of the verb V and the noun $N1$ that give at least one semantic relationship between V and $N1$ is represented in the auxiliary array *arrelvbdep* with the indices of the columns

linenoun, linevb, trole, example.

For arbitrary row of the array *arrelvbdep*, the column *linenoun* contains $c1$ – the number of such row of the array *Arls* that $Arls[c1, ord] = posn1$ (position of the noun $n1$).

For example, for Q1 = "When and where three aluminum containers with ceramics have been delivered from?" $Arls[c1, sem] = container$.

The column *linevb* contains $c2$ – the number of the row of the array *Arls* for which $Arls[c2, ord] = posvb$, i.e. the row $c2$ indicates a certain meaning of the verb V in the position *posvb*.

For example, for Q1 = "When and where three aluminum containers with ceramics have been delivered from?" the column $Arls[c2, sem] = delivery2$. The column *role* is designed to represent the possible semantic relationships between the verb V and the noun $N1$.

If *nrelvbdep* = 0 then the semantic relationships have not been found. Let's assume that this is not possible for the considered input language.

If *nrelvbdep* = 1 then the following meanings have been clearly defined: the meaning of the noun $N1$ (by the row $c1$), the meaning of the verb V (by the row $c2$), and the semantic relationship *arrelvbdep*[*nrelvbdep, role*].

For example, for the question Q1 the following relationships are true: $V =$ "*delivered*," $N1 =$ "*containers*," *nrelvbdep* = 1, *arrelvbdep*[*nrelvbdep, role*] = *Object*1.

If *nrelvbdep* > 1 then it is required to apply the procedure that addresses clarifying questions to the user and to form these questions based on the examples from the column *example*.

9.5.3 Description of the Algorithm Processing the Constructs

Purpose of the algorithm "Find − set − relations − verb − construct"

The algorithm is used for establishing a thematic role connecting a verbal form in the position *posvb* with a construct in the position *posdep*, taking into account a possible preposition before this construct. As a consequence, the algorithm selects one of the several possible values of a verbal form. In order to do this, two concentric loops are required: (1) with the parameter enumerating a possible meaning of a verbal form; (2) with the parameter enumerating a possible verbal-prepositional frame associated with this verbal form.

External specification of the algorithm

Input:

posvb – integer – position of a verbal form (a verb in indicative mood or imperative mood or in infinitive form);

posdep (abbreviation of "position of dependent word") – integer – position of a construct (an expression representing a numerical value of a parameter);

*subclass*1 – string – designation of a construct's sort;

Matr – MSSR of the text;

Arls – projection of the lexico-semantic dictionary on the input text;

Arvfr – projection of the dictionary of verbal-prepositional frames on the input text;

*prep*1 – string – preposition related to a construct or a blank preposition *nil*.

Output:

arrelvbdep – two-dimensional array designed to represent the information about (a) a meaning of a verbal form and (b) a semantic relationship between the verbal form in the position *posvb* and the construct in the position *posdep*;

nrelvbdep – integer – the quantity of meaningful rows in the array *arrelvbdep*.

Algorithm "Find − set − relation − verb − construct"

Begin

      $startrow := Matr[posvb, locunit] - 1$

      $row1 := startrow$

      $numbvalvb := Matr[posvb, nval]$

Comment

The quantity of possible meanings of a verbal form

End-of-comment

      loop − until

      $row1 := row1 + 1$

      $Current - pred := Arls[row1, sem]$

      $K1 := 0; log1 := false$

      loop − until

      $k1 := k1 + 1$

      if $(Arvfr[k1, semsit] = current - pred)$

      then if $(Arvfr[k1, str] = subclass1)$ *and* $(Arvfr[k1, sprep] = prep1)$

         then

Comment

A relationship exists

End-of-comment

          $nrelvbdep := 1$

          $arrelvbdep[1, linevb] := row1$

          $arrelvbdep[1, role] := Arvfr[k1, trole]$

          $log1 := true$

        end − if

$$end - if$$
$$Exit - when \ (log1 \ = \ true)$$
$$End - of - loop$$
$$Exit - when \ ((log1 \ = \ true) \ or \ (row1 \ = \ startrow \ + \ numbvalvb))$$
$$end - of - loop$$

9.5.4 Description of the Algorithm "Find-set-thematic-roles"

Purpose of the algorithm

The algorithm is used for finding a set of thematic roles connecting a verbal form in the position *posvb* with a word (noun, connective word, construct) in the position *posdep*, taking into account a possible preposition before this word. In order to do this, three concentric loops are required: (1) with the parameter corresponding to a possible meaning of the word in the position *posdep*; (2) with the parameter corresponding to a possible meaning of a verbal form; (3) with the parameter corresponding to a verbal-prepositional frame associated with this verbal form.

External specification

Input:

Rc – classifying representation of a text;

nt – integer – quantity of the text units in *Rc*, i.e., quantity of the rows in *Rc*;

Rm – morphological representation of the lexic units of *Rc*;

posvb – integer – position of a verbal form (verb in a personal or infinitive form);

posdep (abbreviation of "position of dependent word") – integer – position of a dependent word (noun, interrogative pronoun, construct – an expression denoting a numerical value of a parameter);

Matr – string-numeric matrix – initial MSSR of a text;

Arls – array – projection of the lexico-semantic dictionary *Lsdic* on the input text *T*;

Arvfr – array – projection of the verbal-prepositional frame dictionary *Vfr* on the input text *T*.

Output:

*class*1 – string – designation of a class of a text unit in the position *posdep*;

*subclass*1 – designation of a subclass of a text unit in the position *posdep*;

arrelvbdep – two-dimensional array designed to represent the information about (a) a meaning of a dependent word, (b) a meaning of a verbal form, and (c) a semantic relationship between the verbal form in the position *posvb* and the dependent word in the position *posdep*;

nrelvbdep – integer – the quantity of meaningful elements in the array *arrelvbdep*.

$$Algorithm\ "Find - set - thematic - roles"$$

Begin

Fill in with 0 all numerical positions of the array *arrelvbdep* and fill in with the void element *nil* all string positions of the array *arrelvbdep*

$$class1 := Rc\,[posdep, tclass]$$
$$subclass1 := Rc\,[posdep, subclass]$$
$$prep1 := Matr\,[posdep, prep]$$
$$if\ (class1 = noun)$$
$$then\ Find - set - relations - verb - noun$$
$$(Rc, Rm, posvb, posdep, prep1, Arls, Arvfr, Matr,$$
$$nrelvbdep, arrelvbdep)$$
$$end - if$$
$$if\ (class1 = constr)$$
$$then\ Find - set - relations - verb - construct$$
$$(posvb, posdep, prep1, Arls, Arvfr, Matr,$$
$$nrelvbdep, arrelvbdep)$$
$$end - if$$

end

9.5.5 An Algorithm of Searching for a Semantic Relationship Between a Verb and a Dependent Expression

$$Purpose\ of\ the\ algorithm\ "Semantic - relation - verbal - form"$$

The algorithm is used (a) for finding a thematic role connecting a verbal form in the position *posvb* with an expression (noun or construct) in the position *posdep*, taking into account a possible preposition before this expression. As a consequence, (b) for selecting one of several possible values of a verbal form and one of several possible values of a word in the position *posdep*; (c) for entering the obtained information about a value of a verbal form, a value of a dependent text unit and a semantic relationship (i.e. a thematic role) to the MSSR *Matr*.

$$External\ specification\ of\ the\ algorithm$$

Input:
Rc – classifying representation;

nt – integer – quantity of the text units in Rc, i.e., a quantity of the rows in Rc;

Rm – morphological representation of the lexic units of Rc;

$posvb$ – integer – position of a verbal form (a verb in a personal or infinitive form);

$posdep$ – integer – position of a noun or a construct;

$Arls$ – array – projection of the lexico-semantic dictionary $Lsdic$ on the input text T;

$Arvfr$ – array – projection of the dictionary of verbal-prepositional frames Vfr on the input text T;

$Matr$ – initial configuration of MSSR of a text.

Output:

$Matr$ – string-numeric matrix – transformed configuration of an initial matrix $Matr$.

External specification of the algorithm "Select – thematic – roles"

Input:

$posvb$ – integer – position of a verbal form;

$posdep$ – integer – position of a dependent text unit (noun or construct);

$arrelvbdep$ – two-dimensional array representing the information about the possible combinations of (a) a meaning of a verbal form in the position $posvb$, (b) a meaning of a dependent unit in the position $posdep$, and (c) a thematic role rel realized in this combination;

$nrelvbdep$ – integer – the quantity of meaningful rows in the array $arrelvbdep$, i.e., the quantity of possible semantic relationships between the considered verbal form and the unit depending on it.

Output:

$m1$ – integer – the ordered number of a certain meaningful row of the array $arrelvbdep$.

Parameter $m1$ takes a non-zero value as a result of processing the user's answer to the clarifying question of the linguistic processor. A user is asked to indicate what relationship among several semantic relationships is realized in the combination "a verbal form in the position $posvb$ + a dependent unit in the position $posdep$."

With the help of the column *example* a user is given the examples of combinations with the semantic relationships that could be potentially realized between the text units in the positions $posvb$ and $posdep$.

Algorithm "Semantic – relation – verbal – form"

Begin

Find – set – thematic – roles

($Rc, Rm, posvb, posdep, Arls, Arvfr, Matr, nrelvbdep, arrelvbdep$)

Comment

Find the quantity of the elements of the array *nrelvbdep* and form the array *arrelvbdep* describing the possible semantic relationships between a verbal form and a text unit depending on it.

End-of-comment

$$if\ nrelvbdep = 1\ then\ m1 := 1$$
$$else\ Select-thematic-roles$$
$$(posvb, posdep, nrelvbdep, arrelvbdep, m1)$$

Comment

*m*1 – number of row of the array *arrelvbdep* that gives a combination of a value of a verbal form, a value of a dependent text unit, and a semantic relationship between a verbal form in the position *posvb* and a text unit in the position *posdep* taking into account a preposition that may be related to the position *posdep*.

End-of-comment

$$end-if$$
$$rel1 := arrelvbdep\,[m1, role]$$
$$locvb := arrelvbdep\,[m1, linevb]$$

Comment

A row of *Arls*

End-of-comment

$$if\ (class1 = noun)\ then\ locnoun := arrelvbdep\,[m1, linenoun]$$

Comment

A row of *Arls*

End-of-comment

$$end-if$$

Comment

$\Rightarrow$ Entering the information to *Matr* (see a description of *Matr* in Chap. 8)

End-of-comment

$$Matr\,[posvb, posdir] := 0$$
$$Matr\,[posvb, locunit] := locvb$$
$$Matr\,[posvb, nval] := 1$$
$$if\ (class1 = noun)\ then\ Matr\,[posdep, locunit] := locnoun$$
$$end-if$$
$$Matr\,[posdep, nval] := 1$$
$$Matr\,[posdep1, posdir] := posvb$$
$$Matr\,[posdep, reldir] := rel1$$

End

Commentary on the algorithm "Semantic − relation − verbal − form"

If *nrelvbdep* > 1 then it is required to address the clarifying questions to a user with the help of the column *example* of the array *arrelvbdep*. It would allow finding the following data:

locnoun – the row of *Arls* indicating a meaning of a text unit in the position *posdep* if this text unit is a noun;

locvb – row of *Arls* indicating a meaning of a verbal form in the position *posvb*;
*rel*1 – semantic relationship (thematic role) between the text units in the positions *posvb* and *posdep*.

If *nrelvbdep* = 1 then *m*1 := 1. Therefore,

$$locvb = arrelvbn\,[1, linevb];$$

$$locnoun = arrelvbn\,[1, linenoun];$$

$$rel1 = arrelvbn\,[1, role].$$

Then the obtained information is being entered to the matrix *Matr* :

$$Matr\,[posdep, locunit] := locnoun;$$

$$Matr\,[posdep, posdir] := posvb;$$

$$Matr\,[posndep, reldir] := rel1.$$

As a result of the conducted analysis, the meanings of both a verbal form in the position *posvb* and a noun in the position *posdep* are unambiguously defined. Therefore,

$$Matr\,[posvb, nval] := 1,$$

$$Matr\,[posdep, nval] := 1.$$

9.5.6 Final Part of a Description of an Algorithm of Processing Verbs

External specification

of the algorithm "Processing – verbal – form"

Input:
Rc – classifying representation;
nt – integer – the quantity of the text units in *Rc*, i.e. a quantity of the rows in *Rc*;
Rm – morphological representation of the lexic units of *Rc*;
Arls – array – projection of the lexico-semantic dictionary *Lsdic* on the input text *T*;

Arvfr – array – projection of the dictionary of verbal – prepositional frames *Vfr* on the input text *T*;

Matr – initial configuration of a matrix semantic-syntactic representation (MSSR) of the considered text;

pos – integer – the position of the considered verbal form;

posqswd [1 : *nt*] – one-dimensional array storing the positions of interrogative words;

pos – *free* – *dep* [1 : *nt*] – one-dimensional array storing the positions of the text units without discovered governing lexical units;

numb – *free* – *dep* – integer – the quantity of lexical units without discovered governing lexical units;

nmasters [1 : *nt*] – array representing the quantity of governing words for each text unit;

numbqswd – integer – the quantity of found interrogative words;

numbsit – integer – the quantity of situations mentioned in the analyzed text;

class – string representing the part of speech of the verbal form in the position *pos*.

Output:

Transformed configuration of MSSR *Matr.*

Algorithm "Processing – verbal – form"

Begin

     nsit := *nsit* + 1

Comment

nsit - quantity of the situations already mentioned in a text

End-of-comment

     $Matr[pos, mark] := Var('e', nsit)$

     verbpos := *pos*

     if ((*class* = *verb*) *and* (*numbqswd* > 0))

     then

Comment

The governing arrows with the marks of the semantic relationships (thematic roles) are being drawn from the position of the verb in the main clause to the positions of the interrogative words

End-of-comment

     loop for k1 from 1 *to numbqswd*

     $p1 := posqswd[k1]$

     $Matr[p1, posdir, 1] := pos$

     end – *of* – *loop*

     numbqswd := 0

     end – *if*

     if numb – *free* – *dep* > 0

Comment

There are free text units, i.e., such units for which a semantic-syntactic governance by another unit is not yet found

End-of-comment

     then loop for m1 from 1 *to numb* – *free* – *dep*

$$Semantic - relation - verbal - form$$
$$(Rc, nt, Rm, pos, pos - free - dep\,[m1],$$
$$Arls, Arvfr, Matr)$$
$$end - of - loop$$
$$\quad end - if$$
$$end$$

Example. Let Qs1 be the following marked-up representation of the question "When (1) and (2) where (3) the next (4) international (5) scientific (6) conference (7) 'COLING' (8) will be held (9) ? (10)"

If $pos = 9$ then the marked arrows from the position 9 to the positions 1 and 3 will be drawn (in the loop with the parameter $k1$ taking the values from 1 to *numbqswd*).

9.6 Processing of Adjectives, Prepositions, Cardinal Numerals, Names and Nouns

9.6.1 Processing of Adjectives

Description of the algorithm "Processing – adjective"

External specification

Input:
pos – integer – ordered number of a text unit being an adjective;
Matr – MSSR of a text;
Arls – array – projection of the lexico-semantic dictionary *Lsdic* on the input text
T.

Output:
nattr – integer – the quantity of consecutive adjectives;
Attributes – a two-dimensional array with the columns

place, property,

where for every row k, *Attributes*[k, *place*] is the position of an adjective in the classifying representation Rc, and *Attributes*[k, *property*] is a compound semantic item corresponding to this adjective and taken from the array *Arls* – the projection of the lexico-semantic dictionary *Ldic* on the input text.

Example 1. Let Q1 = "Where (1) two (2) green (3) aluminum (4) containers (5) came (6) from (7) ? (8)" Then *nattr* := 2 and the array *Attributes* has the following structure:

place	property
3	$Color\,(z1,\,green)$
4	$Material\,(z1,\,aluminum)$
5	nil

Comment

In the end of the algorithm *Processing − noun* the following operations, in particular, are performed:

- $nattr := 0$;
- the column *place* is being set to 0;
- the column *property* is being filled in with the string *nil* designating the void (empty) string.

$$Algorithm\ ``Processing-adjective''$$

Begin
$\quad\quad nattr := nattr + 1$
$\quad\quad k1 := Matr\,[pos,\,locunit]$
$\quad\quad semprop := Arls\,[k1,\,sem]$
$\quad\quad Attributes\,[nattr,\,place] := pos$
$\quad\quad Attributes\,[nattr,\,prop] := semprop$
end

9.6.2 Processing of Prepositions, Cardinal Numbers, and Names

The algorithms "*Processing − preposition*" and "*Processing − cardinal − numeral*" are very simple. The first algorithm is designed to remember a preposition in the considered position *pos* with the help of the variable *leftprep* ("preposition on the left"). The second algorithm transforms a lexical unit belonging to the class of cardinal numerals to the integer represented by this lexical unit.

For example, the words "three" and "twenty three" correspond to the numbers 3 and 23. The variable *leftnumber* ("number on the left") is designed to remember a number. Input parameters of these algorithms are the classifying representation of a text *Rc* and the variable *pos* (number of a row in *Rc*).

$$Algorithm\ ``Processing-preposition''$$

Begin
$\quad\quad leftprep := Rc\,[pos,\,unit]$
end

$$Algorithm\ ``Processing-cardinal-number''$$

Begin

$\qquad$ *Leftnumber* := *Number* (*Rc* [*pos, unit*])

end

Description of the algorithm "Processing − name"

External specification of the algorithm

Input:

Rc − classifying representation of a text;

pos − position of an expression in inverted commas or apostrophes;

Matr − MSSR of a text

Output:

Transformed value of *Matr*.

Algorithm

Begin

$\qquad$ *Matr* [*pos, posdir,* 1] := *pos* − 1

$\qquad$ *Matr* [*pos, reldir,* 1] := '*Name*'

Comment

Meaning of these operations: a governing arrow with the mark "Name" is being drawn to an expression in inverted commas or apostrophes from a noun standing to the left from this expression.

End-of-comment

end

9.6.3 An Algorithm Searching for Possible Semantic Connections Between Two Nouns with Respect to a Preposition

Purpose of the algorithm

"Find − set − relations − noun1 − noun2"

Algorithm *"Find − set − relations − noun1 − noun2"* ("Find a set of semantic relationships between the noun 1 and the noun 2") allows to establish the semantic relationships that could exist between the noun in the position *posn*1 (going forward referred to as *noun*1) and the noun in the position *posn*2 (going forward referred to as *noun*2) under the condition that a certain preposition in the position between *posn*1 and *posn*2 is related to the second noon.

To do this, three loops are required: (1) with the parameter being a possible value of the word in the position *posn*1, (2) with the parameter being a possible value of the word in the position *posn2*, (3) with the parameter being a prepositional frame connected with the considered preposition.

External specification of the algorithm

"Find − set − relations − noun1 − noun2"

Input:

Rc − classifying representation;

nt − integer − the quantity of the text units in the classifying representation *R1*, i.e., the quantity of rows in *Rc*;

Rm − morphological representation of the lexical units of *R1*;

Posn1 − integer − position of the first noun;

Posn2 − integer − position of the second noun;

Matr − MSSR of a text;

Arls − array − projection of the lexico-semantic dictionary *Lsdic* on the input text *T*;

Arfrp − array − projection of the dictionary of prepositional frames *Frp* on the input text *T*.

Output:

arrelvbdep − two-dimensional array designed to represent the information about a meaning of the first noun, a meaning of the second noun, and a semantic relationship between the word in the position *posn1* and the dependent word in the position *posn2*; this array contains the columns *locn1*, *locn2*, *relname*, *example*;

nreln1n2 − integer − the quantity of meaningful rows in the array *arrelvbdep*.

Algorithm "Find − set − relations − noun1 − noun2"

Begin
$\qquad$ *nreln1n2* := 0
Comment
A preposition is being defined
End-of-comment
$\qquad$ *prep1* := *Matr* [*posn2, prep*]
Comment
A set of grammatic cases is being defined
End-of-comment
$\qquad$ *p1* := *Rc* [*posn2, mcoord*]
$\qquad$ *Grcases* := *Cases* (*Rm* [*p1, morph*])
$\qquad$ *line1* := *Matr* [*posn1, locunit*]
$\qquad$ *numb1* := *Matr* [*posn1, nval*]
Comment
The quantity of the rows with the meanings of Noun 1 in *Arls*
End-of-comment
$\qquad$ *loop for n1 from line1 to line1* + *numb1* − 1
Comment

A loop with the parameter being the row of the array *Arls* corresponding to a noun in the position *posn*1

End-of-comment

> *Set*1 := *empty set*
> *loop for j from* 1 *to m*

Comment

m – semantic dimension of the sort system $S(B(Cb(Lingb)))$, i.e., the maximal quantity of incomparable sorts that may characterize one entity.

End-of-comment

> *current* − *sort* := *Arls* [*n*1, *st*$_j$]
> *if current* − *sort* ≠ *nil*
> *then Range* − *of* − *sort* (*current* − *sort, spectrum*)
>   *Set*1 := *Set*1 ∪ *spectrum*
> *end* − *if*

Comment

For an arbitrary sort *z spectrum* (*z*) is the set of all sorts being the generalizations of the sort *z* including the sort *z* itself.

End-of-comment

> *end* − *of* − *loop*

Comment

End of the loop with the parameter *j*

End-of-comment

Comment

Example:

If u = *dyn.phis.ob*
then spectrum (*u*) = {*dyn.phys.ob, phys.ob, space.ob*}

End-of-comment

> *line*2 := *Matr* [*posn*2, *locunit*]
> *numb*2 := *Matr* [*posn*2, *nval*]

Comment

The quantity of the rows of *Arls* with the meanings of *noun*2

End-of-comment

> *loop for n*2 *from line*2 *to line*2 + *numb*2 − 1

Comment

A loop with the parameter being the number of a row of the array *Arls* corresponding to the noun in the position *posn*2

End-of-comment

> *Set*2 := *empty set*
> *loop for q from* 1 *to m*

Comment

m – semantic dimension of the sort system $S(B(Cb(Lingb)))$, i.e. the maximal quantity of incomparable sorts that may characterize one essence.

End-of-comment

> *current* − *sort* := *Arls* [*n*2, *st*$_q$]
> *if current* − *sort* ≠ *nil*

$$then\ Range - of - sort\ (current - sort,\ spectrum)$$
$$Set2 := Set2 \cup spectrum$$
$$end - if$$
$$end - of - loop$$

Comment

End of the loop with the parameter q

End-of-comment

$$loop\ for\ k1\ from\ 1\ to\ narfrp$$

Comment

Quantity of rows in the array $Arfrp$ – projection of the dictionary of prepositional frames Frp on the input text

End-of-comment

$$if\ Arfrp\,[k1,\ prep] = prep1$$

Comment

The required preposition is found

End-of-comment

$$then\ begin\ s1 := Arfrp\,[k1,\ sr1]$$
$$s2 := Arfrp\,[k1,\ sr2]$$
$$if\ (s1 \in Set1)\ and\ (s2 \in Set2)$$
$$then\ if\ Arfrp\,[k1,\ grc] \in Grcases$$
$$then$$

Comment

Relationship exists

End-of-comment

$$nreln1n2 := nreln1n2 + 1$$
$$arreln1n2\,[nreln1n2,\ locn1] := n1$$
$$arreln1n2\,[nreln1n2,\ locn2] := n2$$
$$arreln1n2\,[nreln1n2,\ relname] := arfrp\,[k1,\ rel]$$
$$end - if$$
$$end - if$$
$$end$$
$$end - if$$
$$end - of - loop$$
$$end - of - loop$$
$$end - of - loop$$

Comment

End of loops with the parameters $n1$, $n2$, $k1$

End-of-comment

$$end$$

$$Commentary\ on\ the\ algorithm$$

$$\text{``}Find - set - relations - noun1 - noun2\text{''}$$

The quantity $nreln1n2$ of semantic relationships between the nouns in the positions $posn1$ and $posn2$ is found. The information about such combinations of the meanings of the first and second nouns that give at least one semantic relationship between the elements in the positions $posn1$ and $posn2$ is represented in the auxiliary array $arreln1n2$ with the indices of the columns

$$locn1, \ locn2, \ relname, \ example.$$

For instance, the following relationships may take place for a certain row k :

$$arreln1n2[k, locn1] = p1,$$

$$arreln1n2[k, locn2] = p2,$$

$$arreln1n2[k, relname] = Against2,$$

$$arreln1n2[k, example] = \text{``}a \ remedy \ for \ asthma.\text{''}$$

The interpretation of the columns is as follows:

- The column $locn1$ contains $p1$ – the ordered number of a row of the array $Arls$ defining a possible meaning of the noun in the position $posn1$.
- The column $locn2$ contains $p2$ – the ordered number of a row of the array $Arls$ defining a possible meaning of the noun in the position $posn2$.
- The column $relname$ is intended to represent the possible relationships between the nouns in the positions $posn1$ and $posn2$.

If $nreln1n2 = 0$ then semantic relationships are not found. Let's assume that this is impossible for the considered input language.

If $nreln1n2 = 1$ then the following meanings have been found: a meaning of the noun in the position $posn1$ (in the row $p1$), a meaning of the noun in the position $posn2$ (in the row $p2$), and a meaning of the semantic relationship $arreln1n2 [nreln1n2, relname]$.

If $nreln1n2 > 1$ then it is required to apply a procedure that addresses clarifying questions to a user and to form these questions on the basis of the examples from the column $example$.

9.7 Finding the Connections of a Noun with Other Text Units

Plan of the algorithm "Processing – noun"

Begin

Entering to *Matr* the information about a number (or cardinal numeral) and adjectives possibly standing to the left by applying the algorithm "Recording-attributes"

Generating a mark of an element and the type of a mark (applying the algorithm "Generating-mark")

If Rc [pos + 1, tclass] = proper − noun then Processing − proper − noun
end − if

Searching a semantic dependence from the noun being closest to the left and governed by the verb in the position *verbpos*

If there is no such dependence

Then in case when *verbpos* ≠ 0 searching a semantic dependence from a verbal form in the position *verbpos*

else (i.e. in case when *verbpos* = 0) the position number *pos* is being entered to the array *pos − free − dep* containing free text units.

end

External specification

of the algorithm "Processing − noun"

Input:

Rc − classifying representation;

nt − integer − quantity of text units in the classifying representation *Rc*, i.e., the quantity of rows in *Rc*;

Rm − morphological representation of the lexical units of *Rc*;

pos − integer − position of a noun;

Matr − initial value of MSSR of a text;

Arls − array − projection of the lexico-semantic dictionary *Lsdic* on the input text *T*;

Arvfr − array − projection of the dictionary of verbal − prepositional frames *Vfr* on the input text *T*;

Arfrp − array − projection of the dictionary of prepositional frames *Frp* on the input text *T*.

Output:

pos − integer − position of a text unit;

Matr − transformed value of the initial matrix *Matr*.

External specifications of auxiliary algorithms

Specification of the algorithm "Find − noun − left"

Input:

pos − integer − position of a noun.

Output:

posleftnoun − integer − position of the noun satisfying two conditions: (a) it is closest from the left to the position *pos*, (b) it may serve as a governing word for the noun in the position *pos* (see below the subsection "Description of the auxiliary algorithms").

Specification of the algorithm

"Processing — proper — noun"

Input:

pos – integer – position of a common noun or a proper noun followed by at least one proper noun;

Arls – projection of the lexico-semantic dictionary *Lsdic* on the input text;

Matr – initial configuration of MSSR of a text.

Output:

Matr – transformed configuration of MSSR of a text (see below the subsection "Description of the auxiliary algorithms").

Specification of the algorithm "Processing — name"

Input:

pos – position of a common noun followed by an expression in inverted commas or apostrophes;

Matr – initial configuration of MSSR of a text.

Output:

Matr – transformed value of MSSR of a text.

Specification of the algorithm "Find — set — thematic — roles"

The specification and the algorithm itself are described above.

Specification of the algorithm

"Semantic — relation — verbal — form"

The specification and the algorithm itself are described above.

Specification of the algorithm

"Find — set — relations — noun1 — noun2"

The specification and the algorithm itself are described above.

Specification of the algorithm

"Select — governance — verb — noun"

Input:

pos – integer – position of a text unit;

posvb – integer – position of a verbal form;

posleftnoun – integer – position of a noun on the left;

prep – string – value of a preposition related to the position *pos*.

Output:

res – string – takes the value 1 or 2 as a result of a clarifying dialogue with a user;

if a noun in the position *pos* directly depends on a verbal form in the position *posvb*, then *res* := 1;

if a noun in the position *pos* (taking into account a preposition) directly depends on the noun standing to the left in the position *posleftnoun*, then *res* := 2.

Specification of the algorithm

"Select − relation − between − nouns"

Input:

posleftnoun – integer – position of the noun 1;

pos – integer – position of the noun 2 standing to the right from the noun 1;

prep – string – value of a preposition (it could also be the void (empty) preposition *nil*) related to the noun 2;

arreln1n2 – two-dimensional array representing the information about possible combinations of the meanings of Noun 1 and Noun 2 and a semantic relationship between them, taking into account the preposition *prep*;

nreln1n2 – integer – the quantity of meaningful rows in the array *arreln1n2*, i.e., the quantity of possible semantic relationships between the considered nouns.

Output:

m2 – integer – ordered number of a certain meaningful row of the array *arreln1n2*.

The parameter *m2* takes a non-zero value as a result of processing the user's answer to the clarifying question of the linguistic processor. A user is asked to indicate which of the several semantic relationships is realized in the combination

"Noun 1 in the position *posleftnoun* + Dependent Noun 2 in the position *pos*" taking into account the preposition *prep*.

With the help of the column *example* a user is given the examples of combinations with the semantic relationships that could be potentially realized between the text units in the positions *posleftnoun* and *pos*.

Specification of the algorithm

"Select − thematic − roles"

External specification of this algorithm is described above.

Algorithm "Processing − noun"

Begin

    if leftnumber > 0 *then Matr* [*pos, qt*] := *leftnumber*
    end − if
    if nattr > 0
    then loop for m from 1 *to nattr*
        *p*1 := *Attributes* [*m, place*]
        Matr [*p*1, *posdir*, 1] := *pos*
        Semprop := *Attributes* [*m, prop*]

$Matr[p1, reldir, 1] := semprop$
$end - of - loop$
$end - if$
$leftnumber := 0$
$nattr := 0$
$Matr[pos, prep] := leftprep$
$leftprep := nil$
$Linenoun := Matr[pos, locunit]$

Comment
Number of the row in *Arls* containing an initial value of a noun
End-of-comment

$Sort1 := Arls[linenoun, st1]$
$if\ Sort1 \neq sit$

Comment
Situation
End-of-comment

$then\ numbent := numbent + 1$

Comment
Quantity of entities mentioned in the looked-through part of a text
End-of-comment

$end - if$
$gramnumber := Number(Rc[pos, mcoord])$
$if\ gramnumber = 1$
$then\ var1 := Varstring('x', numbent)$
$end - if$
$if\ (gramnumber = 2)\ or\ (gramnumber = 3)$
$then\ var1 := Varstring('S', numbent)$
$end - if$
$Matr[pos, mark] := var1$
$Find - noun - left(pos, posleftnoun)$
$if\ posleftnoun = 0$

Comment
To the left from the position *pos* there are no nouns that may govern a noun in
the position *pos*
End-of-comment

$then\ if\ verbpos = 0$
$then\ numb - free - dep := numb - free - dep + 1$
$K1 := numb - free - dep$
$pos - free - dep[k1] := pos$
$else\ posvb := verbpos$
$Semantic - relation - verbal - form(posvb, pos, Matr)$
$else$

Comment
In the case when *posleftnoun* > 0
End-of-comment

$Find - set - relations - noun1 - noun2$
$(posleftnoun, Matr[pos, prep], pos, Matr, nreln1n2, arreln1n2)$
Comment
Possible semantic connections (and their quantity) between the considered noun
in the position pos and the closest from the left noun in the position $postleftnoun$
are being found
End-of-comment
$\quad if\ (nreln1n2 = 0)$
Comment
There is no semantic-syntactic governance by the preceding noun
End-of-comment
$\quad then\ posvb := verbpos$
$\quad\quad if\ posvb > 0$
$\quad\quad then\ Semantic - relation - verbal - form\,(posvb, pos, Matr)$
$\quad\quad else$
Comment
In the case when $posvb = 0$
End-of-comment
$\quad\quad\quad numb - free - dep := numb - free - dep + 1$
$\quad\quad\quad K1 := numb - free - dep$
$\quad\quad\quad pos - free - dep\,[k1] := pos$
$\quad\quad end - if$
$\quad end - if$
Comment
The case when $nreln1n2 = 0$ is considered
End-of-comment
$\quad if\ (nreln1n2 > 0)$
Comment
There is an opportunity of semantic-syntactic governance by the preceding noun
End-of-comment
$\quad then\ posvb := verbpos$
$\quad\quad if\ posvb > 0$
$\quad\quad then\ Find - set - thematic - roles$
$(posvb, pos, class1, subclass1, Matr, nrelvbdep, arrelvbdep)$
$\quad\quad\quad if\ (nrelvbdep = 0)$
Comment
There is no semantic connection with a verbal form
End-of-comment
$\quad\quad\quad then\ if\ (nreln1n2 = 1)$
$\quad\quad\quad\quad then\ m2 := 1$
Comment
$m2$ – number of an element of the array $arreln1n2$ which provides the informa-
tion about the connection between $posleftnoun$ and pos for $Matr$
End-of-comment
$\quad\quad\quad\quad else\ Select - relation - between - nouns$

$$(posleftnoun, Mate[pos, prep], pos, nreln1n2, arreln1n2, m2)$$
$$end - if$$

Adding the information about a connection between the text units in the positions *posleftnoun* and *pos* to *Matr*; this information is taken from the position *m2* of the array *arreln1n2*;

$$end - if$$

Comment
A case when $nrelvbdep = 0$
End-of-comment

$$if \ (nrelvbdep > 0)$$

Comment
A connection with a verb is possible
End-of-comment

$$then \ if \ (nreln1n2 > 0)$$

Comment
A connection with a preceding noun is also possible
End-of-comment

$$then \ Select - governance - verb - noun$$
$$(posvb, Matr[pos, prep], posleftnoun, pos, res)$$

Comment
$res = 1 \Rightarrow$ a connection with a verb;
$res = 2 \Rightarrow$ a connection with a noun in the position *posn*1
End-of-comment

$$if \ (res = 1)$$
$$then \ if \ (nrelvbdep = 1) \ then \ m1 := 1$$
$$else \ Select - thematic - roles$$
$$(posvb, Matr[pos, prep], pos, nrelvbdep, arrelvbdep, m1)$$
$$end - if$$

Comment
Recording an information about a connection between a verbal form in the position *posvb* and a noun in the position *pos* to *Matr*; this information is taken from the row *m1* of the array *arrelvbdep*
End-of-comment

$$nmasters[pos] := nmasters[pos] + 1$$

Comment
A new governing arrow to the position *pos* is found
End-of-comment

$$d := nmasters[pos]$$
$$Matr[pos, posdir, d] := posvb$$
$$Matr[pos, reldir, d] := arrelvbdep[m1, role]$$
$$Matr[posvb, locunit] := arrelvbdep[m1, linevb]$$
$$Matr[posvb, nval] := 1$$
$$Matr[pos, locunit] := arrelvbdep[m1, linenoun]$$
$$Matr[pos, nval] := 1$$
$$end - if$$
$$if \ (res = 2)$$

Comment

There is no connection with a verbal form but there is a connection with the noun in the position *posleftnoun*

End-of-comment

$$\text{then } if \ (nreln1n2 = 1) \ then \ m2 := 1$$
$$else$$
$$Select - relation - between - nouns$$
$$(posleftnoun, prep, pos, nreln1n2, arreln1n2, m2)$$
$$end - if$$

Comment

Recording the information about a semantic connection between the nouns in the positions *posleftnoun* and *pos* to *Matr* taking into account the preposition *prep* (which could also be a blank preposition *nil*); this information is taken from the row *m2* of the array *arreln1n2*

End-of-comment

$$Matr \ [posleftnoun, locunit] := arreln1n2 \ [m2, locn1]$$
$$Matr \ [posleftnoun, nval] := 1$$
$$Matr \ [pos, locunit] := arreln1n2 \ [m2, locn2]$$
$$Matr \ [pos, nval] := 1$$
$$Matr \ [pos, posdir, 1] := posleftnoun$$
$$Matr \ [pos, reldir, 1] := arreln1n2 \ [m2, role]$$
$$end - if$$
$$end - if$$
$$end - if$$
$$end - if$$
$$end - if$$
$$end - if$$
$$if \ Rc \ [pos + 1, subclass] = proper - noun$$

then logname :=(words in the positions *pos* and *pos* + 1 may be connected with the same grammatical case) and (semantic units corresponding to these words in the array *Arls* have the same set of sorts in *Arls*);

$$if \ logname = True \ then \ Processing - proper - noun \ (pos)$$
$$end - if$$
$$end - if$$
$$if \ Rc \ [pos + 1, subclass] = Name \ then \ Processing - name \ (pos)$$
$$end - if$$
$$leftprep := nil$$
$$leftnumber := 0$$
$$nattr := 0$$

Set to nil the column *place* of the array *Attributes*;

Fill in with the string *nil* – the designation of the void string – the column *prop* of the array *Attributes*.

End

Comment

End of the algorithm "*Processing – noun*"

End-of-comment

Description of auxiliary algorithms
Description of the algorithm *"Find − noun − left"*
External specification (see above)
Algorithm
Begin

> $posleftnoun := 0$
> $p1 := pos$
> $loop − until\ p1 := p1 − 1$
> $classleft := Rc\,[p1, tclass]$
> $if\ classleft = noun\ then\ posleftnoun := p1$
> $end − if$
>> $exit − when\ (p1 = 1)\ or\ (posleftnoun > 0)\ or\ (classleft \in \{verb,$

participle, adverb, pronoun, construct, marker})

> $end − of − loop$

end

Example 1. Let Q1 = "How many containers with Indian ceramics came from Novorossiysk?"

Let's transform the question Q1 into the following marked-up representation: "How many (1) containers (2) with (3) Indian (4) ceramics (5) came (6) from (7) Novorossiysk (8) ? (9)" Let *pos* = 5 (position of the word "ceramics"). Then after the termination of the algorithm, *posleftnoun* = 2 (the position of the word "containers").

Description of the algorithm *"Processing − proper − noun"*
External specification (see above)
Algorithm
Begin

> $k1 := pos + 1$
> $while\ Rc\,[k1, tclass] = proper − noun\ loop$
> $m1 := Matr\,[k1, locunit]$

Comment
The first and the only row of the array *Arls* containing the information about the unit $Rc\,[k1, unit]$ is found
End-of-comment

> $Matr\,[k1, posdir, 1] := pos$

Comment
A governing arrow is drawn from the element in the position *pos* to the element in the position $k1$
End-of-comment

> $sem1 := Arls\,[m1, sem]$
> $Matr\,[k1, reldir, 1] := sem1$
> $k1 := k1 + 1$
> $end − of − loop$
> $pos := k1 − 1$

end

Example 2. Let Q2 = "How many articles by professor Igor Pavlovich Nosov were published in 2008?" Then, as a result of applying the algorithm "*Processing – proper – noun*" with the parameter *pos* = 4 (position of the word "professor"), the governing arrows will be drawn (by transforming MSSR *Matr*) from the position *pos* to the positions *pos* + 1, *pos* + 2, *pos* + 3 corresponding to the fragment "Igor Pavlovich Nosov."

9.8 Final Step of Developing an Algorithm Building a Matrix Semantic-Syntactic Representation of the Input Text

9.8.1 Description of the Head Module of the Algorithm

In order to facilitate the understanding of the head module of the algorithm building an MSSR of the input NL-text, its external specification (developed in the Sect. 9.2) is given below.

External specification of the algorithm BuildMatr1

Input:
Lingb – linguistic basis;
T – input text;
lang – string variable with the values *English, German, Russian*.
Output:
nt – integer – the quantity of elementary meaningful text units;
Rc – classifying representation of the input text;
Rm – morphological representation of the input text;
Arls – two-dimensional array – projection of the lexico-semantic dictionary *Lsdic* on the input text *T*;
Arvfr – two-dimensional array – projection of the dictionary of verbal – prepositional frames *Vfr* on the input text *T*;
Arfrp – two-dimensional array – projection of the dictionary of prepositional semantic-syntactic frames *Frp* on the input text *T*;
Matr – matrix semantic-syntactic representation (MSSR) of the input text;
numbqswd – the variable representing the quantity of interrogative words in a sentence;
pos – free – dep – one-dimensional array of integers;
posvb, numb – free – dep – integers;
nmasters [1 : *nt*] – one-dimensional array representing the quantity of governing words for each text unit.

9.8.2 *External Specifications of Auxiliary Algorithms*

Specification of the algorithm

"Build − compon − morphol − representation"

Input:

Lingb − linguistic basis;

T − NL-text.

Output:

Rc − classifying representation of the text *T*;

nt − integer − the quantity of elementary meaningful text units in the input text *T*, i.e. the quantity of meaningful rows in the classifying representation *Rc*;

Rm − morphological representation of the text *T*.

Specification of the algorithm

"Build − projection − lexico − semantic − dictionary"

Input:

Rc, nt, Rm;

Lsdic − lexico-semantic dictionary.

Output:

Arls − two-dimensional array − projection of the dictionary *Lsdic* on the input text .

Specification of the algorithm

"Build − projection − verbal − frames − dictionary"

Input:

Rc, nt, Rm, Arls;

Vfr − dictionary of verbal − prepositional semantic-syntactic frames.

Output:

Arvfr − two-dimensional array − projection of the verbal-prepositional frame dictionary *Vfr* on the input text .

Specification of the algorithm

"Build − projection − prepositional − frames − dictionary"

Input:

Rc, nt, Rm, Arls;

Frp − dictionary of prepositional semantic-syntactic frames.

Output:
$Arfrp$ – two-dimensional array – projection of the dictionary of prepositional frames Frp on the input text .

9.8.3 Algorithm of Building an MSSR of the Input Text

Algorithm BuildMatr

Begin
    Build − compon − morphol − representation
(T, Rc, nt, Rm)
    Build − projection − lexico − semantic − dictionary
(Rc, nt, Rm, Lsdic, Arls)
    Build − projection − verbal − frames − dictionary
(Rc, nt, Rm, Arls, Vfr, Arvfr)
    Build − projection − prepositional − frames − dictionary
(Rc, nt, Rm, Arls, Frp, Arfp)
    Forming − initial − values − of − data
    Defining − form − of − text
(nt, Rc, Rm, leftprep, mainpos, kindtext, pos)
       loop − until
       $pos := pos + 1$
       class := Rc [pos, tclass]
       case class of
       preposition : Processing − preposition (Rc, pos, leftprep)
       adjective : Processing − adjective (Rc, pos, nattr, Attributes)
       cardinal − numeral : Processing − cardinal − numeral (Rc, pos, numb)
       noun : Processing − noun
(Rc, Rm, pos, Arls, Arfrp, Matr, leftprep, numb, nattr, Attributes)
       verb : Processing − verbal − form
(Rc, Rm, pos, Arls, Rqs, Arvfr, Matr, leftprep)
       conjunction : Empty operator
       construct : Processing − construct
       name : Processing − name
       marker : Empty operator
       end − case − of
       exit − when (pos = nt)
end

Thus, in this and previous sections of this chapter the algorithm *BuildMatr*1 has been developed. This algorithm finds (a) semantic relationships between the units of NL-text, (b) specific meanings of verbal forms and nouns from the text. This information is represented in the string-numerical matrix *Matr*.

The texts processed by the algorithm may express the statements (the descriptions of various situations), questions of many kinds, and commands. The texts may contain the verbs (in infinitive form, indicative or imperative mood), nouns, adjectives, numerical values of the parameters (constructs), cardinal numerals, digital representations of the numbers, interrogative words (being pronouns or adverbs), and the expressions in inverted commas or apostrophes serving as the names of various objects.

The algorithm *BuldMatr*1 is original and is oriented at directly discovering semantic relationships in NL-texts. It should be underlined that the algorithm *BuldMatr*1 is multilinguial: it carries out a semantic–syntactic analysis of texts from the sublanguages of English, German, and Russian languages being of practical interest.

Chapter 10
An Algorithm of Semantic Representation Assembly

Abstract An algorithm transforming a matrix semantic-syntactic representation *Matr* of a natural language text into a formal expression *Semrepr* $\in Ls(B)$, where *B* is the conceptual basis being the first component of the used marked-up conceptual basis *Cb*, and *Ls(B)* is the SK-language in the basis *B*, was called above *an algorithm of semantic assembly*. This chapter describes (a) an algorithm of semantic assembly *BuildSem*1, (b) an algorithm of semantic-syntactic analysis *SemSynt*1 being the composition of the algorithms *BuildMatr*1 and *BuildSem*1.

10.1 Initial Step of Building Semantic Representations of Input Texts

Let's consider the algorithm "*Preparation − to − constr − SemRepr*" – the first part of the algorithm of semantic assembly described in this chapter. The algorithm "*Preparation − to − constr − SemRepr*" delivers an initial value of the semantic representation (SR) of the input text; it is an initial value of the string *Semrepr* ("Semantic representation"). The form of this string depends on the form of the input text, i.e., on the value of the variable *kindtext* formed by the algorithm *BuildMatr*1.

The choice of the form of a semantic representation of the input text depends on the value of the string variable *kindtext*. To simplify the form of SR of the input texts, the existential quantifiers (when they are needed according to the approach described in Chap. 4) are not included in the formula but are implied.

It should be mentioned that the input language of the algorithm *BuildSem*1 is considerably broader than the output language of the algorithm *BuildMatr*1. The reason is that the algorithm *BuildSem*1 can deal with matrix semantic − syntactic representations corresponding to the NL-texts with participle constructions and attributive clauses. This possibility corresponds to the input language (a sublanguage of the Russian language) of the algorithm *SemSyn* developed by the author and published in [85].

V.A. Fomichov, *Semantics-Oriented Natural Language Processing*, IFSR International Series on Systems Science and Engineering 27, DOI 10.1007/978-0-387-72926-8_10, © Springer Science+Business Media, LLC 2010

10.1.1 Description of the Algorithm
"Preparation-to-constr-SemRepr"

$Externalspecification$

Input:

Rc – array – classifying representation of the input text;

Rm – array – morphological representation of the input text;

kindtext – string characterizing the form of the input text (the possible values of this string are *Stat, Imp, Genqs, Specqs.relat*1, *Specqs – relat*2, *Specqs.role*, *Specqs – quant*1, *Specqs – quant*2 (see Sect. 9.3);

mainpos – integer – the position of the interrogative word in the beginning of the text;

$Matr$ – MSSR of the text.

Output:

Semrepr – string – an initial value of the semantic representation of the input text.

$$Algorithm "Preparation - to - constr - SemRepr"$$

Begin Case of kindtext from

$\qquad$ *Stat* : *Semrepr* := *the empty string*;

Comment

Example 1. Let T1 = "Professor Igor Novikov teaches in Tomsk."
Then initially

$$Semrepr := the\, empty\, string.$$

After the termination of algorithm *BuildSem*1

$$Semrepr = Situation\,(e1, teaching * (Time, \#now\#)$$

$$(Agent1, certn\, person * (Qualif, professor)$$

$$(Name, 'Igor')\,(Surname, 'Novikov') : x2)$$

$$(Place3, certn\, city * (Name1, 'Tomsk') : x3)).$$

End-of-comment

$\qquad$ *Imp* : *Semrepr* = (*Command* (#*Operator*#, #*Executor*#, #*now*#, *e*1)

$\qquad$ Comment

Example 2. Let T2 = "Deliver the container with the details to warehouse No. 3."

$\qquad$ Then initially

$$Semrepr := (Command\,(\#Operator\#, \#Executor\#, \#now\#, e1)).$$

After the termination of algorithm *BuildSem*1

$$Semrepr = (Command\,(\#Operator\#, \#Executor\#, \#now\#, e1)$$

$$\wedge\, Target\, (e1,\, delivery1\, *\, (Object1,\, certn\, container1\, *$$

$$(Content1,\, certn\, set\, *\, (Qual-compos,\, detail))\, :\, x1)$$

$$(Place2,\, certn\, warehouse\, *\, (Number,\, 3)\, :\, x2)))$$

End-of-comment

$$Genqs:\ Semrepr := Question\,(x1,\ (x1 \equiv Truth-value\,($$

Comment

Example 3. Let T3 = "Did the international scientific conference 'COLING' take place in Asia?"

Then initially

$$Semrepr := Question\,(x1,\ (x1 \equiv Truth-value\,(.$$

After the termination of algorithm *BuildSem*1

$$Semrepr = Question\,(x1,\ (x1 \equiv Truth-value$$

$$(Situation\,(e1,\, taking\_place\, *$$

$$(Time,\, certn\, moment\, *\, (Earlier,\, \#now\#)\, :\, t1)$$

$$(Event,\, certn\, conference\, *\, (Type1,\, international)$$

$$(Type2,\, scientific)\,(Name,\,'COLING')\, :\, x2)$$

$$(Place,\, certn\, continent\, *\, (Name,\,'Asia')\, :\, x3))))).$$

End-of-comment

$$Specqs-relat1,\ Specqs-relat2:$$

$$\quad if\ kindtext = Specqs-relat1\ then\ Semrepr := 'Question\,(x1,\,'$$

$$\quad else\ Semrepr := 'Question\,(S1,\, (Compos\,(S1,\,'$$

$$\quad end-if$$

$$\quad end$$

Comment

Example 4. Let T4 = "What publishing house released the novel 'The Winds of Africa'?"

Then initially

$$Semrepr := Question\,(.$$

After the termination of algorithm *BuildSem*1

$$Semrepr = Question\,(x1,\, Situation\,(e1,\, releasing1\, *$$

$$(Time,\, certn\, moment\, *\, (Earlier,\, \#now\#)\, :\, t1)$$

$$(Agent2,\, certn\, publish-house\, :\, x1)$$

$$(Object3,\, certn\, novel1\, *\, (Name1,\,'The\, Winds\, of\, Africa')\, :\, x2))).$$

End-of-comment

Comment

Example 5. Let T5 = "What foreign publishing houses the writer Igor Somov is collaborating with?"
Then initially

$$Semrepr := Question\,(S1, (Qual - compos\,(S1, .$$

After the termination of algorithm *BuildSem*1

$$Semrepr = Question\,(S1, (Qual - compos\,(S1, publish - house *$$

$$(Type - geographic, foreign)) \land Description$$

$$(arbitrary\,publish - house * (Element, S1) : y1,$$

$$Situation\,(e1, collaboration * (Time, \#now\#)$$

$$(Agent1, certn\,person * (Occupation, writer)$$

$$(Name, 'Igor')\,(Surname, 'Somov') : x1)$$

$$(Organization1, y1))))).$$

End-of-comment
　　　　$Specqs - role:\ Semrepr := {}'Question\,('$
Comment
Example 6. Let T6 = "Who produces the medicine 'Zinnat'?"
Then initially

$$Semrepr := Question\,(.$$

After the termination of algorithm *BuildSem*1

$$Semrepr = Question\,(x1, Situation\,(e1, production1 *$$

$$(Time, \#now\#)\,(Agent2, x1)$$

$$(Product2, certn\,medicine1 * (Name1, 'Zinnat') : x2)))$$

End-of-comment
Comment
Example 7. Let T7 = "For whom and where the three-ton aluminum container has been delivered from?"
Then initially

$$Semrepr := Question\,(.$$

After the termination of algorithm *BuildSem*1

$$Semrepr = Question\,((x1 \land x2),$$

$$Situation\,(e1, delivery2 *$$

$$(Time, certn\, moment * (Earlier, \#now\#) : t1)$$

$$(Recipient, x1)\, (Place1, x2)$$

$$(Object1, certn\, container1 * (Weight, 3/ton)$$

$$(Material, aluminum) : x3))).$$

End-of-comment

$Specqs - quant1 : Semrepr := {}'Question\,(x1, ((x1 \equiv Numb\,(' ;$

Comment

Example 8. Let T8 = "How many people did participate in the creation of the text-book on statistics?"

Then initially

$$Semrepr := {}'Question\,(x1, ((x1 \equiv Numb\,(' .$$

After the termination of algorithm *BuildSem*1

$$Semrepr = Question\,(x1, ((x1 \equiv Numb(S1))$$

$$\wedge Qual - compos\,(S1, person) \wedge$$

$$Description\,(arbitrary\, person * (Element, S1) : y1,$$

$$Situation\,(e1, participation1 *$$

$$(Time, certn\, moment * (Earlier, \#now\#) : t1)$$

$$(Agent1, y1)\, (Type - of - activity, creation1 *$$

$$(Product1, certn\, text - book1 * (Field1, statistics) : x2))))).$$

End-of-comment

$Specqs - quant2 :$

$typesit :=$ selected type *sit* ("situation") of the used conceptual basis;

$Semrepr := Question\,(x1, ((x1 \equiv Numb\,(S1)) \wedge Qual - compos\,(S1,$

$+ sortsit +') \wedge$

$Description\,(arbitrary' + sortsit + ' * (Element, S1) : e1, '$

Comment

Example 9. Let T9 = "How many times Mr. Stepan Semenov flew to Mexico?"

Then initially

$$Semrepr := Question\,(x1, ((x1 \equiv Numb\,(S1))$$

$$\wedge Qual - compos\,(S1, sit) \wedge$$

$$Description\,(arbitrary\, sit * (Element, S1) : e1, .$$

After the termination of algorithm *BuildSem*1

$$Semrepr = Question\,(x1, ((x1 \equiv Numb\,(S1))$$

$$\land Qual - compos\,(S1,\,sit)\,\land$$

$$Description\,(arbitrary\,sit\,*\,(Element,\,S1)\,:\,e1,$$

$$Situation\,(e1,\,flight\,*\,(Time,\,certn\,moment\,*$$

$$(Earlier,\,\#now\#)\,:\,t1)\,(Agent1,\,certn\,person\,*$$

$$(Name,\,'Stepan')\,(Surname,\,'Semenov')\,:\,x2)$$

$$(Place2,\,certn\,country\,*\,(Name1,\,'Mexico')\,:\,x3)))))'.$$

End-of-comment
$$end-case-of$$
$$end$$

10.2 Semantic Representations of Short Fragments of the Input Texts

This section describes the algorithm "Begin-of-constr-SemRepr" aimed at forming the one-dimensional arrays *Sembase* ("Semantic Base"), *Semdes* ("Semantic Description"), *Performers* ("Role performers in the situations mentioned in the input text") and an initial configuration of the two-dimensional array *Sitdescr* ("Situation Description").

10.2.1 Description of the Algorithm "Calculation-of-the-kind-of-case"

Let's start to consider such auxiliary algorithms that their interaction allows for forming the arrays *Sembase, Semdes, Performers,* and an initial configuration of the array *Sitdescr*.

External specification

Input:
Rc – array – classifying representation of the input text;
$k1$ – number of the row of the classifying representation of the input text, i.e. the ordered number of the text unit;
$Arls$ – two-dimensional array – the projection of the lexico-semantic dictionary *Lsdic* on the input text T;
$Matr$ – MSSR of the text;
$class1$ – string defining the class of the text unit;
$sem1$ – semantic unit corresponding to $k1$ text unit.
Output:
$casemark$ – string - takes the values $case1 - case7$ depending on the form of the fragment of the classifying representation of the text.

Algorithm

*Begin if class*1 = *adject then casemark* := '*Case*1'
  end − *if*
  *if class*1 = *constr then casemark* := '*Case*2'
  end − *if*
  *if class*1 = *noun*
  then if $Rc[k1 + 1, tclass]$ = *name then casemark* := '*Case*3'
    *else numb*1 := $Matr[k1, qt]$

Comment
The number corresponding to the noun in the position $k1$
End-of-comment
  end − *if*
  ref := *certn*

Comment
Referential quantifier
End-of-comment
  *beg*1 := *sem*1 [1]

Comment
The first symbol of the string sem1, considering each element of the primary informational universe $X(B(Cb(Lingb)))$ and each variable from $V(B)$ as one symbol
End-of-comment
  *setind*1 := 0

Comment
Mark of individual, not the set of individuals
End-of-comment
  *len*1 := *Length* (*sem*1)
  if (*len*1 ≥ 2) *and* (*sem*1 [2] = '*set*')
  *then setind*1 := 1
  end − *if*

Comment
*sem*1 [2] − 2nd symbol of the structured semantic unit *sem*1, if we interpret the elements of primary informational universe $X(B(Cb))$ as symbols, where Cb − the used marked-up conceptual basis (m.c.b.)
End-of-comment
    if ((*numb*1 = 0) *or* (*numb*1 = 1)) *and* (*beg*1 = *ref*) *and* (*setind*1 = 0)
Comment
i.e. $Rc[k1, unit]$ − the designation of the individual, not the set
End-of-comment
    then casemark := '*Case*4'

Comment
Example: "Belgium"
End-of-comment
    end − *if*
    if(*numb*1 = 0) *and* (*beg*1 ≠ *ref*) *and* (*sem*1 is not the designation of
the function from $F(B(Cb))$, where Cb is the used marked-up conceptual basis)

then begin loc1 := Rc [k1, mcoord], md1 := Rm [loc1, morph];
  if (Number (md1) = 1) then casemark := 'Case5'

Comment
Example: "conference"
End-of-comment

else

Comment
i.e., in the case when *Number (md1) = 2*
End-of-comment

casemark := 'Case6'

Comment
Examples: "5 articles", "3 international conferences"
End-of-comment

end − if

  end
  end − if
  if (numb1 = 0) and (setind1 = 1)

Comment
Example: "with Indian ceramics"
End-of-comment

then casemark := 'Case7'
  end − if
end − if

end

10.2.2 External Specification of the Algorithms BuildSemdes1 – BuildSemdes7

Input:
Rc – array – classifying representation of the input text;
k1 – integer – number of a row from *Rc*;
Arls – two-dimensional array – projection of the lexico-semantic dictionary *Lsdic* on the input text *T*;
Matr – MSSR of the text;
sem1 – string – semantic unit, corresponding to the text unit with the number *k1*;
casemark – string taking the values *case1 – case7* depending on the form of the fragment of the classifying representation of the input text;
arrays *Sembase, Semdes, Performers*.
Output:
arrays *Sembase, Semdes, Performers* containing the blocks for forming the final value of the variable *Semrepr* – semantic representation of the input text.

10.2.3 Description of the Algorithms BuildSemdes1 – BuildSemdes7

Description of the auxiliary algorithms

*Function Transform*1

Arguments:

s – string of either the form $r(z, b)$, where r – the designation of the binary relation and b – the second attribute of the relation, or of the form $(f(z) \equiv b)$, where f – name of function with one variable,

b – string designating the value of the function,

z – letter "z" interpreted as variable.

Value:

String t of the form (r, b) in the first case and of the form (f, b) in the second case.

Example 1. Let T1 = "How many two-ton aluminum containers came from Penza?" Then the linguistic basis can be defined in the following way:

$$for\ k1 = 2\ sem1 := (Weight\,(z) \equiv 2/ton),$$

$$Transform1\,(sem1) = (Weight, 2/ton),$$

$$for\ k1 = 3\ sem1 := Material\,(z, aluminum),$$

$$Transform1\,(sem1) = (Material, aluminum).$$

*Algorithm BuildSemdes*1

Begin

Comment

The reflection of semantics of adjectives in the array *Sembase*

End-of-comment

$$if\ Matr\,[k1 - 1, nattr] = 0$$

Comment

To the immediate left from the $k1$ position there are no adjectives, i.e., on the $k1$ position there is the first adjective from the group of consequent adjectives

End-of-comment

$$then\ Sembase\,[k1] := Transform1\,(sem1)$$

$$else$$

Comment

To the immediate left from the $k1$ position there is an adjective

End-of-comment

$$Sembase\,[k1] := Sembase\,[k1 - 1] + Transform1\,(sem1)$$

Comment

The sign "+" here designates the concatenation, i.e., the operation of addition of the string to the right

End-of-comment

$$end - if$$

end

Example 2. During processing the question T1 = "How many 2-ton aluminum containers came from Penza?" by the algorithm *BuildSemdes*1 the following operators will be used:

$$Sembase\,[2] := (Weight,\,2/ton),$$

$$Sembase\,[3] := (Weight,\,2/ton)\,(Material,\,aluminum).$$

$$Algorithm\,BuildSemdes2$$

Begin

Comment

Processing of a construct

End-of-comment

$$Sembase\,[k1] := sem1;\; Performers\,[k1] := Rc\,[k1,\,unit]$$

Comment

Example: $Performers\,[k1] := {}'720/km'$

End-of-comment

end

Description of the algorithm *BuildSemdes*3 (**"Processing of the names"**)

Purpose: building the semantic representation (SR) of the fragment of the text *T*, which is a combination of the form "noun + expression in quotation marks or apostrophes."

Application condition: in the $k1$ position there is a noun, in the $k1 + 1$ position there is an expression in quotation marks or apostrophes.

Example 3. Let T2 = "Who produces the medicine 'Zinnat'?" Then applying this algorithm will result in the following operator:

$$Performers\,[k1] := certn\,medicine1 * (Name, 'Zinnat').$$

$$Algorithm$$

$Begin\ name := Rc\,[k1 + 1,\,unit];$

    $if\ (Performers\,[k1]$ does not include the symbol $*)$

    $then\ Performers\,[k1] := Performers\,[k1] + {}' * (Name, {}' + name + {}')'$

    $else\ Performers\,[k1] := Performers\,[k1] + {}'(Name, {}' + name + {}')'$

    $end - if$

end

$$Algorithm\,BuildSemdes4$$

Begin

Comment

Processing of the proper names
Example of the context – published in Belgium
End-of-comment
$\qquad$ $Sembase\,[k1] := sem1;\; Semdes\,[k1] := Sembase\,[k1]$
$\qquad$ $Var1 := Matr\,[k1, mark];\; Performers\,[k1] := Semdes\,[k1] + '\,:\,' + var1$
Comment
Example:

$$Performers\,[k1] := '\,certn\,country * (Name, 'Belgium') : x2'$$

End-of-comment
End

$$Algorithm\,BuildSemdes5$$

$Begin$
Comment
Processing of the common nouns
Example of the context: "published a monograph"
End-of-comment
$\qquad$ $if\; Matr\,[k1, nattr] \geq 1$
Comment
There are adjectives to the left
End-of-comment
$\qquad$ $then\; Sembase\,[k1] := sem1 + '\,*\,' + Sembase\,[k1 - 1]$
$\qquad$ $else\; Sembase\,[k1] := sem1$
$\qquad$ $end - if$
$\qquad$ $Ref := '\,certn';\; Semdes\,[k1] := ref\; sem1$
$\qquad$ $Var1 := Matr\,[k1, mark];\; Performers\,[k1] := Semdes\,[k1] + '\,:\,' + var1$
Comment
Example 1: $\qquad Performers\,[k1] := '\,certn\,monograph : x3'$

End-of-comment
Comment
Example 2:

$$Performers\,[k1] := '\,certn\,printer * (Form, ink - jet) : x4'$$

End-of-comment
end

$$Algorithm\,BuildSemdes6$$

$Begin$
Comment
Processing of a combination of nouns designating a set of objects.
Example of the context – "5 three-ton containers are delivered"

End-of-comment

$numb1 := Matr[k1, qt]; Sembase[k1] := sem1$

$if\ numb1 > 0\ then\ Semdes[k1] := {'certn\,set} * (Quant,{'} + numb1 +$
$'){(Compos,'} + Sembase[k1] + ')'$

$else\ Semdes[k1] := {'certn\,set} * (Compos,{'} + Sembase[k1] + ')'$

$end - if$

$beg1 := sem1[1]$

Comment

First symbol of the string $sem1$ considering elements of the primary informational universe $X(B(Cb(Lingb)))$ and variables as symbols

End-of-comment

$Var1 := Matr[k1, mark]; Var2 := Varsetmember(var1);$

Comment

The variable $var2$ designates an arbitrary element of the set with the mark $var1$.

Example: $If\ var1 = S2\ then\ var2 = y2$

End-of-comment

$Performers[k1] := {'arbitrary'} + beg1 + {'} * Elem({'} + Semdes[k1] + {'} :$
$' + var1 + ') : {'} + var2$

Comment

Example:

$$Performers[k1] := {'arbitrary\,container}1 *$$

$$(Elem, certn\,set * (Quant, 5)$$

$$(Compos, container1 * (Weight, 3/ton)) : S1) : y1'$$

End-of-comment

end

Description of the algorithm $BuildSemdes7$ (**"Processing of the collective nouns"**)

Purpose: Building the fragment of the semantic representation (SR) of the text T which includes collective nouns ("Indian ceramics," "Italian shoes," etc.). Application condition: there is a collective noun on the $k1$ position.

Example 4. Let T3 = "Where three containers with Indian ceramics came from?" The word "ceramics" in the question T3 has the ordered number 6. The linguistic basis may be defined in a way that would lead to the execution of the following operators as a result of algorithm $BuildSemdes7$ application:

$$Semdes[6] := certn\,set * (Compos, article\_of\_ceramics *$$

$$(Geographic\_localization, certn\,country * (Name, {'India'}))),$$

$$Performers[6] := certn\,set * (Compos, article\_of\_ceramics *$$

$$(Geographic\_localization, certn\,country * (Name, {'India'}))) : S1.$$

$$The\, description\, of\, auxiliary\, algorithms$$

$$Function\, Transform2$$

Arguments:

s – string representing the semantics of an adjective or sequence of adjectives. For example, s may represent the semantics of an adjective "Indian" and be the string $(Geographic\_localization, certn\, country * (Name, 'India'))$.

String t – the structured semantic unit corresponding to the collective noun and including the substring $(Qual - compos,$ (for example, t may correspond to the noun "ceramics" and be the string $certn\, set * (Qual - compos, article\_of\_ceramics)))$.

Value:

String u formed as follows.

Let $pos1$ – position of the first left bracket "(" in the substring $(Qual - compos$ of the string s, and let $pos2$ – position of the right bracket ")" closing the bracket opened in the position $pos1$.

Let h – substring of the string t placed between the substring $(Qual - compos$ and the right bracket in the position $pos2$.

Then u can be derived from the string t by replacing the substring h with the string $h * s$.

Example 5. In the context of the question T3 = "Where three containers with Indian ceramics came from?" let

$$s = (Geographic\_localization, certn\, country * (Name1, 'India')),$$

$$t = certn\, set * (Qual - compos, article\_of\_ceramics).$$

$$Then\, h = article\_of\_ceramics,$$

$$u = Transform2\,(s, t) = certn\, set *$$

$$(Qual - compos, article\_of\_ceramics *$$

$$(Geographic\_localization, certn\, country * (Name, 'India'))).$$

$$Algorithm\, Build\, Semdes7$$

$Begin\ if\ Rc\,[k1 - 1, tclass] \neq adjective$
$\quad then\ Semdes\,[k1] := sem1$
$\quad else\ prop1 := Sembase\,[k1 - 1]$
$\qquad Semdes\,[k1] := Transform2\,(prop1, sem1)$
Comment
Example:

$$Semdes\,[k1] := certn\, set * (Qual - compos, article\_of\_ceramics *$$

$$(Place - of - production, certn\, country * (Name1, 'India')))$$

End-of-comment

$end-if$

end

10.2.4 Description of the Algorithm ProcessSit

The algorithm *ProcessSit* is designed for representation of the structured units of conceptual level (in other words, semantic units), corresponding to the situations mentioned in the input text by means verbs or participles, in the array *Sitdescr*.

External specification

Input:
Rc – array – classifying representation of the input text T;

$k1$ – number of the row of the classifying representation of the text T, i.e., the ordered number of the verbal text unit;

Rm – array – morphological representation of the text T;

$kindtext$ – string designating the form of the text T;

$Arls$ – projection of the lexico-semantic dictionary $Lsdic$ on the input text T;

$Matr$ – MSSR of the text;

$Sitdescr$ – initial configuration of the array which includes descriptions of the situations mentioned in the text;

$timevarnumb$ – maximal number of the variable designating the moment of time.

Output:
$Sitdescr$ – transformed configuration of the array used for describing the situations mentioned in the text.

Description of the auxiliary algorithms

Function Numb1

Argument:
v – the string of the form $R\,S$, where R – letter of Latin alphabet and S – string representing an integer.

Value:

N – the integer associated with the string S.

Example 6. For the string $e3$ $Numb1(e3)$ is the number 3.

Function Stringvar

Arguments:
R – letter of Latin alphabet,

N – an integer

Value:

String of the form $R\,S$, where S – string representing the integer N.

Example 7. If R - letter "t", N – number 2, then *Stringvar* (R, N) is the string $t2$.

Function Time

Argument:
M – set of the values of morphological properties related to the arbitrary verbal form *vbform* (the verb or participle)
Value:
Figure "1" if *vbform* corresponds to the past; figure "2" if *vbform* corresponds to the present; figure "3" if *vbform* corresponds to the future.

Algorithm ProcessSit

*Begin pos*1 $:= Rc\,[k1, mcoord]$
        $armorph := Rm\,[pos1, morph]$
Comment
The set of the values of morphological properties related to the verbal form in the position $k1$
End-of-comment
        $timevarnumb := timevarnumb + 1$
        $Vartime := Stringvar\,('t', timevarnumb)$
        $time1 := Time\,(armorph)$
        *Case of time*1 *from*
    $'1'$: $timesit := '(Time, certn moment * (Earlier, \#now\#) : ' + vartime + ')'$
        $'2'$: $timesit := '(Time, \#now\#)'$
        $'3'$: $timesit := '(Time, certn moment * (Later, \#now\#) : ' + vartime + ')'$
        $End-case-of$
        $linesit := Matr\,[k1, locunit];\ concsit := Arls\,[linesit, sem]$
        $var1 := Matr\,[k1, mark];\ numbsit := Numb1\,(var1)$
        $if\ (kindtext = Imp)\ and\ (numbsit = 1)$
        $then\ Sitdescr\,[numbsit, expr] := 'Target\,(' + var1 + ',' + concsit + ' * '$
        $else\ Sitdescr\,[numbsit, expr] := 'Situation\,(' + var1 + ',' + concsit + ' *$
$' + timesit$
        $end-if$
Comment
Example 1:

$$Sitdescr\,[1, expr] := 'Situation\,(e1, releasing1 * '$$

$$(Time, certn moment * (Earlier, \#now\#) : t1)'$$

Example 2:

$$Sitdescr\,[1, expr] := 'Target\,(e1, delivery1 *$$

$$(Object1, certn\,container1 : x1)$$

$$(Place2, certn\,warehouse * (Number, 4) : x2))'$$

End-of-comment
end

10.2.5 Description of the Algorithm "Begin-of-constr-SemRepr"

External specification

Input:
Rc – array – classifying representation of the input text T;

Rm – array – morphological representation of the text T;

$Arls$ – projection of the lexico-semantic dictionary $Lsdic$ on the text T;

$kindtext$ – string representing the form of the input text T;

$mainpos$ – integer – position of the interrogative word in the beginning of the text;

$Matr$ – MSSR of the text.

Output:
$Semrepr$ – string – initial value of the semantic representation of the input text;

$Performers$ – one-dimensional array containing semantic representations of the short fragments of the input text.

Algorithm

Begin

    Preparation − *to* − *constr* − *SemRepr*

    $(Rc, Rm, Matr, kindtext, mainpos, Semrepr)$

Comment
Example:

    if $kindtext = genqs$

(closed question, i.e., the question with the answer "Yes/No")

    then $Semrepr := {}'Question\,(x1, (x1 \equiv Truth-value\,('$

End-of-comment

    loop for $k1$ *from* 1 *to* nt

Comment
Formation of the arrays *Sembase, Semdes, Performers*, and the initial configuration of the array *Sitdescr*

End-of-comment

    $class1 := Rc\,[k1, tclass]$

    if $(class1 \neq construct)$ *and* $(class1 \neq name)$ *and* $(class1 \neq marker)$

    then $loc1 := Matr\,[k1, locunit]$; $sem1 := Arls\,[loc1, sem]$

    end − *if*

$if\ (class1\ =\ verb)\ or\ (class1\ =\ participle)$
$then\ ProcessSit\ (Rc,\ Rm,\ Arls,\ Matr,\ Sitdescr,\ timevarnumb)$
$end - if$
$if\ (class1 - of\ the\ set\ \{adject,\ constz,\ noun\})\ element$
$then\ Calculation - of\ the\ form - of\ case$
$(Rc,\ k1,\ Arls,\ Matr,\ class1,\ sem1,\ casemark1);$
$end - if$
$Case\ of\ casemark1\ from$
$'Case1':\ BuildSemdes1\ (List1),$
where $List1$ - the list of parameters
$Rc,\ k1,\ Arls,\ Matr,\ Sembase,\ Semdes,\ Performers,\ casemark1;$
$'Case2':\ BuildSemdes2\ (List1);$
$'Case3':\ BuildSemdes3\ (List1);$
$'Case4':\ BuildSemdes4\ (List1);$
$'Case5':\ BuildSemdes5\ (List1);$
$'Case6':\ BuildSemdes6\ (List1);$
$'Case7':\ BuildSemdes7\ (List1)$
$End - case - of$
$End - of - loop$
End

10.3 Development of the Algorithm "Representation-of-situations"

Let's develop an algorithm constructing separate semantic representations (being K-representations) of the situations mentioned by the verbs and participles in the input NL-text. The principal sourse of data for this algorithm is a matrix semantic – syntactic representation (MSSR) of the input text.

10.3.1 The Key Ideas of the Algorithm

By the time this algorithm is applied, the array *Performers* and the initial configuration of the situation description array *Sitdescr* are already formed. The array *Performers* includes semantic units (primary and compound) corresponding to the constructs (digital values of the parameters), nouns and combinations of the forms "Group of adjectives + Noun," "Number + Noun," "Number + Group of adjectives + Noun," "Cardinal number + Noun," "Cardinal number + Group of adjectives + Noun."

The number of filled-in rows of the array *Sitdescr* equals the number of verbs and participles in the text. The mark of the situation is placed in the column mark (the link to the MSSR *Matr* is realized through the elements of this column). The column *expr* (abbreviation of "the expression") is designed for keeping the semantic descriptions of the situations (events) mentioned in the text.

In the considered algorithm *Representation − of − situations*, the transformation of the information is carried out in two consequent stages. The first stage represents a loop for m from 1 to nt, where m is number of a row of the classifying representation Rc, nt – quantity of the text elements. In this loop the information about the semantic-syntactic relations in the combinations "Verbal form (verb or participle two or gerund) + Dependent fragment of the sentence" is represented in the elements of the column *expr* of the array *Sitdescr* (each of these elements is a description of the certain situation mentioned in the input text).

Here the dependent fragment of the sentence means a construct or a noun or a combination of one of the following forms: "Group of adjectives + Noun," "Number + Noun," "Number + Group of adjectives + Noun," "Cardinal number + Noun," "Cardinal number + Group of adjectives + Noun."

The examples of the dependent fragments of the sentence could be the expressions "in 2002nd year," "European scientific publishers," "two-ton containers," "5 containers," "12 personal computers."

Example 1. Let Qs1 = "How many two-ton containers with Indian ceramics delivered from Novorossiysk were shipped to the firm 'Sail'?"

Then during the first stage of applying the algorithm, the expressions that are the elements of the column *expr* of the array *Sitdescr* are supplemented with the information about the semantic-syntactic relations in the combinations "delivered + two-ton containers," "delivered + Novorossiysk," "shipped + two-ton containers," "shipped + firm "Sail."

As a result of this stage of processing the matrix semantic-syntactic representation of the question Qs1 by the algorithm, the array *Sitdescr* acquires the following configuration:

Table 10.1 The structure of the array *Sitdescr* at the intermediate stage of constructing a semantic representation of the input text

mrk	expr
$e1$	*Description1*
$e2$	*Description2*

where

$$Description1 = Situation\,(e1, delivery2 * (Time,$$

$$certn\,moment * (Earlier, \#now\#) : t1)$$

$$(Object1, arbitrary\,container1 * (Element, certn\,set *$$

$$(Qual - compos, container1 * (Weight, 2/ton) : S1) : y1)$$

$$(Place1, certn\,city * (Name1, 'Novorossiysk') : x2)),$$

$$Description2 = Situation\,(e2, shipment1 * (Time,$$

$$certn\,moment * (Earlier, \#now\#) : t2) (Object1, y1)$$

$$(Recipient, certn\,firm * (Name1, 'Sail') : x3)).$$

The auxiliary array $Used$ of the length nt helps to avoid repeating the semantic representations of the same expression (designating an object or a set of objects) in the column $expr$ of the array $Sitdescr$. Initially for each m from 1 to nt, $Used[m] = 0$. If for a certain k, the string $Performers[k]$ is a fragment of a certain row of the array $Sitdescr$, then $Used[k] := 1$.

That is why not the string $Performers[k]$ but the variable being the ending of the string $Performers[k]$ is being added to other rows of the array $Sitdescr$, if needed.

For example, the first row of the array $Sitdescr$ (Table 10.1) includes the expression

$$arbitrary\,container1 * (Element, certn\,set *$$

$$(Qual - compos, container1 * (Weight, 2/ton) : S1) : y1,$$

being an element of the one-dimensional array $Performers[3]$. That is why the second row of the array $Sitdescr$ uses the variable $y1$ instead of this expression.

Let's call the fragments of the sentence directly governed by the verbal forms (verbs or participles) the dependent elements of the first level.

The second stage of applying the algorithm "*Representation − of − situations*" is searching in the MSSR $Matr$ for such constructs or combinations with noun that are directly governed by the dependent elements of the first level, i.e., searching for the dependent elements of the second level.

For example, in the question Qs1, the fragment "two-ton containers" is governed by the verbal form "were shipped," therefore, this fragment is a dependent element of the first level. At the same time, the combination "two-ton containers" governs the noun group "with Indian ceramics."

The formal representation $descr1$ of the information conveyed by the combination of the form "Dependent element of the first level X + Dependent element of the second level Y" is being added from right to the element $Sitdescr[k, expr]$ with the help of conjunction. Here k here is the ordered number of the situation, its participant is designated by the dependent element X of the first level.

The auxiliary array $Conj$ is initially filled in with zeros. If, as a result of the second stage of applying the algorithm, a certain expression is being added from the right to the element $Sitdescr[k, expr]$ with the help of conjunction, then $Conj[k] := 1$. The value 1 serves as the signal to put the element $Sitdescr[k, expr]$ in the brackets before including this element into the semantic representation of the input text.

Example 2. As a result of the second stage of applying the algorithm to the question Qs1, the following expression will be added from the right to the element $Sitdescr[1, expr]$ via conjunction:

$Content1 (y1, certn set * (Qual − compos, ceramic\_article *$

$(Geographic\_localization, certn country * (Name1, 'India')))).$

As the conjunction was used, the element $Conj\,[1]$ will get the value 1. Finally the array $Sitdescr$ will acquire the following configuration:

Table 10.2 The structure of the array $Sitdescr$ at the final stage of constructing a semantic representation of the input text

mrk	expr
e1	SemRepr1
e2	SemRepr2

where

$SemRepr1 = Situation\,(e1, delivery2 * (Time,$

$certn moment * (Earlier, \#now\#) : t1)$

$(Object1, arbitrary container1 * (Element, certn set *$

$(Qual − compos, container1 * (Weight, 2/ton) : S1) : y1)$

$(Place1, certn city * (Name1, 'Novorossiysk') : x2)),$

$\vee Content1, (y1, certn set *$

$(Qual − compos, ceramic\_article *$

$(Geographic\_localization, certn country * (Name, 'India'))),$

$SemRepr2 = Situation\,(e2, shipment1 * (Time,$

$certn moment * (Earlier, \#now\#) : t2) (Object1, y1)$

$(Recipient, certn firm * (Name1, 'Sail') : x3)).$

10.3.2 Description of the Algorithm "Representation-of-situations"

The algorithm considered below is developed for representing the information about the situations (events) mentioned in the text in the array $Sitdescr$.

External specification of the algorithm

Input:
Rc – array – classifying representation of the input text;

nt – integer – length of the input text (the number of rows in *Rc* and *Matr*);

kindtext – string – designation of the form of the input text;

Matr – MSSR of the text;

Performers – two-dimensional array including the semantic images of the situation paricipants;

maxnumbsit – integer – number of the situations mentioned in the text.

Output:

Sitdescr – array representing the information about the situations mentioned in the input text;

Used [1 : *nt*] – one-dimensional array for keeping the signs of multiple usage of the structured semantic unit in the rows of the array *Sitdescr*;

Conj [1 : *maxnumbsit*] – one-dimensional array for keeping the signs of conjunction usage in the rows of the array *Sitdescr*.

$$Algorithm "Representation - of - situations"$$

Begin

Comment

The processing of direct semantic connections between the verbal form and a noun or a construct

End-of-comment

> *Loop for j1 from 1 to nt Used* [*j1*] := 0
> *End − of − loop*
> *Loop for j2 from 1 to maxnumbsit Conj* [*j2*] := 0
> *End − of − loop*
> *Loop for m from 1 to nt*

Comment

The first passage of the strings in *Rc*

End-of-comment

> *Class*1 := *Rc* [*m, tclass*]
> *If* (*class*1 = *noun*) *or* (*class*1 = *constr*)
> *Then begin d* := *nmasters* [*m*];

Comment

The quantity of the text units governing the text unit *m* is found

End-of-comment

> > *If d* > 0
> > *Then loop for q from 1 to d*
> > > *Begin p*1 := *Matr* [*m, posdir, q*]; *class*2 := *Rc* [*p*1, *tclass*]
> > > *If* (*class*2 = *verb*) *or* (*class*2 = *participle*)
> > > *Then var*2 := *Matr* [*p*1, *mark*];

Comment

Mark of the situation designated by the governing verbal form

End-of-comment

> > > > *numbsit* := *Numb* (*var*2);
> > > > *role* := *Matr* [*m, reldir, q*]

$if\ (class1 = noun)$

$then\ if\ kindtext$ is not included in the set $\{specqs - relat2, specqs - quant1\}$

$then\ if\ Used\,[m] = 0\ then\ actant := Performers\,[m]$

$else\ if\ (class1 = noun)$

$then\ actant := Varbuilt\,(Matr\,[m, mark])$

$end - if$

$if\ (class1 = constr)\ then\ actant := Rc\,[m, unit])$

$end - if$

$end - if$

$else$

Comment

If $kindtext$ is included in the set $\{specqs - relat2, specqs - quant1\}$

End-of-comment

$if\ ((Used\,[m] = 1)\ or\ ((Used\,[m] = 0)\ and\ (Matr\,[m, mark] = S1))$

$then\ actant := Varbuilt\,(Matr\,[m, mark])$

$else\ actant := Performers\,[m]$

$end - if$

$end - if$

Comment

$Kindtext$

End-of-comment

$end - if$

Comment

$class1 = noun$

End-of-comment

Comment

$Varbuilt\,(x_j) = x_j;\ Varbuilt\,(S_j) = y_j$

End-of-comment

$Sitdescr\,[numbsit, expr] := Sitdescr\,[numbsit, expr] + \,'(' + role + \,',+ actant + \,')'\ end - if$

end

$end - of - loop\ \{on\,q\}$

$end - if$

end

$end - if$

$End - of - loop$

Comment

End of a loop on m

End-of-comment

Comment

Example 3. Let's consider the question Qs1 = "Where five aluminum two-ton containers came from?" Let's assume that before applying the algorithm "$Representation - of - situations$", the following relationship took place:

$$Sitdescr\,[1, expr] = Situation\,(e1, delivery2 *$$

$$(Time, certn\,moment * (Earlier, \#now\#) : t1).$$

Then after applying the algorithm "*Representation − of − situations*" with a certain choice of the marked-up conceptual basis the following relationship will be true:

$$Sitdescr\,[1, expr] = Situation\,(e1, delivery2 *$$

$$(Time, certn\,moment * (Earlier, \#now\#) : t1)$$

$$(Place1, x1)\,(Object1, arbitrary\,container1 *$$

$$(Element, certn\,set * (Numb, 5)$$

$$(Qual − compos, container1 * (Weight, 2/ton)$$

$$(Material, aluminum)) : S1) : y1$$

End-of-comment

  *loop for k*1 *from* 1 *to nt*

Comment

Filling in of *Sitdescr* – the second passage of *Rc* – the processing of the text units governed by nouns depending on a verbal form

End-of-comment

  *class*1 := $R_c\,[k1, tclass]$

  if (*class*1 = *noun*) *or* (*class*1 = *constr*)

Comment

The examples of combinations: "containers with Indian ceramics" (class1 = noun), "with the lamps of 60 watt" (class1 = constr)

End-of-comment

  then posmaster := *Matr*[k1, *posdir*, 1]

Comment

The position of the word "containers" or "lamps" for above combinations

End-of-comment

    if posmaster = 0 *then output* ("*Wrong text*")

    *else class*2 := *Rc*[*posmaster*, *tclass*]

     *if class*2 = *noun*

     *then rel*1 := *Matr*[k1, *reldir*, 1]

      varmaster := *Matr*[*posmaster*, *mark*]

      if (*class*1 = *constr*) *then arg*2 := *Sembase*[k1]

      end − if

      if (*class*1 = *noun*)

      then vardep := *Matr*[k1, *mark*]

       if Used[k1] = 0 *then arg*2 := *Performers*[k1]

       *else arg*2 := *vardep*

       end − if

      end − if

      letter := the first symbol of the string *varmaster*

Comment

The variable *varmaster* corresponds to the noun governing the text unit in the position *k*1, moreover this noun is governed by the verbal form – the verb or participle
End-of-comment

$$if\ letter = 'x'\ then\ descr1 := rel1\ (varmaster, arg2)$$
$$else$$

Comment
i.e. if letter = "S"
End-of-comment

$$semhead :=\text{the first element of the } Sembase\,[posmaster]$$
$$descr1 := rel1 + '(arbitrary' + semhead + ' * (Element,' +$$
$$varmaster + '),' + arg2 + ')'$$
$$end-if$$

Comment
Then the expression *descr1* is being added with the help of conjunction $\wedge$ to the right of the expression *Sitdescr* [*numbsit, expr*] characterizing the considered situation with the number *numbsit*.
End-of-comment

Example 4. Let Qs2 = "What writers from Tomsk did participate in the conference?" and after the first loop (for *m* from 1 to *nt*) the array *Sitdescr* has the following configuration:

Table 10.3 The structure of the array *Sitdescr* after the loop with *m* changing from 1 to *nt*

mrk	expr
e1	SemRepr1

where

$$SemRepr1 = Situation\,(e1, participation * (Time,$$
$$certn\,moment\,[*(Earlier, \#now\#) : t1)\,(Agent1,$$
$$arbitrary\,writer\,[*(Element, certn\,set *$$
$$(Qual-compos, writer) : S1) : y1)$$
$$(Event1, certn\,conference : x1)).$$

In the second loop (for *k*1 from 1 to *nt*) the operator

$$Descr1 := Geographic\_localization$$

$$(y1, certn\,city * (Name, 'Tomsk') : x2)$$

is being executed , and then the array *Sitdescr* acquires the new configuration

mrk	expr
e1	SemRepr2

where

$$SemRepr2 = Situation\,(e1,\,participation * (Time,$$

$$certn\,moment\,[* (Earlier,\,\#now\#) : t1)\,(Agent1,$$

$$arbitrary\,writer * (Element,\,certn\,set *$$

$$(Qual - compos,\,writer) : S1) : y1)$$

$$(Event1,\,certn\,conference : x1)) \wedge$$

$$Geographic\_localization\,(y1,\,certn\,city *$$

$$(Name1,\,'Tomsk') : x2).$$

Comment
End of example
End-of-comment

$$possit := Matr\,[posmaster,\,posdir,\,1]$$
$$varsit := Matr\,[possit,\,mark];\;\;numbsit := Numb\,(varsit)$$
$$Sitdescr\,[numbsit,\,expr] := Sitdescr\,[numbsit,\,expr] + {'}\wedge{'} + descr1$$
$$If\;Conj\,[numbsit] = 0\;then\;Conj\,[numbsit] := 1$$
$$end - if$$

Comment
The sign of using the conjunction in the string $Sitdescr\,[numbsit,\,expr]$
End-of-comment

$$end - if$$
$$end - if$$
$$end - if$$
$$end - of - loop$$

$$end$$

Comment
End of the algorithm
End-of-comment

10.4 Final Stages of Developing an Algorithm of Semantic Representation Assembly

Let's consider the final stages of developing the algorithm *BuildSem*1 constructing a K-representation of an input NL-text, proceeding mainly from its matrix semantic-syntactic representation *Matr*.

10.4.1 External Specification of the Algorithm "Final-operations"

The considered algorithm is designed for reflecting the information conveyed by the situations description array *Sitdescr* in the final value of the string *Semrepr* (semantic representation of the input text).

Input and Output Data

Input:

Rc – array – classifying representation of the input text;

Rm – array – morphological representation of the input text;

kindtext – string characterizing the form of the input text;

mainpos – integer – position of the interrogative word in the beginning of the text;

Matr – MSSR of the text;

Performers – two-dimensional array containing the semantic images of the situation's participants;

numbsit – integer – number of the situations mentioned in the input text, i.e. the number of filled-in rows of the array *Sitdescr*;

numbqswd – integer – number of interrogative words in the input text;

Sitdescr – array representing the information about the situations mentioned in the input text;

Semrepr – string – initial value of the semantic representation of the text.

Output:

Semrepr – string – final value of the semantic representation (SR) of the input text.

Descriptions of auxiliary algorithms

Function "Right"

Arguments:

pos1 – the number of a row of the classifying representation *Rc* of the input text, i.e., the position number of the input text unit;

class1 – string – designation of the class of the text unit.

Value:

pos2 – position of the closest from right (to the position *pos1*) text unit of the class *class1*.

Function "Stringvar"

Arguments:

R – letter of Latin alphabet,

N – an integer.

Value:

String of the form RS, where S is the string denoting the integer N.

Example 1. If R is the letter "t," and N is the number 2, then *Stringvar* (R, N) is the string $t2$.

Algorithm

*Begin pos*2 := *pos*1
　　　　loop − *until pos*2 := *pos*2 + 1; *class*2 := *Rc* [*pos*2, *tclass*]
　　　　exit − *when* (*class*2 = *class*1)
　　end

10.4.2 Algorithm "Final-operations"

Algorithm

begin loop for k from 1 *to maxnumbsit*
　　　　event := *Sitdescr* [*k, expr*]
　　　　if Conj [*k*] = 1 *then event* := ′(′ + *event* + ′)′
　　　　end − *if*
　　　　if k = 1 *then situations* := *event*
　　　　else situations := *situations* + ′ ∧′ + *event*
　　　　end − *if*
　　　　end − *of* − *loop*
　　　　if maxnumbsit > 1 *then situations* := ′(′ + *situations* + ′)′
　　　　end − *if*
Comment
The string *situations* describes the situations mentioned in the input text
End-of-comment
　　　　Case of kindtext from
　　　　Stat : *Semrepr* := *Situations*
Comment
Example 2. Let T1 = "Professor Igor Novikov teaches in Tomsk."
Then initially

Semrepr := *the empty string*.

After the termination of the algorithm *BuildSem*1

Semrepr = *Situation* (*e*1, *teaching* ∗ (*Time*, #*now*#)

(*Agent*1, *certn person* ∗ (*Qualif, professor*)

(*Name*, ′*Igor*′) (*Surname*, ′*Novikov*′) : *x*2)

(*Place*3, *certn city* ∗ (*Name*1, ′*Tomsk*′) : *x*3)).

End-of-comment
　　　　Imp : *Semrepr* := *Semrepr* + ′ ∧′ + *Situations* + ′)′
Comment

Example 3. Let T2 = "Deliver the case with details to the warehouse 3."
Then initially

$$Semrepr := {}'(Command\,(\#Operator\#,\#Executor\#,\#now\#,e1){}'.$$

After the termination of the algorithm *BuildSem*

$$Semrepr = (Command\,(\#Operator\#,\#Executor\#,\#now\#,e1)$$

$$\wedge\,Goal\,(e1,delivery1 * (Object1,certn\,case *$$

$$(Content1,certn\,set * (Qual-compos,detail)) : x1)$$

$$(Place2,certn\,warehouse * (Number,3) : x2)))$$

End-of-comment
$$Genqs: Semrepr := Semrepr + Situations + {}')))';$$
Comment
Example 4. Let T3 = "Did the international scientific conference 'COLING' ever take place in Asia?"
Then initially

$$Semrepr := {}'Question\,(x1,(x1 \equiv Truth-value\,('.$$

After the termination of the algorithm *BuildSem*

$$Semrepr := {}'Question\,(x1,(x1 \equiv Truth-value$$

$$(Situation\,(e1,taking\_place * (Time,certn\,moment$$

$$* (Earlier,\#now\#) : t1)\,(Event,certn\,conference$$

$$* (Type1,international)\,(Type2,scientific)$$

$$(Name,``COLING'') : x2)\,(Place,certn\,continent *$$

$$(Name1,{}'Asia') : x3)))))).$$

End-of-comment
$$Specqs-relat1,Specqs-relat2:$$
$$if\,Semrepr = {}'Question\,(x1,'$$
Comment
Example 5. Let T4 = "What publishing house has released the novel 'The Winds of Africa'?"
End-of-comment
$$then\,Semrepr := Semrepr + Situations + {}')'$$
Comment
Example 6. For the question T4

$$Semrepr := {}'Question\,(x1,Situation\,(e1,releasing1 *$$

$$(Time, certn\, moment * (Earlier, \#now\#) : t1)$$

$$(Agent2, certn\, publish-house : x1)\,(Object3, certn\, novel1$$

$$*(Name1, 'The\,Winds\,of\,Africa') : x2)))'$$

End-of-comment
> *else*

Comment

i.e., if $Semrepr = 'Question\,(S1, (Qual-compos\,(S1, '$
End-of-comment
> $posmainnoun := Right\,(mainpos, noun)$

Comment

Example 7. Let T5 = "What foreign publishing houses the writer Igor Somov is collaborating with?"

Then mainpos $= 1$ (the position of the question word "what"),

> $posmainnoun := 3$ (the position of the word group "publishing houses")

End-of-comment
> $sem1 := Sembase\,[posmainnoun]$
> *if sem*1 does not include the symbol $'*'$
> *then semhead* $:= sem1$
> *else loc*1 $:= Matr\,[posmainnoun, locunit]$
> $semhead := Arls\,[loc1, sem]$
> $end-if$

Comment

Example 8. For the question T5

$sem1 := publish-house * (Type-geographic, foreign)$
$semhead := publish\_house$

End-of-comment
> $end-if$
> $Semrepr := Semrepr + sem1 + ') \wedge Description\,(arbitrary' + semhead +$
$' * (Element, S1) : y1, ' + Situations + '))'$

Comment

Example 9. For the question T5 = "What foreign publishing houses the writer Igor Somov is collaborating with?"

$$Semrepr := Question\,(S1, (Qual-compos\,(S1, publish-house$$

$$*(Type-geographic, foreign)) \wedge Description$$

$$(arbitrary\, publish-house * (Element, S1) : y1,$$

$$Situation\,(e1, collaboration * (Time, \#now\#)$$

$$(Agent1, certn\, person * (Occupation, writer)\,(Name, 'Igor')$$

$$(Surname, 'Somov') : x1)\,(Organization1, y1))))))).$$

End-of-comment

$Specqs - role :$

Comment

Example 10. Let T6 = "Who produces the medicine 'Zinnat'?"

Then initially

$$Semrepr := Question\,(.$$

End-of-comment

$Unknowns := {}'x1{}'$

$If\ numbqswd > 1$

$Then\ loop\ for\ k\ from\ 1\ to\ numbqswd - 1$

$\quad Vrb := Stringvar\,({}'x{}', k)$

$\quad Unknowns := Unknowns + {}'\wedge{}' + vrb$

$\quad end - of - loop$

$\quad Unknowns := {}'(' + Unknowns + ')'$

$end - if$

$Semrepr := Semrepr + Unknowns + {}',' + Situations + ')'$

Comment

End of $Specqs - role$

End-of-comment

Comment

Example 11. After the termination of the algorithm $BuildSem1$ for the question T6 = "Who produces the medicine 'Zinnat'?"

$$Semrepr = Question\,(x1, Situation\,(e1, production1$$

$$* (Time, \#now\#)\,(Agent2, x1)$$

$$(Product2, certn\,medicine1 * (Name1, {}'Zinnat') : x2)))$$

End-of-comment

Comment

Example 12. Let T7 = "For whom and where the three-ton aluminum container has been delivered from?"

Then initially

$$Semrepr := {}'Question\,('.$$

After the termination of the algorithm $BuildSem1$

$$Semrepr = Question\,((x1 \wedge x2), Situation\,(e1, delivery2$$

$$* (Time, certn\,moment * (Earlier, \#now\#) : t1)$$

$$(Recipient, x1)\,(Place1, x2)\,(Object1, certn\,container *$$

$$(Weight, 3\,ton))\,(Material, aluminum) : x3))).$$

End-of-comment

    Specqs − quant 1

    posmainnoun := *Right* (*main pos, noun*)

Comment

Example 13. Let T8 = "How many people did participate in the creation of the textbook on statistics?".

    Then main pos := 1 (the position of the question word group "how many"),

      posmainnoun := 2 (the position of the word "people")

End-of-comment

    sem 1 := *Sembase* [*posmainnoun*]

    if sem 1 does not include the symbol $'*'$

    then semhead := *sem* 1

    else loc 1 := *Matr* [*posmainnoun, locunit*]

      semhead := *Arls* [*loc* 1, *sem*]

Comment

Example 14. For the question T8 *sem* 1 := *person, semhead* := *person*

End-of-comment

    end − if

Semrepr := *Semrepr* + *sem* 1 + $'$) $\wedge$ *Description* (*arbitrary*$'$ + *semhead* + $'$ $*$ (*Element, S1*) : *y*1, $'$ + *Situations* + $'$))$'$

Comment

Example 15. For the question T8 = "How many people did participate in the creation of the textbook on statistics?"

$$Semrepr := {}'Question\,(x1,\,((x1 \equiv Numb\,(S1))$$

$$\wedge\,Qual - compos\,(S1,\,person)\,\wedge\,Description$$

$$(arbitrary\,person * (Element,\,S1)\,:\,y1,$$

$$Situation\,(e1,\,participation1 *$$

$$(Time,\,certn\,moment * (Earlier,\,\#now\#)\,:\,t1)$$

$$(Agent1,\,y1)\,(Type - of - activity,\,creation1 *$$

$$(Product1,\,certn\,textbook1 * (Field1,\,statistics)\,:\,x2)))))\,'.$$

End-of-comment

    Specqs − quant 2 : *Semrepr* := *Semrepr* + *Situations* + $'$)))$'$

Comment

Example 16. Let T9 = "How many times Mr. Stepan Semenov flew to Mexico?" Then initially

$$Semrepr := Question\,(x1,\,((x1 \equiv Numb\,(S1))$$

$$\wedge\,Qual - compos\,(S1,\,sit)\,\wedge$$

$$Description\,(arbitrary\,sit * (Element,\,S1)\,:\,e1,\,.$$

After the termination of the algorithm *BuildSem*

$$Semrepr = Question\,(x1,\,((x1 \equiv Numb\,(S1))$$

$$\wedge\,Qual - compos\,(S1,\,sit)\,\wedge$$

$$Description\,(arbitrary\,sit * (Element,\,S1) : e1,$$

$$Situation\,(e1,\,flight * (Time,\,certn\,moment *$$

$$(Earlier,\,\#now\#) : t1)\,(Agent1,\,certn\,person *$$

$$(Name,\,'Stepan')\,(Surname,\,'Semenov') : x2)$$

$$(Place2,\,certn\,country * (Name,\,'Mexico') : x3))))).$$

End-of-comment
end

10.4.3 An Algorithm of Semantic Representation Assembly

Combining the auxiliary algorithms developed above in this chapter, let's construct the algorithm *BuildSem*1 ("Assembly-of-SemRepr").

*External specification of the algorithm BuildSem*1

Input:
Rc – array – classifying representation of the input text T;
Rm – array – morphological representation of the text T;
nt – integer – length of the text T (the quantity of rows in Rc and $Matr$);
$kindtext$ – string characterizing the form of the input text;
$mainpos$ – integer – position of the interrogative word in the beginning of the text;
$Matr$ – MSSR of the text;
$Arls$ – projection of the lexico-semantic dictionary $Lsdic$ on the input text.
Output:
$Performers$ – array;
$Sitdescr$ – array;
$Semrepr$ – string – representation of the input text (semantic representation of the input text which is an expression in a certain standard K-language).

*Algorithm BuildSem*1

$Begin\ Preparation - to - constr - SemRepr$
$\quad(Rc,\,Rm,\,Matr,\,kindtext,\,mainpos,\,Semrepr)$
$\quad Begin - of - constr - SemRepr$
$\quad(Rc,\,Rm,\,Matr,\,kindtext,\,numbqswd,\,Arls,\,Performers,\,Sitdescr,\,Semrepr)$

$$Representation - of - situations$$
$$(Rc, Matr, Performers, Sitdescr)$$
$$Final - operations$$
$$(Rc, kindtext, numbqswd, Matr, Performers, Sitdescr, Semrepr)$$

end

Thus, the algorithm *BuildSem*1 ("Assembly-of-SemRepr") developed in this chapter transforms an MSSR of a question, command, or statement from the practically significant sublanguages of English, German, and Russian languages into its semantic representation being an expression of the SK-language determined by the considered marked-up conceptual basis – a component of a linguistic basis.

10.5 A Multilingual Algorithm of Semantic-Syntactic Analysis of Natural Language Texts

Let's define the algorithm of semantic-syntactic analysis of NL-texts (the questions, commands, and statements) as the composition of the developed algorithms *BuildMatr*1 and *BuildSem*1.

10.5.1 Description of the Algorithm SemSynt1

*External specification of the algorithm SemSynt*1

Input:

Lingb – linguistic basis;

lang – string variable with the values *English, German, Russian*;

T – text from the set *Texts* $(Tform(Lingb))$, where *Tform* is the text-forming system being one of the components of the linguistic basis *Lingb*.

Output

Semrepr – string – K-representation of the input text (a semantic representation of the input text being an expression of a certain SK-language).

*Algorithm SemSynt*1

Begin

      *BuildMatr*1   (*T, Rc, Rm, Arls, kindtext, Matr, mainpos, numbqswd*)

      *BuildSem*1 (*Rc, Rm, Arls, kindtext, Matr, mainpos,*

      numbqswd, maxnumbsit, Semrepr)

End

Example. Let T1 be the question "What Russian publishing house released in the year 2007 the work on multi-agent systems 'Mathematical Foundations of Representing the Content of Messages Sent by Computer Intelligent Agents' by professor Fomichov?", T1germ be the same question in German "Welcher Russischer Verlag im Jahre 2007 die Arbeit ueber Multi-Agenten Systeme 'Mathematische Grund-lagen der Representierung des Sinnes der Nachrichten Schickt von Intelligenten Computeragenten' von Professor Fomichov veroeffentlicht hat?", and let T1russ be the same question in Russian (we use here a Latin transcription of Russian texts) "V kakom rossiyskom izdatelstve byla opublikovana v 2007-m godu rabota po mnogoagentnym sistemam "Matematicheskie Osnovy Predstavleniya Soderzhaniya Poslaniy Komputernykh Intellektualnykh Agentov?" Professora Fomichova?"

Then a linguistic basis *Lingb* can be defined so that at the different stages of processing T1 or T1germ or T1russ the algorithm *SemSynt*1 will create

- two-dimensional arrays *Rc* and *Rm* – the classifying and morphological repre-sentations of the input text (see the examples for T1 in Sect. 8.1);
- the two-dimensional arrays *Arls* – the projection of the lexico-semantic dictio-nary *Lsdic* on the input text, *Arvfr* – the projection of the dictionary of verbal – prepositional semantic-syntactic frames *Vfr* on the input text, and *Arvfr* – the projection of the dictionary of verbal – prepositional semantic-syntactic frames *Vfr* on the input text (see the examples for T1 in Sect. 8.2);
- a matrix semantic-syntactic representation (MSSR) *Matr* of the input text (see the example in Sect. 8.3).

At the final stage of processing T1, the procedure *BuildSem*1 will assemble a K-representation of the text T1 or T1germ or T1russ, this SR can be the K-string *Semrepr* of the following form (independently on the input sublanguage of NL):

$$Question\,(x1,\,Situation\,(e1,\,releasing1 * (Agent2,$$

$$certn\,publish\_house * (Country,\,Russia)\,:\,x2)$$

$$(Time,\,2007/year)\,(Product1,\,certn\,work2 *$$

$$(Name1,\,Title1)\,(Field1,\,multi\_agent\_systems)$$

$$(Authors,\,certn\,person * (Qualif,\,professor)$$

$$(Surname,\,{}'Fomichov')\,:\,x4)\,:\,x3)))),$$

where *Title*1 = "Mathematical Foundations of Representing the Content of Mes-sages Sent by Computer Intelligent Agents."

10.5.2 Some Possible Directions of Using the Algorithm SemSynt1

An important peculiarity of the algorithm *SemSynt*1 is that it directly discovers the conceptual relationships between the meanings of the text's fragments without con-structing a pure syntactic representation of an input text.

The adaptation to a concrete sublanguage of natural language is achieved with the help of the string variable *lang* with the values *English, German, Russian*, this variable is one of the input data of the algorithm *SemSynt*1.

The formal model of a linguistic database described in Chap. 7 and the algorithm of semantic-syntactic analysis *SemSynt*1 can be used, in particular, as a basis for designing NL-interfaces to

- recommender systems of various companies offering the goods and services by means of Web-portals;
- intelligent databases with the information about the manufactured, exported, and imported products;
- autonomous intelligent transport-loading systems (robots);
- intelligent databases with the information about the objects stored in certain sections of an automated warehouse;
- advanced question-answering systems of digital libraries;
- intelligent databases with the information about the scientists, their publications, and projects with their participation;
- the advanced computer systems transforming the NL-annotations of various documents into their semantic annotations being the expressions of SK-languages;
- intelligent databases with the information about the sportsmen and sport competitions.

Chapter 11
Natural Language Processing Applications

Abstract The principles of applying the theory of K-representations to the design of two semantics-oriented natural language processing systems are set forth. The first one is the computer system Mailagent1; the task to be solved by this system is semantic classification of e-mail messages stored in the user's mailbox for enabling the user to more quickly react to the more important and/or urgent messages. The second system is the linguistic processor NL-OWL1 transforming the descriptions in restricted Russian language of situations (in particular, events) and the definitions of notions first into the K-representations and then into the OWL-expressions.

11.1 The Structure of a Computer Intelligent Agent for Semantic Classification of E-mail Messages

11.1.1 The Problem of Semantic Classification of E-mail Messages

One of the peculiarities of the new information society is a huge number of e-mail messages received every day by intensively working specialists. For instance, it is noted in [169] that a specialist may receive every day approximately 100 e-mail messages. That is why in case of a 4-day business trip he/she will face the necessity of analyzing 500 messages (400 received during his/her trip and 100 received just after the return from the trip). The realization of such an analysis demands to spend a lot of time. Most often, a specialist has no such time. Hence it is very likely that he/she will be unable to answer some important messages in due time. That is why the task was posed of designing an intelligent computer agent (an electronic secretary) being able to classify the e-mail messages in the English language. As a result, the computer intelligent agent Mailagent1 was elaborated [99]. Its functions are as follows:

The system Mailagent1 is intended for automatic classification of the e-mail messages stored in the mailbox of a user. The work of the user with the preliminary

V.A. Fomichov, *Semantics-Oriented Natural Language Processing*, IFSR International Series on Systems Science and Engineering 27, DOI 10.1007/978-0-387-72926-8_11,
© Springer Science+Business Media, LLC 2010

sorted e-mail messages saves time and enables the user to more quickly react to the more important and/or urgent messages.

The elaborated system has two main components: the adaptation subsystem and the subsystem of linguistic analysis. It is assumed that the receipt of the e-mail messages is the function of a usual e-mail program. That is why the described system destined for the classification of e-mail messages has the possibility of adaptation to the format used for saving the messages on a hard disk of the user's computer.

The subsystem of linguistic analysis proceeds from the following parameters in order to classify the e-mail messages: (1) whether the receiver is waiting for the considered message (from a particular specialist, from a particular address, or from an address belonging to a particular group of addresses); (2) what is the indicated deadline for sending a reply; (3) is it a message sent personally to the receiver (an individual letter) or a message sent at all addresses of a mailing list (e.g., DBWORLD). If the deadline for sending a reply isn't indicated in the text of the message but a hyperlink (an URL) to a Web page is given, the program finds the corresponding Web page and analyzes the content of this Web page. If this Web document indicates the deadline for sending a reply (e.g., for submitting a paper to the Program Committee of an international conference), then the program adds this information to the considered e-mail message.

The messages addressed personally to a receiver form the most important part of the correspondence. That is why a semantic analysis of such messages is carried out in order to understand (in general) their meaning. The basis for fulfilling the semantic analysis of e-mail messages is a linguistic database. Its central components are the system of semantic-syntactic patterns and the lexical – semantic dictionary. The result of semantic analysis is the distribution of the messages into the conceptual categories. The program Mailagent1 possesses the means for visually representing the categories of the messages obtained in the course of its work. Besides, this program provides the possibility of viewing the contents of the e-mail messages belonging to each conceptual category. This makes the viewing of the results a quick and convenient process.

All methods of linguistic analysis employed in the system of automatic classification of the e-mail messages have shown in practice their working properties and effectiveness. A background for elaborating these methods was provided by the theory of K-representations. If necessary, the methods of linguistic analysis can be expanded for the work with new situations by means of a modification of the linguistic database of the program. The computer system Mailagent1 is implemented with the help of the programming language Java.

11.1.2 An Outline of the Computer Intelligent Agent Mailagent1

The constructed computer system Mailagent1 functions in the following way. After the user of this system inputs the date of viewing the received e-mail messages, Mailagent1 forms two systems of folders: (1) EXPECTED MESSAGES,

(2) OTHER MESSAGES. The folder 1 contains the e-mail messages received from (a) the persons stored in a special list in the computer's memory. A part of these people is simply important for the end user, and he/she expects to receive a message from other people in this list.

For the formation of the folder 1, the following information is used: (a) the first and last names of people being particularly important for the end user; (b) the list of fixed e-mail addresses; (c) the list of fixed Internet sites, if the end user is waiting for a message from such an e-mail address that an ending of this address shows its association with an Internet site from this list.

For instance, a researcher is expecting a message with the decision of the Organizing Committee of an International Conference to be held at the University of Bergen, Norway, about a grant for attending this conference. This decision can be sent by any of the members of the Organizing Committee from his/her personal address, and the only common feature of these addresses is an ending. So in the considered case the list of stored information about the Internet sites is to include the ending "uib.no."

The folder 2, called OTHER MESSAGES, is destined for storing all other received e-mail messages, i.e., the e-mail messages not included in the folder EXPECTED MESSAGES. Both the folders have subfolders called (1) UNDEFINED, (2) OVERDUE, (3) 1 WEEK, (4) 1 MONTH, (5) OTHER MESSAGES. The subfolder UNDEFINED contains all messages with indefinite last date of a reply. All such received messages that the deadline for returning a reply has been over are stored in the subfolder OVERDUE. The subfolders 1 WEEK and 1 MONTH are destined for the messages that are to be answered in one week and one month, respectively.

Each of the folders 1–5 contains the subfolders (1) PERSONAL MESSAGES and (2) COLLECTIVE MESSAGES. If an e-mail message is addressed just to a particular person, then this message is included in the subfolder (1), otherwise (e.g., in case of a message from a mailing list, such as DBWORLD) a message is included in the folder (2).

Personal messages may be most interesting to the end user. Such messages may contain the proposals to participate in an international scientific project, to prepare an article for a special issue of an international journal or to write a chapter for a book, to become a member of the Program Committee of an international scientific conference or a member of the Editorial Board of an international scientific journal, etc. Let's say that this part of personal messages encourages the end user to carry out some action. The other part of personal messages may express gratitude for some action fulfilled before by the end user; for sending a hard copy of a paper, for an invitation to take part in an international workshop, etc. Obviously, there are also some other categories of the received e-mail messages.

That is why the elaborated computer system tries to "extract" from an individual message its generalized meaning and, proceeding from this extracted meaning, to associate this message with some conceptual category. The examples of such conceptual categories (or the goals of sending a message) are ACTION ENCOURAGEMENT and THANKS. The messages may express, in particular, the generalized meanings SEND PAPER and COLLOBORATE IN THE PROJECT.

In a message, the generalized meaning "TO SEND PAPER" can be formulated in two possible contexts: (C1) "I would be glad if you send me your recent paper ⟨...⟩" and (C2) "I would be glad to send you the paper ⟨...⟩." So it is very important to reflect in the representation of the goal of sending a message whether it reflects an expected action of the recipient of this message or it reflects an intention of its sender. With this aim, the meaning of the message is completed with the information "Action" for the context C1 or "Sender's intention" for the context C2. So the meaning of a message for the contexts C1 and C2 is represented with the help of the string "Action: SEND PAPER" and the string "Sender's Intention: SEND PAPER."

11.1.3 General Structure of Computer Intelligent Agent Mailagent1

The elaborated system Mailagent1 consists of the following four modules interacting with its main modules:

1. the module of looking for individually oriented messages;
2. the module of finding the deadline for a reply to a message;
3. the module of the analysis whether a message has been expected; item the module of semantic analysis of individual letters.

The joint work of these modules enables the system to write every message in a folder corresponding to this message. In this chapter, the principal attention is paid to describing the work of the last module.

The structure of the folders of the computer system Mailagent1 is reflected in Fig. 11.1. The subfolders of the lower levels of the folder PERSONAL MESSAGES are formed automatically in the process of semantically classifying the stored e-mail messages. For the messages from such subfolders, their generalized meaning is indicated, and a connection of the described action with the sender of a message or its receiver is explained by means of the heading "Action" or "Sender's Intention."

11.1.4 The Structure of Semantic-Syntactic Patterns and Lexico-Semantic Dictionary

In the elaborated computer intelligent agent Mailagent1, the linguistic analysis of the contents of the received e-mail messages is based on the use of special semantic-syntactic patterns (SSPs). The idea is that the employment of rather simple means is able to give a considerable effect as concerns semantic classification of the received messages. A collection of SSPs (stored as textual files) is a part of the Linguistic Database (LDB) of Mailagent1.

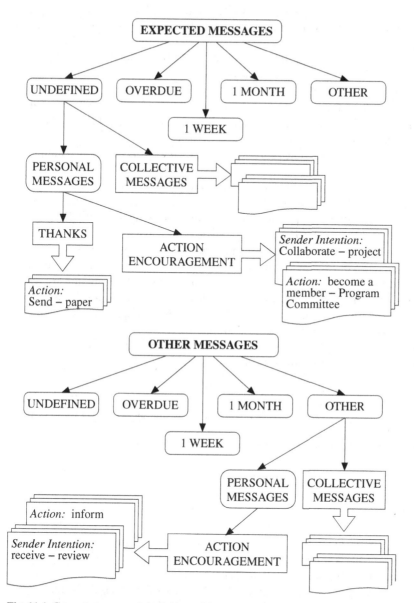

Fig. 11.1 General structure of the folders of the computer system Mailagent1

From the formal point of view, SSPs are strings. Basic components of these strings are substrings called positive indicators of units(PIUs) and negative indicators of units (NIUs). Suppose that arbitrary PIU A1 and NIU B1 are the components of an arbitrary SSP Pt1. Then A1 shows that every NL-text T1 being "compatible" with Pt1 is to include a component of some first definite kind. To the contrary, the meaning of B1 is that such text T1 is not to include the components of some second definite kind.

Let's consider the distinguished classes of PIU (the number of such classes is four). The PIUs of the first class have the beginning $'1\#'$ and show that some English words or short word combinations (for instance, "would," 'grateful," etc.) are to occur in the analyzed text. If for expressing some meaning, any of the synonimical words $d_1, \ldots, d_n$, where $n > 1$, can be used, then a corresponding PIU is to include the fragment $d_1/, d_2/ \ldots,/ d_n$..

For instance, the expression $1\#grateful/thankful/appreciate$ is an example of a PIU.

The PIUs of the *second class* have the beginning $'2\#'$ and say that every text being compatible with the considered semantic-syntactic pattern is to include a word belonging to one of the speech parts indicated after the beginning $'2\#'$. The expressions $'2\#subs't$, $'2\#subst/attr'$ are the examples of the PIUs from the second class, where subst , attr designate the speech parts "substantive" and "attribute."

The *third class* of PIUs is formed by the elements demanding the occurrence in the considered texts of the words and expressions being the designations of some semantic items: "information transmission", "telephone," "a transport means," etc. The PIUs from this class have the beginning $'3\#'$. The examples of such elements are as follows: $'3\#inform.transmis'$, $'3\#telephone'$.

The PIUs of the last, *fourth class* designate such sort or a combination of sorts that this sort s_1 or a combination of sorts

$$s_1 * s_2 * \ldots * s_n,$$

where $n > 1$, is to be associated with some word being necessary for expressing a given meaning.

For instance, a semantic-syntactic pattern may include the element

$$'4\#dynam\_phys\_ob * intel\_syst',$$

where the expressions *dynam_phys_ob* and *intel_syst* are to be interpreted as the sorts "dynamic physical object" and "intelligent system," respectively. This component of an SSP can be used in order to show that any NL-text being compatible with the used SSP is to include a designation of a person (being simultaneously a dynamic physical object and an intelligent system).

The idea of using the concatenations of the sorts stems from the theory of K-representations. Since very many words are associated with several sorts, i.e., general semantic items (speaking metaphorically, the entities denoted by such words have different "coordinates" on different "semantic axes"), different "semantic coordinates" of a word are taken into account in order to find conceptual connections of the words in NL-texts.

Each negative indicator of a unit is a string of the form *%%Expr*, where *Expr* is any PIU. Such components of semantic-syntactic patterns demand the lack in the analyzed NL-texts of the fragments of the four kinds discussed above.

For example, the NIU %%1# would of an arbitrary SSP Pt1 require that any NL-text being compatible with $Pt1$ doesn't include the word "would." Such an element can be used as follows. If a message includes the fragment

"I (or we) would be grateful (or thankful) to you for $\langle Designation\,of\,an\,Action \rangle$",

we understand that the indicated action is a desired action, but it wasn't fulfilled by the moment of sending the message. To the contrary, a similar fragment without the word "would" usually shows that the mentioned action was fulfilled by the moment of sending the message, and one of the purposes of this message is to thank the recipient of the message for some carried out action.

Some special components of the SSPs establish the direction and borders in order to look for the needed fragments of the analyzed texts. Such components are the expressions of the form "M-1," "M-2," "M-3," "M+1," "M+2," etc. If such expression includes the sign "minus," then the needed fragments are to be searched in the sentences preceding the considered sentence. The number k after the sign "minus" ("plus") indicates how many sentences before (after) the considered phrase are to be analyzed.

For instance, if $k = 1$, then only the preceding sentences are to be processed; if $k = +2$, then the considered phrase and next two sentences are to be analyzed. The ending of the zone for the search is the symbol "/".

Every semantic-syntactic pattern (SSP) Pt can be represented in the form

$$A * * * B * * * C,$$

where the fragments A, B, C are as follows: The fragment A is a sequence of positive and (only in some patterns) negative indicators of units (PIUs and NIUs), these indicators are separated by commas. Such a sequence of indicators expresses a system of requirements to be satisfied by each NL-text being compatible with the considered SSP Pt.

The fragment B is a designation of the meaning of every NL-text being compatible with this SSP Pt. The fragment C is a sequence of positive indicators of units enabling to concretize the meaning of the analyzed NL-text and to submit this meaning to the recipient of the e-mail messages. If a component of the fragment C of an SSP isn't destined for including in the meaning submitted to the recipient of the messages, one poses the sign "minus" ("–") before the corresponding component of C.

Example 1. Let $Pt1$ be the expression

$$1\#I/we, \%\%1\# would/in\,advance,$$

$$1\#thankful/grateful/appreciate$$

$$* * *THANKS * * * 4\#phys\_action,$$

$$4\#inf\_object * phys\_object/phys\_object.$$

This expression is a semantic-syntactic pattern (SSP) including both positive and negative indicators of units. If any NL-text T1 is compatible with the SSP $Pt1$, the computer intelligent agent Mailagent1 draws the conclusion that the generalized meaning of T1 is THANKS, i.e., the purpose of sending an e-mail message with T1 was to express gratitude for carrying out some physical action. An important argument in favor of this conclusion is the lack in T1 both of the word "would" and the expression "in advance."

The SSP $Pt1$ helps also to find a fragment of T1 describing such physical action. It is done in the following way. After finding one of the words "thankful," "grateful," or "appreciate" (let's denote it by $d1$), the computer agent analyzes the words to the right from the word $d1$. Suppose that it first discovers the word $d2$ that can be associated with the sort "physical action," and then finds (to the right from $d2$) the word $d3$ such that $d3$ can be associated either both with the sort "informational object" and "physical object" or with the only sort "physical object."

Then the computer agent submits to the mail box user the string $d2\,h\,d3$, where h is either the null string or the substring of T1 separating $d2$ and $d3$. In this case, the submitted string $d2\,h\,d3$ is interpreted as a description of the carried out action (this was the cause of sending an e-mail message with an expression of gratitude).

Example 2. Let's consider the expression $Pt2$ of the form

$$1\#thank,\ 1\#in\,advance,\ M-2/3\#possibility***$$

$$ACTION\,ENCOURAGEMENT***M-3/4\#phys\_action,$$

$$4\#inf\_object*phys*object/phys\_object.$$

The ideas underlying this SSP are as follows. An e-mail letter can implicitly encourage its receiver to carry out some physical action. Let's imagine, for instance, that a letter contains the following fragment T2:

"I would be happy to receive from you a hard copy of your paper published last year in Australia. Would it be possible? If yes, thank you in advance for your time and efforts."

The SSP $Pt2$ will be matched against T2 as follows. The third sentence of T2 contains the word "thank" and the expression "in advance." The element $'M-2/'$ stimulates the computer agent to look for the semantic unit "possibility" as one of the semantic units associated with the words in two sentences before the considered third sentence. Since the second sentence contains the word "possible," Mailagent1 draws the conclusion that the generalized meaning of T2 is ACTION ENCOUR-AGEMENT.

The next question is what action is to be carried out. For answering this question, the SSP $Pt2$ recommends to analyze three sentences to the left from the found word expressing the meaning "possibility" (due to the element "M-3/") and to look in these sentences for (a) a word or word combination that can be associated with the sort "physical action" and (b) for the word or word combination that can be associated with each of two sorts "informational object," "physical object" or with the only sort "physical object". The case (a) takes place, for instance, for the verbs and verbal substantives "to send," "the sending," "to sign," and the case (b) – for the expressions "a hard copy of your article," "CD-ROM," "this contract."

The system of semantic-syntactic patterns is associated with the other component of a linguistic database (LDB); this component is called Lexico-Semantic Dictionary. It is one of the relations of the LDB; its attributes are as follows: (1) word, (2) speech part, (3) semantic unit (SU), (4) sort 1 associated with the semantic unit, (4) sort 2 associated with the SU or the empty sort NIL, (5) sort 3 associated with the SU or the empty sort NIL. This dictionary is stored in the computer's memory (for instance, in the form of a table) and can be modified. For the elaborated computer program, the lexico-semantic dictionary is a textual file.

11.1.5 Implementation Data

The computer intelligent agent Mailagent1 has been implemented in the programming language Java (version JDK 1.1.5). This computer system is intended for automatic semantic classification of the e-mail messages in English stored on the hard disk of a computer. The program Mailagent1 was tested in the environment Windows 95/98/2000/NT. The file with the messages for the classification was represented as a file of the e-mail system that received these messages from the Internet.

11.2 A Transformer of Natural Language Knowledge Descriptions into OWL-Expressions

In the context of transforming step by step the existing Web into Semantic Web (see Sect. 6.10), the need for large Web-based and interrelated collections of formally represented pieces of knowledge covering many fields of professional activity is a weighty ground for increasing the interest of the researchers to the problem of automated formation of ontologies.

It seems that the most obvious and broadly applicable way is to construct a family of NLPSs being able to transform the descriptions of knowledge pieces in NL (in English, Russian, German, Chinese, Japanese, etc.) into the OWL-expressions and later, possibly, into the expressions of an advanced formalism for developing ontologies.

This idea underlay the design of the computer system NL-OWL1, it is a Russian-language interface implementing a modification of the algorithm of semantic-syntactic analysis *SemSyn* stated in [85]. The main directions of expanding the input language of the algorithm *SemSyn* are as follows:

- the definitions of notions in restricted Russian language can be the input texts of the system;
- a mechanism of processing the homogeneous members of the sentence is added to the algorithm of semantic-syntactic analysis;

- a part of input sentences (the descriptions of the events and the definitions of notions) is transformed not only into the K-representations (i.e., into the expressions of a certain SK-language) but also, at the second stage, into the OWL-expressions.

Figure 11.2 illustrates the structure of the computer system NL-OWL1.

Let's consider the examples illustrating the principles of processing NL-texts by the experimental Russian-language interface NL-OWL1, implemented in the Web programming system PHP.

Example 1. Definition: "Carburettor is a device for preparing a gas mixture of petrol and air."

K-representation:

$$ModuleOfKnowledge\,(definition;\,carburetor;\,x1;$$

$$(Definition1\,(certn\,carburetor\,:\,x1,\,certn\,device\,:\,x2\wedge$$

$$Purpose\,(x2,\,certn\,preparation1\,:\,x\_e1)\wedge$$

$$Description\,(preparation1,\,Object1\,(certn\,mixture*$$

$$(Type,\,gas)\,:\,x3))\wedge Product1\,(x3,\,certn\,petrol\,:\,x4)$$

$$\wedge Product1\,(x3,\,air))))$$

OWL-expression:
$\langle owl\,:\,Class\,rdf\,:\,ID = ``ModuleOfKnowledge"/\rangle$
$\langle hasFormModule\,rdf\,:\,resource = ``\#definition"/\rangle$
$\langle hasConcept\,rdf\,:\,resource = ``\#carburetor\_x1"/\rangle$
$\langle owl\,:\,Class\,rdf\,:\,ID = ``Action"/\rangle$
$\langle Action\,rdf\,:\,ID = ``Concept"/\rangle$
$\langle owl\,:\,Class\,rdf\,:\,ID = ``Situation"/\rangle$
$\langle Situation\,rdf\,:\,ID = ``x1"\rangle$
$\langle hasAction\,rdf\,:\,resource = ``\#Concept"/\rangle$
$\langle hasDetermination\,rdf\,:\,resource = ``\#Device\_x2"/\rangle$
$\langle /Situation\rangle$
$\langle owl\,:\,Class\,rdf\,:\,ID = ``Device"/\rangle$
$\langle Device\,rdf\,:\,ID = ``Device\_x2"/\rangle$
$\langle Destination\,rdf\,:\,resource = ``\#preparation\_x\_e1"/\rangle$
$\langle /Device\rangle$
$\langle owl\,:\,Class\,rdf\,:\,ID = ``Destination"/\rangle$
$\langle Destination\,rdf\,:\,ID = ``Destination\_x\_e1"/\rangle$
$\langle Object1\,rdf\,:\,resource = ``\#Mixture\_x3"/\rangle$
$\langle /Destination\rangle$
$\langle owl\,:\,Class\,rdf\,:\,ID = ``Mixture"/\rangle$
$\langle Mixture\,rdf\,:\,ID = ``Mixture\_x3"/\rangle$
$\langle Form\,rdf\,:\,resource = ``\#gas"/\rangle$
$\langle Product1\,rdf\,:\,resource = ``\#Petrol\_x4"/\rangle$

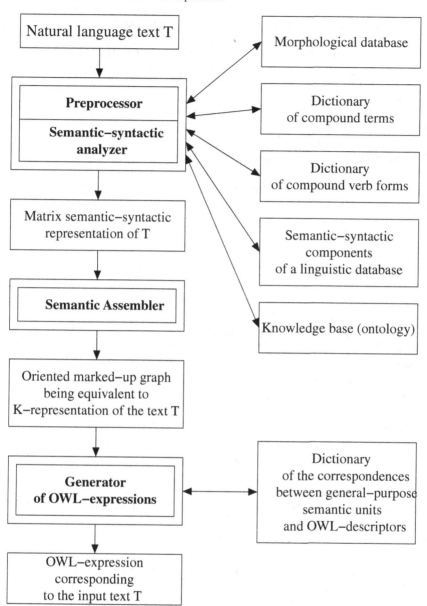

Fig. 11.2 The structure of the computer system NL-OWL1

$\langle Product1\ rdf : resource = \text{``}\#Air\_x5\text{''}/\rangle$
$\langle /Mixture\rangle$
Example 2. Ramp is an inclined surface for the entrance of transport means.
K-representation:

$ModuleOfKnowledge\,(definition;\,ramp;\,x1;\,(Definition1$

$(certn\,ramp\,:\,x1,\,certn\,surface*(Position,\,inclined)\,:\,x2\,\wedge$

$Purpose\,(x2,\,certn\,entrance1\,:\,x\_e1)\,\wedge\,Description\,(entrance1,$

$Object1\,(certn\,set*(Compos,\,mean*(Type,\,transport))\,:\,S1)))))$

OWL-expression:
$\langle owl\,:\,Class\,rdf\,:\,ID\,=\,"ModuleOfKnowledge"/\rangle$
$\langle hasFormModule\,rdf\,:\,resource\,=\,"\#definition"/\rangle$
$\langle hasConcept\,rdf\,:\,resource\,=\,"\#ramp\_x1"/\rangle$
$\langle owl\,:\,Class\,rdf\,:\,ID\,=\,"Action"/\rangle$
$\langle Action\,rdf\,:\,ID\,=\,"Concept"/\rangle$
$\langle owl\,:\,Class\,rdf\,:\,ID\,=\,"Situation"/\rangle$
$\langle Situation\,rdf\,:\,ID\,=\,"x1"\rangle$
$\langle hasAction\,rdf\,:\,resource\,=\,"\#Concept"/\rangle$
$\langle hasDetermination\,rdf\,:\,resource\,=\,"\#Surface\_x2"/\rangle$
$\langle /Situation\rangle$
$\langle owl\,:\,Class\,rdf\,:\,ID\,=\,"Surface"/\rangle$
$\langle Surface\,rdf\,:\,ID\,=\,"Surface\_x2"/\rangle$
$\langle Disposition\,rdf\,:\,resource\,=\,"\#inclined"/\rangle$
$\langle Destination\,rdf\,:\,resource\,=\,"\#entrance\_x\_e1"/\rangle$
$\langle /Surface\rangle$
$\langle owl\,:\,Class\,rdf\,:\,ID\,=\,"Destination"/\rangle$
$\langle Destination\,rdf\,:\,ID\,=\,"Destination\_x\_e1"/\rangle$
$\langle Object1\,rdf\,:\,resource\,=\,"\#Resource\_S1"/\rangle$
$\langle /Destination\rangle$
$\langle owl\,:\,Class\,rdf\,:\,ID\,=\,"Resource"/\rangle$
$\langle Resource\,rdf\,:\,ID\,=\,"Resource\_S1"/\rangle$
$\langle Form\,rdf\,:\,resource\,=\,"\#transport"/\rangle$
$\langle /Resource\rangle$

Example 3. A hand screw is a transportable mechanism for lifting and holding an object at a small height.

K-representation:

$ModuleOfKnowledge\,(definition;\,with\,a\,hand\,screw;\,x1;$

$(Definition1\,(certn\,hand\,screw\,:\,x1,$

$certn\,mechanism*(Feature,\,transportability)\,:\,x2\,\wedge$

$Purpose\,(x2,\,certn\,lifting1\,:\,x\_e1)\,\wedge$

$Purpose\,(x2,\,certn\,holding1\,:\,x\_e2)\,\wedge$

$Description\,(holding1,\,Object1\,(certn\,object\,:\,x3))\,\wedge$

$Place1\,(x3,\,certn\,height*(Degree,\,small)\,:\,x4))))$

OWL-expression:

$\langle owl \; : \; Class \, rdf \; : \; ID \; = \; "ModuleOfKnowledge"/\rangle$

$\langle hasFormModule \, rdf \; : \; resource \; = \; "\#definition"/\rangle$

$\langle hasConcept \, rdf \; : \; resource \; = \; "\#with \, a \, hand \, screw\_x1"/\rangle$

$\langle owl \; : \; Class \, rdf \; : \; ID \; = \; "Action"/\rangle$

$\langle Action \, rdf \; : \; ID \; = \; "Concept"/\rangle$

$\langle owl \; : \; Class \, rdf \; : \; ID \; = \; "Situation"/\rangle$

$\langle Situation \, rdf \; : \; ID \; = \; "x1"\rangle$

$\langle hasAction \, rdf \; : \; resource \; = \; "\#Concept"/\rangle$

$\langle hasTime \, rdf \; : \; resource \; = \; "\#Now"/\rangle$

$\langle hasDetermination \, rdf \; : \; resource \; = \; "\#Mechanism\_x2"/\rangle$

$\langle /Situation\rangle$

$\langle owl \; : \; Class \, rdf \; : \; ID \; = \; "Mechanism"/\rangle$

$\langle Mechanism \, rdf \; : \; ID \; = \; "Mechanism\_x2"/\rangle$

$\langle Property \, rdf \; : \; resource \; = \; "\#transportable"/\rangle$

$\langle Destination \, rdf \; : \; resource \; = \; "\#lifting\_x\_e1"/\rangle$

$\langle Destination \, rdf \; : \; resource \; = \; "\#holding\_x\_e2"/\rangle$

$\langle /Mechanism\rangle$

$\langle owl \; : \; Class \, rdf \; : \; ID \; = \; "Destination"/\rangle$

$\langle Destination \, rdf \; : \; ID \; = \; "Destination\_x\_e2"/\rangle$

$\langle Object1 \, rdf \; : \; resource \; = \; "\#Object\_x3"/\rangle$

$\langle /Destination\rangle$

$\langle owl \; : \; Class \, rdf \; : \; ID \; = \; "Object"/\rangle$

$\langle Object \, rdf \; : \; ID \; = \; "Object\_x3"/\rangle$

$\langle Place1 \, rdf \; : \; resource \; = \; "\#Height\_x4"/\rangle$

$\langle /Object\rangle$

$\langle owl \; : \; Class \, rdf \; : \; ID \; = \; "Height"/\rangle$

$\langle Height \, rdf \; : \; ID \; = \; "Height\_x4"/\rangle$

$\langle Extent \, rdf \; : \; resource \; = \; "\#small"/\rangle$

$\langle /Height\rangle.$

Due to the broad expressive possibilities of SK-languages, the intelligent power of the transformer NL-OWL1 can be considerably enhanced. That is why the formal methods underlying the design of the system NL-OWL1 enrich the theoretical foundations of the Semantic Web project.

Appendix A
A Proofs of Lemmas 1, 2 and Theorem 4.5 from Chapter 4

All's well that ends well

A.1 Proof of Lemma 1

For the sake of convenience, let's repeatedly adduce the basic definition used in Lemma 1 and the formulation of this lemma.

Definition. Let B be an arbitrary conceptual basis, $n \geq 1$, for $i = 1,\ldots,n$, $c_i \in D(B)$, $s = c_1 \ldots c_n$, $1 \leq k \leq n$. Then let the expressions $lt_1(s, k)$ and $lt_2(s, k)$ denote the number of the occurrences of the symbol $'('$ and the symbol $'\langle'$ respectively in the substring $c_1 \ldots, c_k$ of the string $s = c_1 \ldots c_n$.

Let the expressions $rt_1(s, k)$ and $rt_2(s, k)$ designate the number of the occurrences of the symbol $')'$ and the symbol $'\rangle'$ into the substring $c_1 \ldots c_k$ of the string s. If the substring $c_1 \ldots c_k$ doesn't include the symbol $'('$ or the symbol $'\langle'$, then let respectively

$$lt_1(s, k) = 0, \; lt_2(s, k) = 0,$$

$$rt_1(s, k) = 0, \; rt_2(s, k) = 0.$$

Lemma 1. Let B be an arbitrary conceptual basis, $y \in Ls(B)$, $n \geq 1$, for $i = 1,\ldots, n$, $c_i \in D(B)$, $y = c_1 \ldots c_n$. Then
(a) if $n > 1$, then for every $k = 1,\ldots, n-1$ and every $m = 1, 2$

$$lt_m(y, k) \geq rt_m(y, k);$$

$$(b) \; lt_m(y, n) = rt_m(y, n).$$

Proof. Let's agree that for arbitrary sequence $s \in Ds^+(B)$, the length of S (the number of elements from $Ds(B)$) will be denoted as $l(s)$ or $Length(s)$.

Let B be an arbitrary conceptual basis, $y \in Ls(B)$, $n \geq 1$, for $I = 1,\ldots, n$, $c_i \in D(B)$, $y = {}_1 \ldots {}_n$. If $n = 1$, then it immediately follows from the rules $P[0], P[1],\ldots, P[10]$ that $y \in X(B) \cup V(B)$.

According to the definition of conceptual basis, the symbols from the sets $X(B)$ and $V(B)$ are distinct from the symbols $'(', ')', '\langle', '\rangle'$. That is why

V.A. Fomichov, *Semantics-Oriented Natural Language Processing*, IFSR International Series on Systems Science and Engineering 27, DOI 10.1007/978-0-387-72926-8, © Springer Science+Business Media, LLC 2010

$$lt_1(y,n) = rt_1(y,n) = 0,$$

$$lt_2(y,n) = rt_2(y,n) = 0.$$

If $n > 1$, then y has been constructed with the help of some rules from the list $P[1],\ldots,P[10]$ (in addition to the rule $P[0]$). Let's prove the lemma by induction, using the number q of the applications of the rules $P[1],\ldots, : P[10]$ for building y as the parameter of induction.

Part 1. Let's assume that $q = 1$.

Case 1.1. Let a string y be obtained by applying just one time the rule $P[2]$ or $P[4]$. Then

$$y = f(u_1,\ldots,u_m) \text{ or } y = r(u_1,\ldots,u_m),$$

where $m \geq 1$, $f \in F(B)$, $r \in R(B)$,

$$u_1,\ldots,u_m \in X(B) \cup V(B).$$

Obviously, in this situation

$$lt_1(y, 1) = rt_1(y, 1) = 0, lt_1(y, n) = rt_1(y, n) = 1,$$

for $1 < k < n$, $lt_2(y, k) = 1$, $rt_2(y, k) = 0$,
    for $k = 1,\ldots,n$ $lt_2(y, k) = rt_2(y, k) = 0$.

Case 1.2. If y is obtained by applying the rule $P[3]$, then there are such elements $u_1, u_2 \in X(B) \cup V(B)$, that $y = (u_1 \equiv u_2)$. Then for $k = 1,\ldots,n-1$,

$$lt_1(y, k) = 1, rt_1(y, k) = 0,$$

$$lt_1(y, n) = rt_1(y, n) = 1,$$

for $k = 1,\ldots,n$,

$$lt_2(y, k) = rt_2(y, k) = 0.$$

Case 1.3. Let's suppose that the string y is constructed as a result of applying just just one time the rule $P[7]$, and with this aim the binary logical connective $bin \in \{\wedge, \vee\}$ and the elements $d_1,\ldots,d_m \in X(B) \cup V(B)$, where $m > 1$, were used. Then

$$y = (d_1 \, bin \, d_2 \, bin \ldots bin \, d_m).$$

Therefore, for $k = 1,\ldots,n-1$,

$$lt_1(y, k) = 1, rt_1(y, k) = 0,$$

$$lt_1(y, n) = 1, rt_1(y, n) = 1,$$

for $k = 1,\ldots,n$, $lt_2(y, k) = rt_2(y, k) = 0$.

Case 1.4. If y is obtained by applying just one time the rule $P[1]$ or $P[6]$, then y doesn't contain the symbols $'('$, $')'$, $'\langle'$, $'\rangle'$. Then

$$lt_1(y, k) = rt_1(y, k) = lt_2(y, k) = rt_2(y, k) = 0$$

for $k = 1, \ldots, n$.

Case 1.5. The analysis of the rule $P[9$ shows that the string y couldn't be constructed from the elements of the sets $X(B)$ and $V(B)$ by means of applying just one time the rule $P[9]$ (and, possibly, the rule $P[0]$).

Supposing that it is not true we have $y = Qv(concept)\,descr$, where $Q \in \{\forall,\ exists\}$, $v \in V(B)$, $concept \in X(B)$, $descr\,\&\,P \in Ts(B)$, where P is the sort "a meaning of proposition." But the string des is to include the variable v, where $tp(v) = [entity]$. That is why $descr = v$. This relationship contradicts the relationship $descr\,\&\,P \in Ts(B)$.

Case 1.6. If the string y was obtained by applying one time the rule $P[10]$, y doesn't include the symbols $'('$, $')'$, and the first symbol of y is $'\langle'$. Then for $k = 1, \ldots, n$,

$$lt_1(y, k) = 1,\ rt_1(y, k) = 0,$$

for $k = 1, \ldots, n-1$, $lt_2(y, k) = 1$, $rt_2(y, k) = 0$,

$$lt_2(y, n) = rt_2(y, n) = 1.$$

Part 2. Let such $q \geq 1$ exist that the statements (a), (b) of Lemma 1 are valid for every string y constructed out of the elements from the sets $X(B)$ and $V(B)$ by applying not more than q times the rules $P[1], \ldots, P[10]$ (and, possibly, by applying arbitrarily many times the rule $P[0]$).

Let's prove that in this case the statements (a), (b) are true for every string y obtained by applying $q+1$ times the rules from the list $P[1], \ldots, P[10]$. Consider possible cases.

Case 2.1. $y = f(a_1, \ldots, a_m)$ or $y = r(a_1, \ldots, a_m)$, where $m > 1$. For $a_1, \ldots, a_m$ the conditions (a), (b) are satisfied. That is why, obviously,

$$lt_1(y, 1) = rt_1(y, 1) = 0,$$

$$lt_1(y, 2) = 1,\ rt_1(y, 2) = 0;$$

for $i = 3, \ldots, n-1$,

$$lt_1(y, i) > rt_1(y, i),\ lt_1(y, n) = rt_1(y, n).$$

Case 2.2. If y is the string $(a_1 \equiv a_2)$, then it follows from the inductive assumption that

$$lt_1(y, 1) = 1,\ rt_1(y, 1) = 0,$$

for $i = 2, \ldots, n-1$, $lt_1(y, n) = rt_1(y, n)$.

Case 2.3. Let such a binary logical connective $b \in \{\wedge, \vee\}$, such $m > 1$, and such strings $a_1, \ldots, a_m$ exist that y was obtained from $b, a_1, \ldots, a_m$ by applying just one time the rule $P[7]$. Then

$$y = (a_1\,b\,a_2\,b \ldots b\,a_m).$$

Let $Length(a_1) = n_1, \ldots, Length(a_m) = n_m$. Obviously, for $a_1, \ldots, a_m$ the statements (a), (b) of Lemma 1 are valid. Let's notice that the connective b is located in y in the positions

$$n_1 + 2, \, n_1 + n_2 + 3, \, n_1 + \ldots + n_m + (m + 1);$$

then

$$lt_1(y, 1) = 1, \, rt_1(y, 1) = 0;$$

for $p = n_1 + 2, \, n_1 + n_2 + 3, \, n_1 + n_2 + \ldots + n_m + m + 1,$

$$rt_1(y, p) = lt_1(y, p) - 1;$$

for $i = 1, \ldots, m - 1$ and for every such p that

$$n_1 + \ldots + n_i + i + 1 < p < n_1 + \ldots + n_i + n_{i+1} + i + 2.$$

$$rt_1(y, p) \le lt_1(y, p) - 1;$$

$$lt_1(y, n) = rt_1(y, n).$$

A.2 Proof of Lemma 2

For the sake of convenience, let's repeatedly adduce the formulation of Lemma 2 from Sect. 3.8.

Lemma 2. Let B be an arbitrary conceptual basis, $y \in Ls(B), n > 1, y = c_1 \ldots c_n$, where for $i = 1, \ldots, n, \, c_i \in D(B)$, the string y includes the comma or any of the symbols $\equiv, \wedge, vee$, and k be such arbitrary natural number that $1 < k < n$. Then

(a) if c_k is one of the symbols $\equiv, \wedge, \vee$, then

$$lt_1(y, k) > rt_1(y, k) \ge 0;$$

(b) if c_k is the comma, then at least one of the following relationships takes place:

$$lt_1(y, k) > rt_1(y, k) \ge 0,$$

$$lt_2(y, k) > rt_2(y, k) \ge 0.$$

Proof. Let for B and y the assumptions of Lemma 2 are true, and

$$Symb = \{\equiv, \wedge, vee\}.$$

Since y contains at least one of the elements of the set $Symb$, the string y is constructed out of the elements of the set $X(B) \cup V(B)$ and several auxiliary symbols by applying r times, where $r \geq 1$, some rules from the list $P[1], \ldots, P[10]$ (and, possibly, by applying several times the rule $P[0]$).

Let's prove the lemma by induction on r.

Case 1. If $r = 1$, then the truth of the goal statement for y immediately follows from the analysis of all situations considered in Part 1of the proof of Lemma 1.

Case 2 (inductive step). Let such $r \geq 1$ exist that the statement of the lemma is true for arbitrary conceptual basis and for every such $y \in Ls(B)$ that n the process of constructing y, m rules from the list $P[1], \ldots, P[10]$ were used, where $1 \leq m \leq r$. Let's prove that in this case the statement of the lemma is true and for every such $z \in Ls(B)$ that, while constructing y, $r+1$ rules from the list $P[1], \ldots, P[10]$ were used.

Case 2.1a. Let the rule $P[7]$ be used on the last step of constructing the string z. Then the string z can be represented in the form

$$z = (y_1 q y_2 q \cdots q y_m),$$

where $m > 1$, $q \in \{\wedge, \vee\}$, for $i = 1, \ldots, m$, $y_i \in Ls(B)$. According to inductive assumption, for constructing every string from the strings $y_1, \ldots, y_m$, the rules $P[1], \ldots, P[10]$ were employed not more than r times.

Let $2 < k < l(y_1) + 1$. Then, obviously, the symbol c_k is an inner element of the string y_1. According to inductive assumption, if $lt_1(y_1, k-1) > 0$, then $lt_1(y_1, k-1) > rt_1(z, k-1)$; if $lt_2(y_1, k-1) > 0$, then $lt_2(y_1, k-1) > rt_2(y_1, k-1)$.

But $lt_1(z, k) = lt_1(y_1, k-1) + 1$, $rt_1(z, k) = rt_1(y_1, k-1)$, that is why $lt_1(z, k) > rt_1(z, k)$.

Besides, $lt_2(z, k) = lt_2(y_1, k-1)$, $rt_2(z, k) = rt_2(y_1, k-1)$.

That is why, in accordance with inductive assumption, if $lt_2(z, k) > 0$, then $lt_2(z, k) > rt_2(z, k)$.

Case 2.1b. Let's assume, as above, that the rule $P[7]$ was employed at the last step of constructing the string z and that the string z can be represented in the form

$$(y_1 q y_2 q \cdots q y_m),$$

where $m > 1$, $q \in \{\wedge, \vee\}$, for $i = 1, \ldots, m$, $y_i \in Ls(B)$. Besides, let there be such i, $1 \leq i \leq m$, that

$$1 + l(y_1) + 1 + \ldots + l(y_i) + 2 < k < 1 + l(y_1) + 1 + \ldots + l(y_i) + 1 + l(y_{i+1}),$$

where $l(y_h) = Length(y_h)$, $h = 1, \ldots, i+1$. This means that the symbol $_k$ is an inner element of a certain substring $_{i+1} \in Ls(B)$.

Let the k th position of the considered string z coincide with the -th position of the string y_{i+1}. Then

$$lt_1(z, k) = 1 + lt_1(y_1, l(y_1)) + \ldots + lt_1(y_i, l(y_i)) + lt_1(y_{i+1}, p),$$

$$rt_1(z, k) = rt_1(y_1, l(y_1)) + \ldots + rt_1(y_i, l(y_i,)) + rt_1(y_{i+1}, p).$$

In accordance with Lemma 1, for $j = 1, \ldots, i$,

$$lt_1(y_j, l(y_j)) = rt_1(y_j,$$

$$l(y_j)), \; lt_2(y_j, l(y_j)) = rt_2(y_j, l(y_j)),$$

$$lt_1(y_{+1}, p) \geq rt_1(y_{j+1}, p).$$

That is why $lt_1(z, k) > rt_1(z, k)$.

Case 2.1c. Let there exist such $m > 1$, $y_1, \ldots, y_m \in Ls(B)$, $q \in \{\wedge, \vee\}$ that

$$z = (y_1 \, q \, y_2 \, q \ldots q \, y_m),$$

and there is such i, $1 \leq i < m$ that

$$k = 1 + l(y_1) + 1 + \ldots + l(y_i) + 1 = l(y_1) + \ldots + l(y_i) + i + 1.$$

This means that c_k is an occurrence of the logical connective q, separating y_i and $y_i + 1$. It follows from Lemma 1 that for $s = 1, \ldots, i$,

$$lt_1(y_s, l(y_s)) = rt_1(y_s, l(y_s)).$$

That is why

$$lt_1(z, k) = 1 + lt_1(y_1, l(y_1)) + \ldots + lt_1(y_i, l(y_i)),$$

$$rt_1(z, k) = rt_1(y_1, l(y_1)) + \ldots + rt_1(y_i, l(y_i)).$$

Therefore, $lt_1(z, k) = rt_1(z, k) + 1$.

Case 2.2. During the last step of constructing the string z, the rule $P[3]$ has been used. In this situation, there are such $y_0, y_2 \in Ls(B)$ that z can be represented in the form $(y_0 \equiv y_2)$. Obviously, this situation can be considered similar to the Case 2.1 for $m = 2$.

Case 2.3. On the last step of constructing the string z, one of the rules $P[2]$ or $P[4]$ was applied. Then there are such $y_1, \ldots, y_n \in Ls(B)$ that z can be represented in the form $f(y_1, \ldots, y_n)$ or in the form $r(y_1, \ldots, y_n)$ respectively, where f is an n-functional symbol, r is an n-ary relational symbol. Obviously, this case is considered similar to the Case 2.1.

Case 2.4. During the final step of constructing the string z, one of the rules $P[8]$ or $[9]$ or $[10]$ has been used. These cases are to be considered similar to the Case 2.1 too.

A.3 Proof of Theorem 4.5

Let there be an arbitrary conceptual basis, $z \in Ls(B) \setminus (X(B) \cup V(B))$. Then it directly follows from the structure of the rules $P[0]$, $P[1]$, $\ldots$, $P[10]$ and the definition of the set $Ls(B)$ the existence of such k, where $1 \leq k \leq 10$, $n \geq 1$, and $y_0, y_1, \ldots, y_n \in Ls(B)$ that

$$y_0 \& y_1 \& \ldots \& y_n \& z \in Ynr_{10}^k(B).$$

That is why the main attention is to be paid to proving the uniqueness of such $n+3$-tuple $(k, n, y_0, \ldots, y_n)$. Remember that, in accordance with the definition from Sect. 3.8, for arbitrary conceptual basis,

$$D(B) = X(B) \cup V(B) \cup \{',', '(', ')', ':', '*', '\langle', '\rangle'\},$$

$$Ds(B) = D(B) \cup \{'\&'\},$$

$D^+(B)$ and $Ds^+(B)$ are the sets of all non empty finite sequences of the elements from $D(B)$ and $Ds(B)$ respectively.

Suppose that B is an arbitrary conceptual basis, z is an arbitrary formula from $Ls(B) \setminus (X(B) \cup V(B))$, $1 \leq k \leq 10$, $n \geq 1$, and $y_0, y_1, \ldots, y_n$ are such formulas from $Ls(B)$ that

$$y_0 \& y_1 \& \ldots \& y_n \& z \in Ynr_{10}^k(B).$$

Let's prove that in this case the $n+3$-tuple $(k, n, y_0, \ldots, y_n)$ is the only such $n+3$-tuple that for this tuple the above relationship takes place. For achieving this goal, it is necessary to consider rather many possible cases.

The length of arbitrary string $s \in Ds^+(B)$ (the number of elements from $Ds(B)$) will denote, as before, $l(s)$. With respect to our agreement, the elements of arbitrary primary informational universe $X(B)$ and the variables from the set $V(B)$ are considered as symbols, i.e as non structured elements. That is why the length $l(d$ of arbitrary $d \in (X(B) \cup V(B)$ is equal to 1. The number 0 will be considered as the lenth of empty string e.

Case 1. The first symbol of the string z, denoted as $z[1]$, is the left parenthesis "(". Then, obviously, $k = 3$ or $k = 7$. If $k = 3$, then $n = 2$, and z is the string of the form $(y_0 \equiv y_2)$, and y_1 is the symbol $\equiv$.

If $k = 7$, then $n \geq 2$, y_0 is the logical connective $\vee$ or $\wedge$, and z is the string of the form

$$(y_1 \, y_0 \, y_2 \, y_0 \, \ldots \, y_0 \, y_n).$$

That is, z is the string of the form $(y_1 \wedge y_2 \wedge \ldots \wedge y_n)$ in case $y_0 = \wedge$ and the string of the form $(y_1 \vee y_2 \vee \ldots \vee y_n)$ in case $y_0 = \vee$.

Suppose that z is the string of the form $(y_0 \equiv y_2)$, $y_0, y_2 \in D^+(B)$ and prove that it is impossible to present z in a different way. It will be a proof by contradiction.

Case 1a. Suppose that there are such $w_1, w_2 \in Ls(B)$ that $l(y_1) > l(w_1)$, and z can be represented in the form $(w_1 \equiv w_2)$. Then the string $w_1 \equiv$ is the beginning of the string y_0. Let there be such k, m that

$$1 \leq k < m, \; y_1 = c_1 \ldots c_m,$$

$$w_1 = c_1 \ldots c_k,$$

where $c_1 \ldots c_m \in D(B.)$ Then $c_{k+1} = {}'\equiv'$, and, since no one string from $Ls(B)$ has no ending being the symbol ${}'\equiv'$, the inequality $k+1 < m$ is valid.

Since y_0 includes the symbol ${}'\equiv'$, then the string y_1 includes the left parenthesis ${}'('$ in a certain position $j < k + 1$. That is why $lt_1(y_1, k+1) > 0$. But it follows from the relationship $c_{k+1} = {}'\equiv'$ (in accordance with Lemma 2) that

$$lt_1(y_0, k+1) > rt_1(y_0, k+1).$$

Obviously,

$$lt_1(y_0, k+1) = lt_1(w_1, k),$$

$$rt_1(y_0, k+1) = rt_1(w_1, k).$$

That is why $lt_1(w_1, k) > rt_1(w_1, k)$. However, with respect to Lemma 1, it follows from $w_1 \in Ls(B)$ and $k = l(w_1)$ that

$$lt_1(w_1, k) = rt_1(w_1, k).$$

Therefore, a contradiction has been obtained.

Similar speculations can be fulfilled in case of the assumption about the existence of such $w_1, w_2 \in Ls(B)$ that

$$l(w_1) > l(y_0),$$

$$l(w_2) < l(y_2),$$

$$z = (w_1 \equiv w_2).$$

That is why such w_1 and w_2 don't exist.

Case 1b. Let's assume, as before, that the string z can be presented in the form $(y_0 \equiv y_2)$, where $y_0, y_2 \in D^+(B)$. Suppose also that there are such $q \in \{\wedge, \vee\}$, $m > 1$, and $w_1, \ldots, w_m \in Ls(B)$ that

$$z = (w_1 q w_2 q \ldots q w_m).$$

Obviously, $l(w_1) \neq l(y_2)$. Let $l(y_0) < l(w_1)$. Then the string $y_0 \equiv$ is a beginning of the string $w_1 \in Ls(B)$. Taking this into account, we easily receive a contradiction, repeating the speculations of the Case 1a.

If $l(w_1) < l(y_0)$, the string $w_1 q$ is a beginning of the string $y_0 \in Ls(B)$. In this case we again apply the Lemma 2 and the way of reasoning used in the process of considering Case 1a.

Therefore, if $z \in Ls(B)$ and for some $y_0, y_2 \in Ls(B)$, z is a string of the form $(y_0 \equiv y_2)$, where $y_0, y_2 \in D^+(B,)$ then the string z can't be obtained with the help of any rule (on the final step of inference) being different from the rule [3], and only from the "blocks" y_0, y_2.

Case 1c. Let there be such $q \in \{\wedge, \vee\}$, $m > 1$, and $y_1, \ldots, y_m \in Ls(B)$ that

$$z = (y_1 q y_2 q \ldots q y_m).$$

Then it is necessary to consider two situations:

1. there are such $u_1, u_2 \in Ls(B)$ that z is the string of the form $(u_1 \equiv u_2)$;
2. there are such $p > 1, w_1, \ldots, w_p \in Ls(B), d \in \{\wedge, \vee\}$ that

$$z = (w_1 d w_2 d \ldots d w_p),$$

besides, the $m+1$-tuple $(q, y_1, \ldots, y_m)$ is distinct from the $m+1$-tuple $(d, w_1, \ldots, w_p)$.

Consider the situation (1). Let $l(u_1) < l(y_1)$, then $u_1, y_1 \in Ls(B)$, and the string $u_1 \equiv$ is a beginning of the string y_1. This situation was analyzed in Case 1, where we obtained a contradiction.

If $l(y_1) < l(u_1)$, then $u_1, y_1 \in Ls(B)$, and the string $y_1 q$ is a beginning of the string u_1; that is why, obviously, $l(y_1 q) < l(u_1)$.

Let's use the Lemma 2. Suppose that

$$y_1 q = c_1 \ldots c_r c_{r+1},$$

where for $i = 1, \ldots, r + 1, c_i \in D(B), c_{r+1} = q$.

Since c_{r+1} is the symbol $' \equiv'$, it follows from Lemma 2 that

$$lt_1(u_1, r+1) > rt_1(u_1, r+1).$$

But we have the relationships

$$lt_1(u_1, r+1) = lt_1(u_1, r) = lt_1(y_1, r),$$

$$rt_1(u_1, r+1) = rt_1(u_1, r) = rt_1(y_1, r).$$

But, since $y_1 \in Ls(B)$, it follows from Lemma 1 that

$$lt_1(y_1, r) = rt_1(y_1, r).$$

We have obtained a contradiction.

Let's analyze the situation (2). Suppose that $l(w_1) < l(y_1)$. Then $w_1, y_1 \in Ls(B)$, and the string $w_1 d$ is a beginning of y_1, where d is one of the symbols $\wedge, \vee$. But in this case we are able to obtain a contradiction exactly in the same way as while analyzing the situation (1).

Obviously, the case $l(y_1) < l(w_1)$ is symmetric to the just considered case. Let's assume that $q = d$, and there is such $j, 1 \leq j < m$ that

$$y_1 = w_1, \ldots, y_j = w_j,$$

but $w_{j+1} \neq y_{j+1}$. Then either the string $w_{j+1} d$ is a beginning of the string y_{j+1} or $y_{j+1} d$ is a beginning of w_{j+1}. That is why we are in the situation considered above.

Thus, the analysis of the Case 1 has been finished. We have proved the following intermediary statement:

Let B be an arbitrary conceptual basis, $z \in Ls(B)$
$: (X(B)) \cap V(B))$, and the first symbol of the string z is the left parenthesis $'('$.

Then there is such unique $n+3$-tuple $(k, n, y_0, \ldots, y_n)$, where $1 \leq k \leq 10$, $n \geq 1$, $y_0, y_1, \ldots, y_n \in Ls(B)$ that

$$y_0 \& y_1 \& \ldots \& y_n \& z \in Lnr_{10}^k(B);$$

in this case, $k = 3$ or $k = 7$.

Case 2. Let B be an arbitrary conceptual basis, z be an arbitrary formula from $Ls(B)$
$: (X(B) \cup V(B))$, $1 \leq k \leq 10$, and $y_0, y_1, \ldots, y_n$ are such formulas from $Ls(B)$ that the relationship above takes place.

Suppose that z has a beginning being a functional symbol, i.e., $z[1] \in F(B)$, and that the second before the end symbol z is distinct from the symbol $':'$. Then it immediately follows from the structure of the rules $P[0], P[1], \ldots, P[10]$ that the string z is obtained as a result of applying the rule $P[2]$ at the last step of inference. That is why $z = f(y_1, \ldots, y_n)$, where $n \geq 1$, $y_1, \ldots, y_n \in Ls(B)$.

If $f \in F_1(B)$, i.e., f is an unary functional symbol, then $n = 1$, and y_1 is unambiguously determined by z. Suppose that $n > 1$, $f \in F_n(B)$, and there is such a sequence $u_1, \ldots, u_n$ being distinct from the sequence $y_1, \ldots, y_n$ that $u_1, \ldots, u_n \in Ls(B)$, and string $(y_1, \ldots, y_n)$ coincides with the string $(u_1, \ldots, u_n)$.

Applying Lemma 2 similarly to the way of reasoning used while considering Case 1c, it is easy to show that the existence of two different representations $(y_1, \ldots, y_n)$ and $(u_1, \ldots, u_n)$ of the string $z \in Ls(B)$ is impossible due to the same reason as two different representations

$(y_1 \, d \, y_2 \, d \ldots d \, y_n)$ and $(u_1 \, d \, u_2 d \ldots d \, u_n)$, where $d \in \{\wedge, \vee\}$, are impossible.

Case 3. Let the assumptions of the formulation of the Theorem 4.5 be true for the conceptual basis B and the string z; this string z has the beginning $z[1]$ being a relational symbol, and $z[1]$ is not a functional symbol (that is, $z[1] \in R(B) \setminus F(B)$), and the second from the end symbol of the string z is distinct from the symbol $':'$. This situation is similar to the Case 2, only z is constructed as a result of employing the rule $P[4]$ at the last step of inference.

That is why there exists the only (for z) such sequence $(k, n, y_0, \ldots, y_n)$ that $1 \leq k \leq 10$, $n \geq 1$,

$$y_0, y_1, \ldots, y_n \in Ls(B),$$

$$y_0 \& y_1 \& \ldots \& y_n \& z \in Ynr_{10}^k(B),$$

where $k = 4$.

Case 4. Let B be an arbitrary conceptual basis, $z \in Ls(B) \, setminus(X(B) \cup V(B))$, for $i = 1, \ldots, l(z)$, $z[i]$ the i-th symbol of the string z, $z[1] \in X(B)$, $z[2] = '*'$, the next to last symbol z is distinct from the symbol $':'$. Then it directly follows from the rules $P[0], P[1], \ldots, P[10]$ that for constructing the string z, the rule $P[8]$ has been employed during the last step of inference. That is why there are such $p \geq 1$ and a sequence $(r_1, b_1, \ldots, r_p, b_p)$, where for $i = 1, \ldots, p$, $r_i \in R_2(B)$, $b_i \in Ls(B)$, that if $a = z[1]$, then

$$z = a * (r_1, b_1) \ldots (r_p, b_p).$$

We must now prove that for every $n+3$-tuple $(k, n, y_0, \ldots, y_n)$, such that $1 \le k \le 10$, $n \ge 1$, for $i = 0, \ldots, n$, $y_i \in Ls(B)$, it follows from the relationship

$$y_0 \mathbin{\&} y_1 \mathbin{\&} \ldots \mathbin{\&} y_n \mathbin{\&} z \in Ynr_{10}^k(B)$$

that the $n+3$-tuple $(k, n, y_0, \ldots, y_n)$ coincides with the tuple

$$(8, 2p+1, a, r_1, b_1, \ldots, r_p, b_p).$$

The relationship $z = a * (r_1, b_1) \ldots (r_p, b_p)$ implies that it is necessary to consider only the following situation: there are such

$$m \ge 1, h_1, \ldots, h_m \in R_2(B), d_1, \ldots, d_m \in Ls(B)$$

that

$$Z = a * (h_1, d_1) \ldots (h_m, d_m),$$

and this representation is distinct from the representation

$$z = a * (r_1, b_1) \ldots (r_p, b_p).$$

The elements $a, r_1, \ldots, r_p, h_1, \ldots, h_m$ are interpreted as symbols. That is why, obviously, $r_1 = h_1$.

Let $d_1 \ne b_1$ and $l(b_1) < l(d_1)$. Then the string $b_1)$ is a beginning of the string d_1. Using the Lemma 2 and Lemma 1 similar to the way of reasoning in the course of considering the Cases 1a and 1c, we obtain a contradiction. Therefore, $b_1 = d_1$.

Let there be such i that $1 \le i \le minp, m$ and for $j = 1, \ldots, I$,

$$r_j = h_j, b_j = d_j,$$

$$r_{j+1} = h_{j+1}, b_{j+1} \ne d_{j+1}.$$

Then it follows from $l(b_{j+1}) < l(d_{j+1})$ that the string b_{j+1} is a beginning of the string d_{j+1}, but we've just analyzed a similar situation.

If $l(b_{j+1}) < l(d_{j+1})$, the string (d_{j+1}) is a beginning of the string b_{j+1}, and we encounter, in essence, the same situation. Therefore, the representations

$$z = a * (r_1, b_1) \ldots (r_p, b_p),$$

$$Z = a * (h_1, d_1) \ldots (h_m, d_m),$$

coincide.

Case 5. Suppose that for the conceptual basis B and the string z, the assumptions of the formulation of the Theorem 3.5 are true, $z[1]$ is the quantifier $\forall$ or $\exists$, the symbol $z[n-1] \ne ':'$, where $n = Length(z)$. Then, obviously, the rule $P[9]$ was applied at the last step of building the string Z. . That is why z is the string of the form

$$Q \, var \, (conc) \, A,$$

where $Q = z[1]$, $var \in V(B)$, $conc, A \in Ls(B)$, $conc \in X(B)$ or $conc$ is such a string that for constructing it, the rule $P[8]$ was applied at the last step of inference, and $A \& P \in Ts(B)$, where P is the sort "a meaning of proposition" of the basis B.

Let $l(z) = m$ (that is, m is the length of z). If $conc \in X(B)$, then the uniqueness of the representation of z is obvious:

$$Q = z[1], \ v = z[2], \ '(' = z[3],$$

$$conc = z[4], \ ')' = z[5], \ A = z[6] \ldots z[m].$$

If $conc \in Ls(B) \setminus (X(B) \cup V(B))$, then let's use the Lemma 1.
Let $l(conc) = p$, where $p > 1$ then for $i = 1, \ldots, p$,

$$lt_1(conc, i) \geq rt_1(conc, i);$$

$$lt_1(conc, p) = rt_1(conc, p).$$

That is why for $q = 1, \ldots, p+3$,

$$lt_1(z, q) > rt_1(z, q),$$

$$lt_1(z, p+4) = rt_1(z, p+4).$$

Therefore, the position of the right parenthesis $')'$ immediately after the string $conc$ can be defined as such minimal integer s $leq 1$ that

$$lt_1(z, s) > rt_1(z, s),$$

$$lt_1(z, s) = rt_1(z, s).$$

That is why, if $z \in Ls(B)$ and $z[1]$ is the quantifier Q, where $Q = \forall$ or $Q = \exists$, then there exists the only such $n+3$-tuple $(k, n, y_0, \ldots, y_n)$, where $1 \leq k \leq 10$, $n \geq 1$, $y_0, y_1, \ldots, y_n \in Ls(B)$, that

$$y_0 \& y_1 \& \ldots \& y_n \& z \in Ynr_{10}^k(B);$$

besides, $k = 9$, $n = 4$, $y_0 = Q$, $y_1 = v$, $y_3 = conc$, $y_4 = A$ (with respect to the relationship $z = Q var (conc) A$.

Case 6. Suppose that the assumptions of the formulation of the Theorem 3.5 are true for the conceptual basis B and the string z, $z[1] = '\langle'$, and the next to last symbol z is distinct from ':' (colon). Then, obviously, the rule $P[10[$ was used at the last step of constructing z. Assume that there exists such sequence m, $u_1, \ldots, u_n$ and a different sequence n, $y_1, \ldots, y_n$, where $m, n \geq 1$, $u_1, \ldots, u_m, y_1, \ldots, y_n \in Ls(B)$, that z can be represented in two forms

$$\langle u_1, \ldots, u_m \rangle,$$

$$langley_1, \ldots, y_n \rangle.$$

Applying Lemma 2 similar to the way of reasoning in Case 2, we conclude that two different representations of z are impossible due to the same reason as the impossibility of two different representations $f(u_1,\ldots,u_m)$ and $f(y_1,\ldots,y_n)$ of the same string from $Ls(B)$.

Case 7. Let the assumptions of the Theorem 4.5 be true for the conceptual basis B and the string z, $z[1]$ be an intensional quantifier from $Int(B)$, z has no ending of the form : v, where $v \in V(B)$, $m = l(z)$. Then, obviously, z can be obtained with the help of the rule with the number $k = 1$ out of the "blocks" $y_1 = z[1]$ and $y_2 = z[2] \in z[m]$.

Case 8. Let B be an arbitrary conceptual basis, $z \in Ls(B) \setminus (X(B) \setminus V(B))$, $z[1] = \neg$. Then there is such string $y \in Ls(B)$ that $z = \neg y$. The analysis of the structure of the rules $P[0], P[1],\ldots, P[10]$ shows that only two situations are possible: (1) y has no ending : v, where $v \in V(B)$, that is why z was constructed out of the pair $(\neg, y)$ as a result of employing the rule $P[6]$; (2) $y = w : v$, where $v \in V(B)$; in this case, z was built out of the pair of operands $(\neg w, v)$ by applying one time the rule $P[5]$.

Let's pay attention to the fact that the situation when z was constructed out of the pair of operands $(\neg, w : v)$ by means of applying the rule $P[6]$, is impossible; the reason is that the rule $P[6]$ allows for constructing the expressions of the form $\neg a$ only in case when either a $inX(B)$ or no one of the rules $P[2], P[5], P[10]$ was employed at the last step of constructing the string a.

Case 9. Let B be an arbitrary conceptual basis, $z \in Ls(B) \setminus (X(B) \cup V(B))$, z has an ending being the substring of the form : var, where $var \in V(B)$, and z has no beginning $'\neg'$. Let's show that in this case the rule $P[5]$ was applied at the last step of obtaining the string a.

Assume that it is not true, and the rule P_m, where $1 \le m \le 10$, m $neq5$, $m \ne 6$. was used at the last step of constructing the string a. If $m = 1$, then $z = qtr\,des$, where q is an intensional quantifier from $Int(B)$, $des \in X(B)$ or

$$des = cpt * (r_1, b_1) \ldots (r_p, b_p),$$

where $p \ge 1$, $a \in X(B)$, $r_1,\ldots, r_n \in R(B)$, $b_1,\ldots, b_p \in Ls(B)$.

Since z has an ending being the substring of the form : var, where $var \in V(B)$, then the situation $z = qtr\,des$, where $q \in Int(B)$, $des \in X(B)$, is excluded. But $v \ne ')'$; therefore, we obtain a contradiction, and the situation $m = 1$ is impossible. If m is one of the integers 2, 3, 4, 7, 8, the last symbol of z is the right parenthesis $')'$. That is why each of these situations is impossible. If $m = 10$, the last symbol of z is $')'$, and it is impossible too.

Suppose that the rule $P[9]$ was applied at the last step of inference. Then there are such

$$qex \in \{\exists, \forall\}, \ var \in V(B), \ des \in Ls(B), \ A \in Ls(B),$$

that $z = qex\,var\,(des)\,A$.

According to the definition of the rule $P[9]$, the string A has no ending of the form : z , where z is an arbitrary variable from $V(B)$. That is why in Case 9, the rule $P[5]$ was applied at the last step of constructing z. Thus, we've considered all possible cases, and Theorem 4.5 is proved.

Glossary

c.b. conceptual basis

CIA Computer Intelligent Agent

CMR Component-Morphological Representation of an NL-text

DRT Discourse Representation Theory

e-Commerce Electronic Commerce

e-Contracting Electronic Contracting

e-Negotiations Electronic Negotiations

e-Science Electronic Science

EL Episodic Logic

FIPA Foundation for Intelligent Physical Agents

IFS Integral Formal Semantics of Natural Language

K-calculus Knowledge Calculus

KCL-theory Theory of K-calculuses and K-languages

l.b. linguistic basis

LP Linguistic Processor

MAS Multi-Agent System (or Systems)

m.c.b. marked-up conceptual basis

MR Morphological Representation of an NL-text

MSSR Matrix Semantic-Syntactic Representation of an NL-text

NL Natural Language

NLPS Natural Language Processing System

NLPSs Natural Language Processing Systems

OWL Ontology Web Language

SCL-theory Theory of S-calculuses and S-languages

SK-language Standard Knowledge Language

SR Semantic Representation of an NL-text

TCG Theory of Conceptual Graphs

UNL Universal Networking Language

USR Underspecified Semantic Representation of an NL-text

References

1. Abrahams, A. S., Eyers, D. M., Kimbrough,S. O.: On defining and capturing events in business communication systems. In: IDT brownbag (2006) http://opim-sky.wharton.upenn.edu/šok/sokpapers/2007/flbc-implementation-foils-20060422.pdf. Cited 22 Apr 2006

2. Aczel, P., Israel, D., Katagiri, Y., Peters, S.: Situation Theory and Its Applications. CSLI Publications, Stanford (1993)

3. Ahrenberg, L.: On the integration and scope of segment-based models of discourse. the Third Nordic Conf. on Text Comprehension in Man and Machine, Linkoeping Univ., 1–16 (1992)

4. Allen, J.F. , Schubert, L.K., Ferguson, G. M., Heeman, P. A., Hwang, C. H., Kato, T., Light, M., Martin, N. G., Miller, B. W., Poesio, M., Traum, D.R.: The TRAINS project: A case study in building a conversational planning agent. Journal Experimental and Theoretical Artificial Intelligence. **7**, 7–48 (1995)

5. Alshawi, H., van Eijck, J.: Logical forms in the core language engine. 27th Ann. Meeting of the ACL, Vancouver, Canada., 25–32 (1989)

6. Alshawi, H.: Resolving quasi logical forms. Computational Linguistics. **16**, 133–144 (1990)

7. Alshawi, H.: The Core Language Engine. MIT Press, Cambridge, MA (1992)

8. Alshawi, H., Crouch, R.: Monotonic semantic interpretation. In: Proceedings of the 30th Annual Meeting of the Association for Computational Linguistics, University of Delaware, pp. 32–39 (1992)

9. Alshawi, H., van Eijck, J.: Logical forms in the core language engine. 27th Ann. Meeting of the ACL, Vancouver, Canada., 25–32 (1989)

10. Angelova, G.: Language technology meet ontology acquisition. In: Dau, F., Mugnier, M.-L., Stumme, G. (eds.) International Conference on Conceptual Structures 2005. LNCS 3596, pp. 367–380. Springer, Heidelberg (2005)

11. Apresyan, Yu. D.: Lexical Semantics. Synonymic Means of Language. Nauka Publ., Moscow, (in Russian) (1974)

12. Baclawski, K., Niu, T.: Ontologies for Bioinformatics. MIT Press, Cambridge, MA (2005)

13. Baldwin, T.: Making sense of Japanese relative clause constructions. In: Hirst, G., Nirenburg, S. (eds.) Proceedings of the Second Workshop on Text Meaning and Interpretation held in cooperation with ACL-2004, 25–26 July 2004, Barcelona, Spain, pp. 49–56 (2004)

14. Barwise, J., Cooper, R.: Generalized quantifiers and natural language. Linguistics & Philosophy. **4**, 159–219 (1981)

15. Barwise, J., Cooper, R.: Extended kamp notation: a graphical notation for situation theory. Aczel, Israel, Katagiri, fc Peters., 29–53 (1993)

16. Barwise, J., Perry, J.: Situations and Attitudes. The MIT Press, Cambridge, MA (1983)

17. Beale, S., Lavoie, B., McShane, M., Nirenburg, S., Korelsky, T.: Question answering using ontological semantics. In: Hirst, G., Nirenburg, S. (eds.) Proceedings of the Second Work-

shop on Text Meaning and Interpretation held in cooperation with ACL-2004, 25–26 July 2004, Barcelona, Spain, pp. 41–48 (2004)

18. Berners-Lee, T., Hendler, J., Lassila, O.: The semantic web. Scientific American, 34–43 (2001)

19. Cannataro, M., Talia, D.: Semantics and knowledge grids: building the next-generation grid. IEEE Intelligent Systems (Jan/Feb 2004), Special issue on E-Science (2004)

20. Caron, J.: Precis de Psycholinguistique. Presses Universitaires de France, Paris, (1989)

21. Carpenter B., Penn G.: Attribute logic engine (ALE), Version 3.2 http://www.cs.toronto.edu/ gpenn/ale.html.

22. Chafe, W. L.: Meaning and the Structure of Language. The University of Chicago Press Chicago and London, (1971)

23. Chai, J., Lin, J., Zadrozny, W., Ye, Y., Budzikowska, M., Horvath, V., Kambhatla, N., Wolf, C.: The role of a natural language conversational interface in online sales: A case study. International Journal of Speech Technology. 4, 285–295 (2001)

24. Chai, J., Horvath, V., Nicolov, N., Stys, M., Kambhatla, N., Zadrozny, W., Melville, P.: Natural language assistant – a dialog system for online product recommendation. Artificial Intelligence Magazine (AI Magazine). 23, 63–76 (2002)

25. Chen, H., Wang, Y., Wu, Z.: Introduction to semantic e-Science in biomedicine (2007)

26. Chierchia, G.: Structured meanings, thematic roles and control. In Chierchia, G., Partee, B. H., Turner, R. (eds.) Properties, Types and Meaning, Semantic Issues., pp. 131–166. Kluwer, Dordrecht (1989)

27. Clifford, J.: QE-III : a formal approach to natural language querying. In William K. (ed.) Proceedings of the National Conference on Artificial Intelligence (AAAI-83), pp. 79–83. AAAI, Los Altos, California (1983)

28. Clifford, J.: Language querying of historical data bases. Computational Linguistics. 14, 10–34 (1988)

29. Cresswell, M. J.: Structured Meanings: The Semantics of Prepositional Attitudes. MIT Press, Cambridge, MA (1985)

30. Cooper, R.: Three lectures on situation theoretic grammar. In Filgueiras, M. (eds.) Natural Language Processing, pp. 102–140. Springer-Verlag, Berlin (1991)

31. CrossFlow Project. Insurance Requirements. CrossFlow Consortium (1999) http://www.crossflow./public/pubdel/D1b.pdf. Cited 1999

32. Reference description of the DAML + OIL ontology markup language. In DAML plus OIL (2001) http://www.daml.org/2001/03/daml+oil. Cited March 2001

33. Delmonte, R.: Text understanding with GETARUNS for Q/A and summarization. In: Hirst, G., Nirenburg, S. (eds.), Proceedings of the Second Workshop on Text Meaning and Interpretation held in cooperation with ACL-2004, 25–26 July 2004, Barcelona, Spain, pp. 97–104 (2004)

34. Dekker, P.: The scope of negation in discourse: towards a flexible dynamic Montague grammar. In Quantification and Anaphora 1. DYANA. Deliverable R2.2.A. July 1990. Edinburgh, 79–134 (1990)

35. De Roure, D., Hendler, J. A.: E-science: the grid and the semantic web. IEEE Intelligent Systems (Jan/Feb 2004), Special issue on E-Science (2004)

36. van Eijck, D. J. N., Kamp, H.: Representing Discourse in Context. The University of Amsterdam, Amsterdam (1996)

37. Eklund, P. W., Ellis, G., Mann, G.: Conceptual structures; knowledge representation as Interlingua. In Eklund, P. W., Ellis, G., Mann, G. (eds.) Proceedings of the 4th Intern. Conf. on Conceptual Structures, p. 321. ICCS'96, Sydney, Australia, Soringer-Verlag, Berlin, Heidelberg, NewYork (1996)

38. Ershov, A. P.: Machine Fund of Russian Language: An External Task Statement. In Karaulov, Yu. N. (ed.), Machine Fund of Russian Language: Ideas and Opinions. Nauka Publ. Moscow, pp. 7–12, in Russian (1986)

39. Fauconnier, G.: Mental Spaces: Aspects of Meaning Construction in Natural Language. Cambridge University Press, Cambridge (1985)

40. Fellbaum, C., Hahn, U., Smith, B.: Towards new information resources for public health – From WordNet to MedicalWordNet. Journal of Biomedical Informatics **39**, 321–332 (2006)

41. Fensel, D.: OIL in a Nutshell. Knowledge acquisition, modeling, and management, Proceedings of the European Knowledge Acquisition Conference (EKAW-2000). In Dieng, R. (ed.) Lecture Notes in Artificial Intelligence (LNAI). Springer-Verlag, Berlin (2000)

42. Fenstad, J.E., Halvorsen, P. K., Langholm, T. , Van Benthem, J.: Situations, Language and Logic. D. Reidel, Dordrecht (1987)

43. Fenstad, J.E., Langholm, T., Vestre, E.: Representations and interpretations. In Rosner, M., Johnson, R. (eds.) Computational Lingnuistics and Formal Semantics, Cambridge University Press, Cambridge, MA 31–95 (1992)

44. Fenstad, J. E., Lonning, J. T.: Computational semantics: steps towards "Intelligent" Text Processing. In Studer, R. (ed.) Natural Language and Logic. Springer-Verlag, Berlin, 70–93 (1990)

45. Fillmore, C.: Frames and the semantics of understanding. Quaderni di Seman-tica, **6**, 222–253 (1985)

46. Finin, T., Weber, J., Wiederhold, G., Genesereth, M., Fritzson, R., McKay, D., Mc Guire, J., Pelavin, R., Shapiro, S., Beck, C.: Specification of the KQML. Agent-Communication Language. The DARPA Knowledge Sharing Initiative, External Interfaces Working Group (1993)

47. FIPA SL Content Language Specification. In: FIPA SL CL (2002) http://www.fipa.org/specs/fipa00008/SC000081.htm. Cited 2002

48. Fisher, C., Gleitman, H., Gleitman, L. R.: On the semantic content of subcate-gorization frames. Cognitive Psychology. **23**, 331–392 (1991)

49. Fodor, J. A.: The Language of Thought. Crowell New York, (1975)

50. Fodor, J. A.: Psychosemantics: The Problem of Meaning in the Philosophy of Mind. 2nd print. The MIT Press Cambridge, MA, London, (1988)

51. Fomichov, V. A. (1980). Some principles of mathematical describing natural language's subsets. In Ailamazyan, A. K. (ed.) Proceedings of the 1st and 2nd sc.-techn. conf. of young specialists. Informational Processes and Their Automation. Moscow, VNTITSentr, 125–136, in Russian (1980)

52. Fomichov, V. A.: To the theory of logic-algebraic modelling the speech-forming mechanisms of conceptual level. 1. Task statement and the idea of an approach to its resolving. Moscow Institute of Electronic Engineering, 85 pp. – The paper deposited in the All-Union Institute of Scientific and Technical Information; in Russian – Vsesoyuzny Institut Nauchnoy i Teknicheskoy Informatsii, or VINITI (27/10/81, No. 4939-81 Dep. – The abstract 2B1386DEP in the Abstracts Journal "Mathematika" of VINITI), 1982, No. 2; in Russian (1981)

53. Fomichov, V.A.: About elaboration and application of the theory of logic-algebraic modeling of some natural language's text-forming mechanisms of conceptual level. IX All-Union Symposium on Cybernetics (Theses of symposium. Sukhumi, 10–15 Nov. 1981), Vol. 1, Moscow: Sc, Council on complex problem "Cybernetics" of As. Sc USSR, 186–188 (1981)

54. Fomichov, V. A.: Development and application of methods of logic-algebraic modeling the mechanisms of speech-formation semantics. Ph.D. dissertation. Moscow: Moscow Inst. of Electronic Eng. (1982)

55. Fomitchov, V. A. (1983). Formal systems for natural language man-machine interaction modelling. In International Symposium on Artificial Intelligence IFAC, Vol. 1. Leningrad: Ac. Sc. USSR, 223–243 (in two variants – in Russian and in English). Paper presented at the First Symposium of the Int. Federation of Automatic Control on Artificial Intelligence (1983)

56. Fomitchov, V.: Formal systems for natural language man-machine interaction modeling. In Ponomaryov, V. M. (ed.) Artificial Intelligence. Proc. of the IFAC Symposium, Leningrad, USSR (1983), IFAC Proc. Series, pp. 203–209. Oxford, UK; New York (1984)

57. Fomichov, V.A.: Calculuses for modeling the processes of treating the conceptual information. Moscow, 83pp. Unpublished manuscript (1985)

58. Fomichov, V.A.: A mathematical model of linguistic processor and its application to the design of training complexes. In Y. A. Ryazanov (ed.) Automatic Regulation and Control.

Mathematical Modeling of the Non-stationary Processes of Automatic Control Systems, Moscow: All-Union Machine-Building Institute for Tuition by Correspondence, pp. 110–114 (1986)

59. Fomichov, V. A.: About synthesis of tasks formal specifications in the dialogue system TEMP-1. In: Markov, V.N. (ed.) Mathematical Provision and Programming for Calculating and Control Systems, pp. 10–13. Moscow Institute of Electronic Engineering Publishing House, Moscow (1986)

60. Fomichov, V.A.: The experience of mathematical designing a system of communication with the database of a flexible manufacturing system. In: Problems of creating flexible manufacturing systems in machine-building. Theses of papers òf the Republican sc.-techn. Conf. 1–2 October 1987, Kaunas. Vilnius : Lithuanian Sc, -Res. Inst, of Sc. -Techn. Information and Technical-Economical Studies, Kaunas Polytechnic Inst., Sc. Council Ac. Sc. USSR on Problems of Machine-Building and Technological Processes, 90–91 (1987)

61. Fomichov, V.A.: K-calculuses and K-languages as instruments for investigation of some key questions of the fifth generation computer systems theory. In: Informatics-87, II All-Union conf. on actual problems of informatics and computer technics. Theses of papers, Yerevan, 20–22 October 1987, Yerevan: Ac. Sc. Armyan. SSR Publ. House, pp. 182–183 (1987)

62. Fomichov, V. A.: Representing Information by Means of K-calculuses: Textbook. The Moscow Institute of Electronic Engineering (MIEE) Press, Moscow, in Russian (1988)

63. Fomichov, V. A.: About the means of constructing natural-language-communication mathematical theory. In Markov, V. N. (ed.); Mathematical Provision of Calculating, Informational, and Control Systems. The MIEE Press, Moscow; pp. 21–25. in Russian (1988)

64. Fomichov, V.A.: About mathematical models for designing analyzers of discourses. In: Petrov, A.E. (ed.) Scientific-Technical Means of Informatization, Automatization and Intellectualization in National Economy. Materials of a seminar. Moscow: Central Russian Knowledge House Press, pp. 62–70; in Russian (1991)

65. Fomichov, V.: Mathematical models of natural-language-processing systems as cybernetic models of a new kind. Cybernetica (Belgium). **35**, 63–91 (1992)

66. Fomichov, V. A.: Towards a mathematical theory of natural-language communication. Informatica. An International Journal of Computing and Informatics (Slovenia). **17**, 21–34 (1993)

67. Fomichov, V. A.: K-calculuses and K-languages as powerful formal means to design intelligent systems processing medical texts. Cybernetica (Belgium). **36**, 161–182 (1993)

68. Fomichov, V. A.: Integral formal semantics and the design of legal full-text databases. Cybernetica (Belgium). **37**, 145–17 (1994)

69. Fomichov, V. A.: A variant of a universal metagrammar of conceptual structures. Algebraic systems of conceptual syntax (Invited talk). In Nijholt, A., Scollo, G., Steetskamp, R. (eds.) Algebraic Methods in Language Processing. Proceedings of the Tenth Twente Workshop on Language Technology joint with First AMAST Workshop on Language Processing, pp. 195–210. University of Twente, Enschede, The Netherlands, December 6–8, 1995 (1995)

70. Fomichov, V. A.: A mathematical model for describing structured items of conceptual level. Informatica (Slovenia). **20**, 5–32 (1996)

71. Fomichov, V. A.: An outline of a formal metagrammar for describing structured meanings of complicated discourses. In II Intern. Conf. on Mathematical Linguistics (ICML'96), Abstracts, Tarragona, Grup de Recerca en Linguistica Matematica i Enginyeria del Llenguatge (GRLMC), Report 7/96, pp. 31–32. Universitat Rovira i Virgili (1996).

72. Fomichov, V. A.: K-calculuses and the problem of conceptual information retrieval in textual data bases. Knowledge Transfer (Volume II). In Behrooz, A. (ed.) International Conference "Knowledge Transfer – 1997 (KT97)", Symposium "Information Technology", pp. 52–58. London, University of London (1997)

73. Fomichov, V. A.: Formal studies of anticipative systems as a basis of effective humanitarian and computer applications. In Lasker, G. E., Dubois, D., Teiling, B. (eds.) Modelling of Anticipative Systems, Vol. II, pp. 73–77. University of Windsor, Canada (1997)

74. Fomichov, V. A.: K-calculuses and matrix grammars as powerful tools for designing natural language processing systems// Abstracts of Papers, Interdisciplinary Symposium "Artificial

Intelligence, Cognitive Science, and Philosophy for Social Progress" (President – Chairperson: Mr. Vladimir Fomichov, Russia), 15th International Congress on Cybernetics, Namur, Belgium, August 24–28. Namur, International Association for Cybernetics (1998)

75. Fomichov, V. A.: A comprehensive mathematical framework for designing agent communication languages. Proceedings of the International Conference "Information Society (IS'98)", pp. 81–84. Ljubljana, Slovenia, 6–7 October (1998)

76. Fomichov, V. A.: Theory of restricted K-calculuses as a comprehensive framework for constructing agent communication languages; Special Issue on NLP and Multi-Agent Systems. In Fomichov, V. A., Zeleznikar, A.P. (eds.). Informatica. An International Journal of Computing and Informatics (Slovenia). **22(4)**, 451–463 (1998)

77. Fomichov, V. A.: Theory of restricted K-calculuses as a universal informational framework for electronic commerce. Database, Web and Cooperative Systems. The Proceedings of 1999 International Symposium on Database, Web and Cooperative Systems, August 3–4, 1999 in Baden-Baden, Germany – DWACOS'99 / In: Lasker, G. E. (ed.). University of Windsor and Yanchun Zhang, University of Southern Queensland. The International Institute for Advanced Studies in Systems Research and Cybernetics, University of Windsor, Windsor, Ontario, Canada. **1**, pp. 41–46 (1999)

78. Fomichov, V. A.: A universal resources and agents framework for electronic commerce and other applications of multi-agent systems; 7th International Workshop on Computer Aided Systems Theory and Technology 1999 – EUROCAST'99. In: Kopacek, P. (ed.). September 29th – October 2nd, 1999, Vienna, Austria, Vienna University of Technology. pp. 99–102 (1999)

79. Fomichov, V. A.: An ontological mathematical framework for electronic commerce and semantically-structured web. In Zhang, Y., Fomichov, V.A., Zeleznikar, A.P. (eds.) Special Issue on Database, Web, and Cooperative Systems. Informatica. An International Journal of Computing and Informatics (Slovenia, Europe), 24(1), 39–49 (2000)

80. Fomichov, V. A.: New content languages for electronic commerce and digital libraries. In Binder, Z. (ed.), Management and Control of Production and Logistics 2000 (MCPL 2000). A Proceedings volume from the 2nd IFAC/IFIP/IEEE Conference, Grenoble, France, 5–8 July 2000. Pergamon, An Imprint of Elsevier Science, Oxford, UK., New York, Tokyo. **2**, pp. 503–508 (2001)

81. Fomichov, V.A.: The method of constructing the linguistic processor of the animation system AVIAROBOT. Preconference Proceedings "Collaborative Decision-Support Systems" (Focus Symposium in conjunction with the 14th International Conference on Systems Research, Informatics and Cybernetics – InterSymp-2002, July 29 – August 3, 2002, Germany). Focus Symposia Chair: Jens Pohl – CAD Research Center, Cal Poly, San Luis Obispo, CA, USA, pp. 91–102 (2002)

82. Fomichov, V. A.: Theory of K-calculuses as a powerful and flexible mathematical framework for building ontologies and designing natural language processing systems. In Andreasen, T., Motro, A., Christiansen, H., Larsen, H.L. (eds.) Flexible Query Answering Systems. 5th International Conference, FQAS 2002, Copenhagen, Denmark, October 27–29, 2002. Proceedings; LNAI 2522 (Lecture Notes in Artificial Intelligence, **2522**), Springer: Berlin, Heidelberg, New York, Barcelona, Hong Kong, London, Milan, Paris, Tokyo. pp. 183–196 (2002)

83. Fomichov, V.A.: Mathematical foundations of representing texts' meanings for the development of linguistic informational technologies. Part 1. A model of the system of primary units of conceptual level. Part 2. A system of the rules for constructing semantic representations of phrases and complex discourses. Informational Technologies (Moscow) **2002/10–11**, 16–25, 34–45 (2002)

84. Fomichov, V.A.: Theory of standard K-languages as a model of a universal semantic networking language. Preconference Proceedings "Intelligent Software Systems for the New Infostructure" (Focus Symposium in conjunction with the 16th International Conference on Systems Research, Informatics and Cybernetics – InterSymp-2004, July 29 – August 5, 2004,

Germany). Focus Symposia Chair: Jens Pohl – CAD Research Center, Cal Poly, San Luis Obispo, CA, USA. pp. 51–61 (2004)

85. Fomichov, V.A.: The Formalization of Designing Natural Language Processing Systems. MAX Press, Moscow; in Russian (2005)

86. Fomichov, V.A.: A new method of transforming natural language texts into semantic representations. Informational Technologies (Moscow) **2005/10**, 25–35 (2005)

87. Fomichov, V.A.: Standard K-languages as a powerful and flexible tool for building contracts and representing contents of arbitrary E-negotiations. In Bauknecht, K., Proell, B., Werthner, H. (eds.) The 6th Intern. Conf. on Electronic Commerce and Web Technologies "EC-Web 2005", Copenhagen, Denmark, Aug. 23–26, Proceedings. Lecture Notes in Computer Science. **3590**. Springer Verlag. pp. 138–147 (2005)

88. Fomichov, V.A.: A class of formal languages and an algorithm for constructing semantic annotations of Web-documents. Vestnik of N.E. Bauman State Technical University. Series "Instrument-Making". **60**, 73–86 (2005)

89. Fomichov, V.A.: The theory of SK-languages as a powerful framework for constructing formal languages for electronic business communication. Preconference Proceedings "Advances in Intelligent Software Systems" (Focus Symposium in conjunction with the 18th International Conference on Systems Research, Informatics and Cybernetics – InterSymp-2006, August 7–12, 2006, Germany). Focus Symposia Chair: Jens Pohl – CAD Research Center, Cal Poly, San Luis Obispo, CA, USA. pp. 123–130 (2006)

90. Fomichov, V.A.: Theory of K-representations as a framework for constructing a semantic networking language of a new generation. Preconference proceedings "Representation of Context in Software: Data => Information => Knowledge" (Focus Symposium in conjunction with the 19th International Conference on Systems Research, Informatics and Cybernetics – InterSymp-2007, July 30 – August 4, 2007, Germany). Focus Symposia Chair: Jens Pohl – CAD Research Center, Cal Poly, San Luis Obispo, CA, USA. pp. 61–70 (2007)

91. Fomichov, V. A.: Mathematical Foundations of representing the Content of Messages Sent by Computer Intelligent Agents. The Publishing House "TEIS", Moscow (2007)

92. Fomichov, V.A.: The approach of the theory of K-representations to reflecting in a formal way the content of contracts and records of negotiations in the field of electronic commerce. Business-informatics (Moscow), **2007/2**, 34–38 (2007)

93. Fomichov, V. A.: Theory of K-representations as a powerful framework for eliminating the language barrier between the users of internet. Robert Trappl (ed.) Cybernetics and Systems 2008. Proceedings of the Ninetieth European Meeting on Cybernetics and Systems Research, held at the University of Vienna, Austria, 25–28 March 2008. The Austrian Cybernetic Society for Cybernetic Studies, Vienna. 2008. Vol. 2. pp. 537–542 (2008)

94. Fomichov, V. A.: A comprehensive mathematical framework for bridging a gap between two approaches to creating a meaning-understanding Web. International Journal of Intelligent Computing and Cybernetics (Emerald Group Publishing Limited, UK). 2008. **1**(1), 143–163. (2008)

95. Fomichov, V. A.: Towards Joining Systems Science and Engineering of Semantics-Oriented Natural Language Processing Systems. – Preconference Proceedings "Intelligent Software Tools and Services" (Focus Symposium in conjunction with the 20th International Conference on Systems Research, Informatics and Cybernetics – InterSymp-2008, July 24–30, 2008, Germany). Focus Symposia Chair: Jens Pohl. CAD Research Center, Cal Poly, San Luis Obispo, CA, USA. pp. 55–66. (2008)

96. Fomichov, V. A., Akhromov, Ya. V.: Standard K-languages as a Powerful and Flexible Tool for Constructing Ontological Intelligent Agents. Preconference Proceedings "Collaborative Decision-Support Systems" (Focus Symposium in conjunction with the 15th International Conference on Systems Research, Informatics and Cybernetics – InterSymp-2003, July 28–August 01, 2003, Germany). Focus Symposia Chair: Jens Pohl – CAD Research Center, Cal Poly, San Luis Obispo, CA, USA, pp. 167–176 (2003)

97. Fomichov, V. A., Chuykov, A. V.: An intelligent agent with linguistic skills for the Web-search of production manufacturers. InterSymp-2000, 12th International Conference on

Systems research, Informatics and Cybernetics (July 31–August 4, 2000, Germany). Pre-conference Proceedings "Advances in Computer-Based and Web-Based Collaborative Systems (Focus Symposia, August 1 and 2, 2000)", Focus Symposia Chairs Jens Pohl and Thomas Fowler, IV. – Collaborative Agent Design (CAD) Research Center, Cal Poly, San Luis Obispo, CA, USA. pp. 137–147. (2000)

98. Fomichov, V. A., Kochanov, A. A.: Principles of Semantic Search for Information on the Web by the Intelligent Agent LingSearch-1. Preconference Proceedings "Advances in Computer-Based and Web-Based Collaborative Systems" (Focus Symposia in conjunction with the 13th International Conference on Systems Research, Informatics and Cybernetics – InterSymp-2001, July 31–August 1, 2001, Germany). Focus Symposia Chairs: Jens Pohl and Thomas Fowler, IV. – Collaborative Agent Design (CAD) Research Center, Cal Poly, San Luis Obispo, CA, USA. pp. 121–131 (2001)

99. Fomichov, V. A., Lustig, I. V.: A Computer Intelligent Agent for Semantic Classification of E-mail Messages. Preconference Proceedings "Advances in Computer-Based and Web-Based Collaborative Systems" (Focus Symposia in conjunction with the 13th International Conference on Systems Research, Informatics and Cybernetics – InterSymp-2001, July 31 – August 1, 2001, Germany). Focus Symposia Chairs: Jens Pohl and Thomas Fowler, IV. – Collaborative Agent Design (CAD) Research Center, Cal Poly, San Luis Obispo, CA, USA. pp. 29–37. (2001)

100. Fomichova O. S., Fomichov, V. A.: (2004). A new approach to designing children-oriented web-sites of art museums. In Cybernetics and Systems 2004. **2**. Proceedings of the Seventeenth European Meeting on Cybernetics and Systems Research (University of Vienna, Austria, 13–16 April 2004). In Trappl, R. (ed.) University of Vienna and Austrian Society for Cybernetic Studies. pp. 757–762 (2004)

101. Gaerdenfors, P. (ed): Generalised Quantifiers. Linguistic and Logical Approaches. Reidel, Dordrecht (1987)

102. Gazdar, G., Klein, E., Pullum, G., Sag, I.: Generalized Phrase Structure Grammar. Blackwell, Oxford, (1985)

103. Geeraerts, D.: Editorial Statement. Cognitive Linguistics. **1** (1), 1–3 (1990)

104. Genesereth, M. R., Fikes, R. E.: Knowledge Interchange Format Version 3 Reference Manual. Technical Report Logic-92-1, Computer Science Department, Stanford University (1992)

105. Grishman, R.: Computational Linguistics : An Introduction. Cambridge University Press Cambridge, (1986)

106. Groenendijuk, J., Stokhof, M.: Dynamic montague grammar. In Kalman, L. Polos L. (eds.) Papers from the Second Symposim on Logic and Language. Akademiai Kiado Budapest, – 1990. – pp. 3–48. (1990)

107. Groenendijuk, J., Stokhof, M.: Two theories of dynamic semantics. In J. van Eijck (ed.) Logics in Artificial Intelligence. Springer-Verlag, Berlin, 55–64 (1991)

108. Groenendijuk, J., Stokhof, M.: Dynamic predicate logic. Linguistics and Philosophy, **14(1)**, pp. 39–101. (1991)

109. Gruber, T. R.: A translation approach to portable ontology specifications. Knowledge Acquisition Journal. **5** 199–220 (1993)

110. Guarino, N.: Formal ontology and information systems. In: Proceedings of FOIS'98, Trento, Italy, pp. 3–15. IOS Press, Amsterdam (1998)

111. An accommodation recommender system based on associative networks (2003) Available online: http://ispaces.ec3.at/papers/DITTENBACH.EA_2003_a-natural-language.pdf

112. How to analyze free text descriptions for recommending TV programmes (2006) Available online. http://www8.informatik.uni-erlangen.de/IMMD8/Lectures/EINFKI/ecai_2006.pdf

113. Habel: Kognitionswissenschaft als Grundlage der Computerlinguistik. In Computerlinguistik und ihre theoretischen Grundlagen. Berlin etc.: Springer-Verlag, 204–209 (1988)

114. Hahn, U.: Making understands out of parsers: semantically driven parsing as a key concept for realistic text understanding applications. International Journal of Intelligent Systems. **4** (3), 345–385 (1989)

115. Hahn, U., Schulz, S.: Ontological foundations for biomedical sciences. Artificial Intelligence in Medicine. **39**, 179–182 (2007)

116. Hahn, U., Valencia, A.: Semantic Mining in Biomedicine (Introduction to the papers selected from the SMBM 2005 Symposium, Hinxton, U.K., April 2005). Bioinformatics. **22**, 643–644 (2005)

117. Hunter, J., Drennan, J., Little, S.: Realizing the hydrogen economy through semantic web technologies. IEEE Intelligent Systems (Jan/Feb 2004), Special issue on E-Science (2004)

118. Hasselbring, W., Weigand, H.: Languages for electronic business communication: state of the art. Industrial Management and Data Systems. **1001**, 217–226 (2001)

119. Hausser, R.: Database semantics for natural language. Artificial Intelligence (AIJ). **130**, (1), Elsevier, Dordrecht (2001)

120. Hausser, R: Foundations of Computational Linguistics, Human-Computer Communication in Natural Language. 2nd Edition. Springer, Berlin, New York (2001)

121. Hausser, R.: A Computational Model of Natural Language Communication; Interpretation, Inference, and Production in Database Semantics. Springer, Berlin, Heidelberg, New York (2006)

122. Heim, I.: E-Type pronouns and donkey anaphora. Linguistics & Philosophy. **13**(2), 137–177 (1990)

123. Hendler, J., Shadbolt, N., Hall, W., Berners-Lee, T., Weitzner, D.: Web Science: an interdisciplinary approach to understanding the web. Communications of the ACM. **51** , 60–69 (2008)

124. Herzog, O. Rollinger, C.-R. (eds.): Text Understanding in LILOG. Integrating Computational Linguistics and Artificial Intelligence. Final Report on the IBM Germany LILOG-Project. Springer-Verlag Berlin, (1991)

125. Hirst, G.: Semantic Interpretation and Ambiguity. Artificial Intelligence. **34**, 131–177 (1988)

126. Hobbs, J.R., Stickel, M.E., Appelt, D.E., Martin, P.: Interpretation as abduction. Artificial Intelligence. **63**, 69–142 (1993)

127. Horrocks, I.: A denotational semantics for Standart OIL and Instance OIL. Department of Computer Science, University of Manchester, UK, (2000) http://www.ontoknowledge.org/oil/downl/semantics.pdf

128. Horrocks, I, van Harmelen, F., Patel-Schneider, P. DAML + OIL Release (March 2001) http://www.daml.org/2001/03/daml+oil+index.

129. Hwang, C.H. A Logical Approach to Narrative Understanding. Ph.D. Dissertation, U. of Alberta, Edmonton, Canada. (1992)

130. Hwang, C. H., Schubert, L.K.: Meeting the interlocking needs of LF-computation, deindexing, and inference: An organic approach to general NLU. // Proc. 13th Int. Joint Conf. on AI (IJCAI'93). Chambery, France, 28 August–3 September (1993)

131. Hwang, C. H., Schubert, L.K, Episodic Logic: a situational logic for natural language processing. In Aczel, P., Israel, D., Katagiri, Y., Peters, S. (eds.) Situation Theory and Its Application, CSLT Publication, Stanford, 303–338 (1993)

132. Hwang, C.H., Schubert, L.K.: Episodic Logic: a comprehensive, natural represen-tation for language understanding. Minds and Machines. 1993b, **3**, 381–419 (1993)

133. Hwang, C. H., Schubert, L.K.: Tense trees as the 'fine structure' of discourse. In: Proceedings of the 30th Annual Meeting of the ACL, Newark, DE, pp. 232–240 (1992)

134. Hwang, C. H., Schubert, L.K.: Interpreting tense, aspect, and time adverbials: a compositional, unified approach. In: Gabbay, D.M., Ohlbach, H.J. (eds.) Proceedings of the 1st Int. Conf. on Temporal Logic, July 11–14, Bonn, Germany, pp. 238–264. Springer, Heidelberg (1994)

135. Jackendoff, R.S.: Semantic Structures. MIT Press, Cambridge, MA (1990)

136. Java, A., Finin, T., Nirenburg, S.: Text Understanding Agents and the Semantic Web. In: Proceedings of the 39th Hawaii International Conference on System Sciences, (2006) http://ebiquity.umbc.edu/file directory/papers/205.pdf.

137. Johnson, M.: The Body in the Mind: The Bodily Basis of Meaning, Imagination, and Reason. The University of Chicago Press Chicago, (1987)
138. Johnson, M.: Attribute-Value Logic and the Theory of Grammar. CSLI, Stanford (1988)
139. Johnson-Laird, P.: Mental Models. Harvard University Press Cambridge, MA (1983)
140. Johnsonbaugh, R.: Discrete Mathematics. Fifth edition. Prentic Hall, Upper Saddle River, NJ 621–621 (2001)
141. Jowsey, E. Montague Grammar and First Order Logic. Edinburgh Working Papers in Cognitive Science., 1, Univ. of Edinburgh (1987)
142. Kamp, H.: A theory of truth and semantic representation. In Groenendijk, J., Janssen, T., Stokhof, M. (eds.) Formal Methods in the Study of Natural Language. Part 1. Mathematical Centre, Amsterdam, 227–322. (1981)
143. Kamp, H., Reyle, U.: Introduction to Modeltheoretic Semantics of Natural Language, Formal Logic and Discourse Representation Theory. Kluwer Academic Publishers, Dordrecht (1993)
144. Kamp, H., Reyle, U. A calculus for first order discourse representation structures. Journal for Logic, Language and Information (JOLLI). 5, 297–348 (1996)
145. Kimbrough, S.O. A Note on Getting Started with FLBC: Towards a User Guide for FLBC. August 25, (2001) http://grace.wharton.upenn.edu/sok/sokpapers/2001-2/lap/lap-acm-questions-flbc-20010825.pdf.
146. Kimbrough, S.O.: A Note on the FLBC Treatment of an Example X12 Purchase Order (2001) http://opim-sun.wharton.upenn.edu/ sok/flbcinfo/span-fullEDI-sok-sam.pdf.
147. Kimbrough, S.O.: Reasoning about the Objects of Attitudes and Operators: Towards a Disquotation Theory for Representation of Propositional Content //Proceedings of the Tenth International Conference on Artificial Intelligence and Law- ICAIL'2001, (2001) http://grace.wharton.upenn.edu/sok/sokpapers/2000-1/icail/sok-icail01.pdf.
148. Kimbrough, S.O., Moore, S.A.: On automated message processing in electronic commerce and work support systems: speech act theory and expressive felicity. ACM Transactions on Information Systems, 15(4), 321–367. (1997)
149. Lakoff, G.: Women, Fire and Dangerous Things: What Categories Reveal about the Mind. The University of Chicago Press, Chicago (1987)
150. Lakoff, G., Johnson, M.: Metaphors We Live By. The University of Chicago Press, Chicago (1980)
151. Langacker, R. W.: Foundations of Cognitive Grammar. Vol. 1. Theoretical Prerequisites. Stanford University Press, Stanford (1987)
152. Langacker, R. W.: Concept, Image, and Symbol. The Cognitive Basis of Grammar. Mouton de Gruyter, Berlin, New York (1990)
153. Lee, C.H., Na, J.C., Khoo C.: (Singapore), In: ICADL 2003, LNCS 2911 (2003)
154. McShane, M., Zabludowski, M., Nirenburg, S., Beale, S.: OntoSem and SIMPLE: Two multilingual world views. In: Hirst, G., Nirenburg, S. (eds.) Proceedings of the Second Workshop on Text Meaning and Interpretation held in cooperation with ACL-2004, 25–26 July 2004, Barcelona, Spain, pp. 25–32 (2004)
155. Mel'cuk, I., Zolkovskij, A.: Towards a functioning meaning-text model of language. Linguistics 57, 10–47 (1970)
156. Minsky, M.: Framework for representing knowledge. In Winston, P. H. (ed.), The Psychology of Computer Vision, Mc Graw-Hill book comp., New York (1975)
157. Meyer, R.: Probleme von Zwei-Ebenen-Semantiken. Kognitionswissenschaft, 4, 32–46 (1994)
158. Montague, R.: Universal Grammar. Theoria. 36, 373–398 (1970)
159. Montague, R.: English as a formal language. In Thompson, R.H. (ed.) Formal Philosophy. Selected papers of Richard Montague. Yale University Press, New Haven and London 188–221 (1974)
160. Montague, R.: The proper treatment of quantification in ordinary english. In Thompson, R.H. (ed.) Formal Philosophy, Yale University Press, London, 247–270. (1974)

161. Navigli, R., Velardi, P.: Through automatic semantic annotation of on-line glossaries. In Proc. of European Knowledge Acquisition Workshop (EKAW)-2006, LNAI 4248, pp. 126–140 (2006)

162. OWL Web Ontology Language Semantics and Abstract Syntax, Patel-Schneider, P. F., Hayes, P., Horrocks, I. (eds.) W3C Recommendation, 10 February 2004 (2004) http://www.w3.org/TR/2004/REC-owl-semantics-20040210/. Latest version available at http://www.w3.org/TR/owl-semantics/

163. OWL Web Ontology Language Overview, Deborah L. McGuinness and Frank van Harmelen, Editors. W3C Recommendation, 10 February 2004 (2004) http://www.w3.org/TR/2004/REC-owl-features-20040210/. Latest version available at http://www.w3.org/TR/owl-features/

164. OWL Web Ontology Language Reference, Mike Dean and Guus Schreiber, Editors. W3C Recommendation, 10 February 2004 (2004) http://www.w3.org/TR/2004/REC-owl-ref-20040210/. Latest version available at http://www.w3.org/TR/owl-ref/

165. Panedes-Frigolett, H.: Interpretation in a cognitive architecture. In: Hirst, G., Nirenburg, S. (eds.) Proceedings of the Second Workshop on Text Meaning and Interpretation held in co-operation with ACL-2004, 25–26 July 2004, Barcelona, Spain, pp. 1–8 (2004)

166. Partee, B. H. (ed.) Montague Grammar. Academia Press, New York (1976)

167. Partee, B. H., ter Meulen, A., Wall, R.: Mathematical Methods in Linguistics. Kluwer, Dordrecht (1990)

168. Peregrin, J.: On a logical formalization of natural language. Kybernetika. **26**, 327–341 (1990)

169. Pohl, J.: Adapting to the information age. In: Preconference Proceedings "Advances in Computer-Based and Web-Based Collaborative Systems (Focus Symposia, August 1 and 2, 2000)", Focus Symposia Chairs Jens Pohl and Thomas Fowler, IV. 12th International Conference on Systems Research, Informatics and Cybernetics – InterSymp-2000 (July 31–August 4, 2000, Baden-Baden, Germany), pp. 9–25. Collaborative Agent Design (CAD) Research Center, Cal Poly, San Luis Obispo, CA, USA (2000)

170. Putnam, H.: Reason, Truth, and History. Cambridge University Press, Cambridge (1981)

171. Resource Description Framework (RDF) Model and Syntax Specification, Ora Lassila, Ralph R. Swick, Editors. World Wide Web Consortium Recommendation, (1999) http://www.w3.org/TR/1999/REC-rdf-syntax-19990222/. Latest version available at http://www.w3.org/TR/REC-rdf-syntax/.

172. RDF/XML Syntax Specification (Revised), Dave Beckett, Editor. W3C Recommendation, 10 February 2004 (2004) http://www.w3.org/TR/2004/REC-rdf-syntax-grammar-20040210/. Latest version available at http://www.w3.org/TR/rdf-syntax-grammar

173. Resource Description Framework (RDF): Concepts and Abstract Syntax, Graham Klyne and Jeremy J. Carroll, Editors. W3C Recommendation, 10 February 2004 (2004) http://www.w3.org/TR/2004/REC-rdf-concepts-20040210/. Latest version available at http://www.w3.org/TR/rdf-concepts/

174. Resource Description Framework (RDF) Schema Specification. W3C Recommendation, March 1999 (1999) http://vvww.w3.org/TR/WD-rdf-schema. Cited 1999

175. RDF Vocabulary Description Language 1.0: RDF Schema, Dan Brickley and R.V. Guha, Editors. W3C Recommendation, 10 February 2004 (2004) http://www.w3.org/TR/2004/REC-rdf-schema-20040210/. Latest version available at http://www.w3.org/TR/rdf-schema/. Cited 2004

176. Reeve, L., Han, H.: Survey of semantic annotation platforms. Proc. of the 20th Annual ACM Symposium on Applied Computing and Web Technologies (2005)

177. Saint-Dizier, P.: An approach to natural-language semantics in Logic Programming. Journal of Logic Programming. **3** (4), 329–356 (1986)

178. Schank, R.C.: Conceptual dependency: a theory of natural language understanding. Cognitive Psychology. 1972. **3**(4), 552–631 (1972)

179. Schank, R.C., Goldman, N.M., Rieger, C.J., Riesbeck, C.K.: Conceptual Information Processing. North-Holland Publ. Company; Amsterdam, Oxford American Elsevier Publ. Comp., Inc., New York (1975)

180. Schank, R., Birnbaum, L., Mey, J.: Integrating semantics and pragmatics. Quaderni di Semantica. VI(2) (1985)

181. Schubert, L.K.: Dynamic skolemization. In Bunt, H. Muskens, R. (eds.) Computing Meaning, vol. 1, Studies in Linguistics and Philosophy Series, Kluwer Academic Press, Dortrecht (also Boston, London), 219–253 (1999)

182. Schubert, L.K.: The situations we talk about. In Minker, J. (ed.) Logic-Based Artificial Intelligence. Kluwer, Dortrecht, 407–439 (2000)

183. Schubert, L. K., Hwang, C. H.: An episodic knowledge representation for narrative texts. Proceedings of the First Int. Conf. on Principles of Knowledge Representation and Reasoning (KR'89), Toronto, Canada. pp. 444–458. (1989)

184. Schubert, L.K., Hwang, C.H.: Episodic logic meets little red riding hood: a comprehensive, natural representation for language understanding. In Iwanska, L. Shapiro, S.C. (eds.) Natural Language Processing and Knowledge Representation: Language for Knowledge and Knowledge for Language, MIT/AAAI Press, Menlo Park, CA, and Cambridge, MA, 2000, 111–174 (2000)

185. Sembock, T.M.T., Van Rijsbergen, C.J.: SILOL : A Simple Logical-linguistic Document Retrieval System. Information Processing and Management, 26(1), 111–134 (1990)

186. Semantic Web Activity Statement. W3C, (2001) http://www.w3.org/2001/sw/activity

187. Seuren, P. A. M.: Discourse Semantics. Basil Blackwell, Oxford (1985)

188. Saint-Dizier, P.: An approach to natural-language semantics in Logic Programming. Journal of Logic Programming. 3, 329–356 (1986)

189. Sgall, P., Hajicova, E., Panevova, J.: The Meaning of the Sentence in Its Semantic and Pragmatic Aspects. Academia and D. Reidel, Prague and Dordrecht (1986)

190. Sgall, P.: The tasks of semantics and the perspectives of computers. Computers and Artificial Intelligence. 8 (5), 403–421 (1989)

191. Shadbolt, N., Hall, W., Berners-Lee, T.: The Semantic Web Revisited. IEEE Intelligent Systems. 21, 96–101 (2006)

192. Shapiro S.C.: Formalizing English. International Journal of Expert Systems. 9 (1), 151–171. (1996)

193. Shapiro, S.C.: SnepPS: a logic for natural language understanding and commonsense reasoning. In: Iwanska, L.M., Shapiro, S.C. (eds.) Natural Language Processing and Knowledge Representation: Language for Knowledge and Knowledge for Language, pp. 175–195. AAAI Press/The MIT Press, Menlo Park, CA (2000)

194. Shoenfield, J.L.: Mathematical Logic. Addison-Wesley Publ. Co., Reading, MA (1967)

195. Sowa, J.F.: Conceptual Structures: Information Processing in Mind and Machine. Addison-Wesley Publ. Comp, Reading (1984)

196. Sowa, J.F.: Toward the expressive power of natural language. In Sowa, J.F. (ed.) Principles of Semantic Networks. Explorations in the Representation of Knowledge. Morgan Kaufman Publ., Inc. San Mateo pp. 157–189. (1991)

197. Sowa, J.F.: Knowledge Representation: Logical, Philosophical, and Computational Foundations. Brooks/Cole, Pacific Grove, CA (2000)

198. Thibadeau, R., Just, M. A., Carpenter, P. A.: A model of the time course and content of reading. Cognitive Science. 6, 157–203 (1982)

199. Uchida, H., Zhu, M., Della Senta T.: A Gift for a Millennium, A book published by the United Nations University, (1999) http://www.unl.ias.unu.edu/publications/gm/index.htm. Cited 1999

200. Universal Networking Language (UNL) Specifications, Version 2005 (2005) www.undl.org/unlsys/unl/unl2005. Cited 2005

201. Universal Networking Language (UNL) Specifications, Version 2005, Edition 2006, 30 August 2006, UNL Center of UNDL Foundation (2005) http://www.undl.org/unlsys/unl/unl2005-e2006. Cited 2006

202. Wermter, J., Hahn, U.: Massive Biomedical Term Discovery. Discovery Science, 281–293 (2005)

203. Wierzbicka, A.: Lingua Mentalis. The Semantics of Natural Language. Academic Press, Sydney, (1980)

204. "Voice Browser" Activity – Voice enabling the Web! W3C paper. (2001) http://www.w3.org/Voice/. Cited 2001

205. Wilks, Y.: An artificial intelligence approach to machine translation. In Computer Models of Thought and Language. Schank, R. Colby, K. (eds.) Freeman, San Francisco pp. 114–151. (1973)

206. Wilks, Y.: Form and content in semantics. Synthese, **82**(3), 329–351 (1990)

207. Winograd, T.: Understanding Natural Language. Academic Press, New York, Edinburgh Univ. Press Edinburgh (1972)

208. Wolska, M., Kruijff-Korbayova, I., Horacek, H.: Lexical-semantic interpretation of language input in mathematical dialogs. In: Hirst, G., Nirenburg, S. (eds.) Proceedings of the Second Workshop on Text Meaning and Interpretation held in cooperation with ACL-2004, 25–26 July 2004, Barcelona, Spain, pp. 81–88 (2004)

209. Wroe, C., Goble, C., Greenwood, M., Lord, P., Miles, S., Papay, J., Payne, T., Moreau, L.: Automating Experiments Using Semantic Data on a Bioinformatics Grid. IEEE Intelligent Systems (Jan/Feb 2004), Special issue on E-Science (2004)

210. Xu, L., Jeusfeld, M.A.: A concept for monitoring of electronic contracts. Tilburg University, The Netherlands (2003) http://infolab.uvt.nl/pub/itrs010.pdf. 19 p. Cited 2003

211. Zhu, M., Uchida, H.: Universal Word and UNL Knowledge Base. In:Proceedings of the International Conference on Universal Knowledge Language (ICUKL-2002), 25-29 November 2002, Goa of India (1999) http://www.unl.ias.edu/publications/UW%20and%20UNLKB.htm. Cited 1999

212. Zhuge, H.: China's E-Science Knowledge Grid Environment. IEEE Intelligent Systems (Jan/Feb 2004), Special issue on E-Science (2004)

213. Yasukawa, H., Suzuki, H., Noguchi, N.: Knowledge representation language based on situation theory. In Fuehi, K., Kott, L. (eds.) Programming of Future Generation Computers II: Proceedings of the 2nd Franco-Japanese Symp. Amsterdam: North-Holland, pp. 431–460 (1988)

Index